A Laboratory Guide to the Tight Junction

A Laboratory Guide to the Tight Junction

Jianghui Hou
Division of Nephrology, Washington University Saint Louis,
St. Louis, MO, United States

ELSEVIER

ACADEMIC PRESS
An imprint of Elsevier

Academic Press is an imprint of Elsevier
125 London Wall, London EC2Y 5AS, United Kingdom
525 B Street, Suite 1650, San Diego, CA 92101, United States
50 Hampshire Street, 5th Floor, Cambridge, MA 02139, United States
The Boulevard, Langford Lane, Kidlington, Oxford OX5 1GB, United Kingdom

British Library Cataloguing-in-Publication Data
A catalogue record for this book is available from the British Library

Library of Congress Cataloging-in-Publication Data
A catalog record for this book is available from the Library of Congress

ISBN: 978-0-12-818647-3

For Information on all Academic Press publications
visit our website at https://www.elsevier.com/books-and-journals

Publisher: Andre Gerhard Wolff
Acquisitions Editor: Andre Gerhard Wolff
Editorial Project Manager: Susan Ikeda
Production Project Manager: Swapna Srinivasan
Cover Designer: Victoria Pearson

Typeset by MPS Limited, Chennai, India

Dedication

To my parents and my wife
who have consistently supported my scientific inquiry.

To my son Harvey
who will pursue his curiosity.

Contents

Author biography xix
Preface xxi
Acknowledgments xxiii

1 Introduction

1.1 Cell junction

1.1.1 Junction category 1
1.1.2 Tight junction 1
1.1.3 Adherens junction 3
1.1.4 Desmosome 3
1.1.5 Gap junction 4
References 5

1.2 Cell adhesion

1.2.1 Cadherin interaction 6
1.2.2 Claudin interaction 6
1.2.3 Connexin interaction 8
1.2.4 Cadherin compatibility and tissue morphogenesis 8
1.2.5 Claudin compatibility and tissue barrier 10
1.2.6 Connexin compatibility and electric synapse 10
References 10

1.3 Paracellular channel

1.3.1 Ion channel in tight junction 12
1.3.2 Electric conductance of paracellular channel 12
1.3.3 Ion selectivity of paracellular channel 14
1.3.4 Size selectivity of paracellular channel 16
1.3.5 Paracellular water channel 16
References 17

1.4 Perijunctional cytoskeleton

1.4.1 Actin polymerization 20
1.4.2 Actin reorganization 20
1.4.3 Contractile apparatus 23

1.4.4 Mechanosensitive signal transduction 24
1.4.5 Contractility and paracellular permeability 24
References 25

1.5 Junction signaling

1.5.1 Catenin signaling 26
1.5.2 Small GTPases 26
1.5.3 Receptor tyrosine kinases 26
1.5.4 Canonical Wnt signaling 28
1.5.5 Hippo signaling 28
References 31

2 Biochemical approaches for tight junction

2.1 Biochemistry of tight junction

2.1.1 Biochemical organization of tight junction 33
2.1.2 Tight junction enriched protein fraction 33
2.1.3 Tight junction integral protein fraction 34
2.1.4 Molecular structure of claudin protein 35
2.1.5 Models of claudin interaction 36
2.1.6 Claudin interaction with tight junction plaque proteins 37
2.1.7 Tight junction anchorage onto cytoskeleton 38
References 39

2.2 Tight junction isolation by subcellular fractionation

2.2.1 Background knowledge 41
2.2.2 Materials and reagents 42
2.2.3 Experimental procedure 44
2.2.4 Data analysis 45
2.2.5 Troubleshooting 45
2.2.6 Concluding remarks 47
References 47

2.3 Immunoprecipitation of *cis* and *trans* claudin interactions

2.3.1 Background knowledge 49
2.3.2 Materials and reagents 50
2.3.3 Experimental procedure 51
2.3.4 Data analysis 56
2.3.5 Troubleshooting 57
2.3.6 Concluding remarks 58
References 58

2.4 Isolation of claudin oligomer by chemical cross-linking

2.4.1 Background knowledge 59
2.4.2 Materials and reagents 59
2.4.3 Experimental procedure 63
2.4.4 Data analysis 65
2.4.5 Troubleshooting 66
2.4.6 Concluding remarks 68
References 69

2.5 Yeast two-hybrid assay of claudin interaction

2.5.1 Background knowledge 70
2.5.2 Materials and reagents 72
2.5.3 Experimental procedure 72
2.5.4 Data analysis 74
2.5.5 Troubleshooting 76
2.5.6 Closing remarks 77
References 77

2.6 Recombinant claudin protein production in *Pichia pastoris*

2.6.1 Background knowledge 78
2.6.2 Materials and reagents 79
2.6.3 Experimental procedure 81
2.6.4 Data analysis 86
2.6.5 Troubleshooting 86
2.6.6 Closing remarks 88
References 88

3 Biophysical approaches for tight junction

3.1 Electrophysiology of epithelial transport

3.1.1 Electric potential, resistance, and capacitance of cell membrane 89
3.1.2 Basic principles of cell membrane electrophysiology 91
3.1.3 Electrophysiology of an epithelium 94
3.1.4 Noise prevention and signal conditioning 96
3.1.5 Data acquisition and digitization 97
References 99

3.2 Epithelial cell cultures in Ussing chamber

3.2.1 Background knowledge 100
3.2.2 Materials and instrumentation 105
3.2.3 Experimental procedure 107

3.2.4 Data analysis 112
3.2.5 Troubleshooting 114
3.2.6 Closing remarks 115
References 116

3.3 Epithelial tissues in Ussing chamber

3.3.1 Background knowledge 117
3.3.2 Materials and instrumentation 118
3.3.3 Experimental procedure 121
3.3.4 Data analysis 123
3.3.5 Troubleshooting 125
3.3.6 Closing remarks 126
References 126

3.4 Epithelial ohmmeter

3.4.1 Background knowledge 128
3.4.2 Materials and instrumentation 128
3.4.3 Experimental procedure 129
3.4.4 Data analysis 129
3.4.5 Troubleshooting 130
3.4.6 Closing remarks 131
References 132

3.5 Impedance measurement in Ussing chamber

3.5.1 Background knowledge 133
3.5.2 Materials and instrumentation 136
3.5.3 Experimental procedure 136
3.5.4 Data analysis 137
3.5.5 Troubleshooting 137
3.5.6 Closing remarks 139
References 139

3.6 Flux assay in Ussing chamber

3.6.1 Background knowledge 140
3.6.2 Materials and instrumentation 141
3.6.3 Experimental procedure 142
3.6.4 Data analysis 142
3.6.5 Troubleshooting 143
3.6.6 Closing remarks 143
References 144

3.7 Measurement of water permeability in Ussing chamber

3.7.1 Background knowledge 145
3.7.2 Materials and instrumentation 146
3.7.3 Experimental procedure 148
3.7.4 Data analysis 148
3.7.5 Troubleshooting 149
3.7.6 Closing remarks 150
References 150

4 Histological approaches for tight junction

4.1 Fixation and fixatives

4.1.1 Classification of fixatives 153
4.1.2 Mechanism of fixation 153
4.1.3 Concentration of fixatives 155
4.1.4 Osmolality of fixative solution 155
4.1.5 Penetration of fixatives 156
4.1.6 Temperature of fixation 156
4.1.7 Duration of fixation 157
4.1.8 Fixation artifacts 157
References 157

4.2 Fixation

4.2.1 Introduction 158
4.2.2 Materials and reagents 158
4.2.3 Experimental procedure 158
4.2.4 Data analysis 160
4.2.5 Troubleshooting 160
4.2.6 Concluding remarks 160
References 161

4.3 Tight junction atlas

4.3.1 Introduction 161
4.3.2 Survey of tight junction in organ systems 162
4.3.3 Concluding remarks 172
References 174

5 Light microscopy for tight junction

5.1 Theory of light microscopy

5.1.1 Lateral resolution in light microscopy 175
5.1.2 Axial resolution in light microscopy 176
5.1.3 Depth of field in light microscopy 177

5.1.4 Fluorescence microscopy 177
5.1.5 Fluorescent labels 178
5.1.6 Autofluorescence 179
5.1.7 Photobleaching 180
References 180

5.2 Wide-field fluorescence microscopy for cells on cover glass

5.2.1 Background knowledge 181
5.2.2 Materials and reagents 182
5.2.3 Experimental procedure 183
5.2.4 Data analysis 184
5.2.5 Troubleshooting 184
5.2.6 Concluding remarks 187
References 187

5.3 Wide-field fluorescence microscopy for thin tissue section

5.3.1 Background knowledge 189
5.3.2 Materials and reagents 192
5.3.3 Experimental procedure 193
5.3.4 Data analysis 194
5.3.5 Troubleshooting 196
5.3.6 Concluding remarks 197
References 197

5.4 Confocal microscopy for cells on Transwell

5.4.1 Background knowledge 199
5.4.2 Materials and reagents 201
5.4.3 Experimental procedure 202
5.4.4 Data analysis 203
5.4.5 Troubleshooting 203
5.4.6 Concluding remarks 206
References 206

5.5 Confocal microscopy for thick tissue sections

5.5.1 Background knowledge 207
5.5.2 Materials and reagents 207
5.5.3 Experimental procedure 208
5.5.4 Data analysis 208
5.5.5 Troubleshooting 210
5.5.6 Concluding remarks 210
References 211

6 Electron microscopy for tight junction

6.1 Theory of electron microscopy

6.1.1 Wave-particle duality of electron 213
6.1.2 Electromagnetic lens 213
6.1.3 Specimen preparation 214
6.1.4 Ultramicrotomy 215
6.1.5 Positive staining 216
6.1.6 Negative staining 216
6.1.7 Low temperature methods 217
6.1.8 Immunolabeling techniques 217
References 218

6.2 Transmission electron microscopy for cell culture

6.2.1 Background knowledge 220
6.2.2 Materials and reagents 221
6.2.3 Experimental procedure 222
6.2.4 Data analysis 224
6.2.5 Troubleshooting 225
6.2.6 Concluding remarks 226
References 226

6.3 Transmission electron microscopy for tissue section

6.3.1 Background knowledge 228
6.3.2 Materials and reagents 229
6.3.3 Experimental procedure 230
6.3.4 Data analysis 232
6.3.5 Troubleshooting 232
6.3.6 Concluding remarks 234
References 234

6.4 Transmission electron microscopy for tracer assay

6.4.1 Background knowledge 235
6.4.2 Materials and reagents 235
6.4.3 Experimental procedure 237
6.4.4 Data analysis 237
6.4.5 Troubleshooting 238
6.4.6 Concluding remarks 239
References 239

6.5 Transmission electron microscopy for immunolabeling application

6.5.1 Background knowledge 240
6.5.2 Materials and reagents 240

6.5.3 Experimental procedure 242
6.5.4 Data analysis 243
6.5.5 Troubleshooting 244
6.5.6 Concluding remarks 245
References 245

6.6 Freeze-fracture electron microscopy

6.6.1 Background knowledge 246
6.6.2 Materials and reagents 249
6.6.3 Experimental procedure 251
6.6.4 Data analysis 253
6.6.5 Troubleshooting 254
6.6.6 Concluding remarks 255
References 255

6.7 Freeze-fracture replica immunolabeling technique

6.7.1 Background knowledge 256
6.7.2 Materials and reagents 256
6.7.3 Experimental procedure 258
6.7.4 Data analysis 258
6.7.5 Troubleshooting 260
6.7.6 Concluding remarks 260
References 261

7 Cell models of tight junction biology

7.1 Cell culture

7.1.1 Primary culture and cell transformation 263
7.1.2 Subculture and propagation 263
7.1.3 Anchorage independence 264
7.1.4 Cloning and selection 265
7.1.5 Gene transfer 265
7.1.6 Cryopreservation 265
7.1.7 Contamination 266
References 267

7.2 Culture of epithelial cells

7.2.1 Background knowledge 269
7.2.2 Materials and reagents 270
7.2.3 Experimental procedure 271
7.2.4 Data analysis 272
7.2.5 Troubleshooting 272
7.2.6 Concluding remarks 273
References 273

7.3 Calcium switch assay

7.3.1 Background knowledge 275
7.3.2 Materials and reagents 275
7.3.3 Experimental procedure 276
7.3.4 Data analysis 277
7.3.5 Troubleshooting 278
7.3.6 Concluding remarks 278
References 278

7.4 Retrovirus-mediated transgene expression

7.4.1 Background knowledge 279
7.4.2 Materials and reagents 282
7.4.3 Experimental procedure 283
7.4.4 Data analysis 285
7.4.5 Troubleshooting 285
7.4.6 Concluding remarks 287
References 287

7.5 Retrovirus-mediated RNA interference

7.5.1 Background knowledge 288
7.5.2 Materials and reagents 289
7.5.3 Experimental procedure 290
7.5.4 Data analysis 291
7.5.5 Troubleshooting 292
7.5.6 Concluding remarks 293
References 300

8 Mouse models of tight junction physiology

8.1 Mouse genetics and transgenics

8.1.1 Laboratory mouse 303
8.1.2 Mouse strain 303
8.1.3 Mouse genome 305
8.1.4 Random mutagenesis in laboratory mouse 305
8.1.5 Transgenesis in laboratory mouse 305
8.1.6 Gene targeting in laboratory mouse 307
References 309

8.2 Transgenic overexpression by DNA injection

8.2.1 Background knowledge 310
8.2.2 Materials and reagents 313
8.2.3 Experimental procedure 313
8.2.4 Data analysis 314
8.2.5 Troubleshooting 315

8.2.6 Concluding remarks 316
References 316

8.3 Lentivirus-mediated gene knockdown

8.3.1 Background knowledge 318
8.3.2 Materials and reagents 318
8.3.3 Experimental procedure 320
8.3.4 Data analysis 322
8.3.5 Troubleshooting 323
8.3.6 Concluding remarks 326
References 328

8.4 Conditional gene knockout by homologous recombination

8.4.1 Background knowledge 329
8.4.2 Materials and reagents 331
8.4.3 Experimental procedure 332
8.4.4 Data analysis 335
8.4.5 Troubleshooting 338
8.4.6 Concluding remarks 340
References 340

9 Perspective

9.1 Scanning ion conductance microscopy

9.1.1 Concept of conductance scanning 341
9.1.2 Practical application 342
9.1.3 Instrumentation 343
9.1.4 Tight junction conductance measurement 343
9.1.5 Limitation and future direction 345
References 346

9.2 Cryo-electron microscopy

9.2.1 Single-particle cryo-electron microscopy 347
9.2.2 Cryo-electron microscopy of vitreous section 349
9.2.3 Limitation and future direction 349
References 351

9.3 Super-resolution microscopy for tight junction

9.3.1 Super-resolution microscopy 352
9.3.2 Spatial separation of tight junction components 352
9.3.3 Architectural alteration in tricellular tight junction 354
9.3.4 Limitation and future direction 355
References 356

9.4 Novel binders to tight junction

9.4.1 *Clostridium perfringens* enterotoxin 357
9.4.2 TJ modulating peptidomimetics 357
9.4.3 Anti-claudin antibodies 360
9.4.4 Limitation and future direction 361
References 363

9.5 De novo assembly of tight junction

9.5.1 Concept of de novo assembly of subcellular organelle 366
9.5.2 Giant unilamellar vesicle 367
9.5.3 Protein incorporation and topological orientation in giant unilamellar vesicle 367
9.5.4 Probing claudin interactions in giant unilamellar vesicle 369
9.5.5 Limitation and future direction 370
References 371

9.6 Organoid model of tight junction biology

9.6.1 Organoid culture 372
9.6.2 Organ on a chip 372
9.6.3 Bioprinting of organ 374
9.6.4 Limitation and future direction 374
References 376

Index 379

Author biography

Dr. Jianghui Hou is a professor of Molecular Medicine and Cell Biology at Washington University in St. Louis, United States. He holds a PhD degree in Molecular Biology from the University of Edinburgh, United Kingdom. He specialized in tight junction biology during his postdoctoral training at Harvard Medical School, United States. He has been studying tight junction biology for the past 15 years and published over 45 peer-reviewed articles on this topic in leading journals. His major research interests include tight junction structure and function, super-resolution measurement of tight junction conductance, pathophysiology of tight junction disease, and pharmacologic development of tight junction modulator.

Preface

My first volume on tight junction, "The Paracellular Channel – Biology, Physiology, and Disease" serves as an introduction to the field of tight junction biology. Among the cellular components, tight junction stands out as a structurally and functionally unique organelle, which warrants independent study. A series of research techniques have been developed specifically for tight junction. This volume provides a comprehensive description of the essential technologies used in biochemical, biophysical, histological, and physiological analyses of tight junction.

I must emphasize that many of today's technologies have roots in the biochemical work of the past, in particular, the elegant work of biochemical purification of tight junction by Dr. Daniel Goodenough. In 1984, Dr. Goodenough described the first biochemical protocol to isolate tight junction membrane from the bile canaliculus in the mouse liver. His protocol has paved the way for the discovery of nearly all tight junction proteins, which include ZO-1, occludin, and claudins. This book is conceptually organized into five major areas of research technique for tight junction, which include biochemical, biophysical, histological, imaging and genetic approaches. Tight junction is a cellular architecture made of proteins and lipids. Biochemical techniques aim to address how molecular interactions within tight junction organize proteins and lipids into such a fine structure. Tight junction creates a paracellular pathway to allow ion and solute permeation on the basis of size and charge. Biophysical techniques aim to reveal the real-time changes in tight junction permeability. Tight junction is found in every type of cell throughout human body. Histological techniques allow in situ mapping of the location of tight junction molecules in cells and tissues. Tight junction is a subcellular entity of nanometer dimension. Direct visualization of tight junction depends upon advanced imaging techniques, which can be classified into two categories based upon the types of microscopy being used, that is, light microscopy and electron microscopy. Electron microscopy was once at the center of tight junction research. Modern super-resolution microscopy is offering novel insights of molecular localization at the resolution approaching electron microscopy. Since the discovery of the genetic linkage of tight junction proteins to human diseases, studying mutations in tight junction genes has become a new research direction to address the causative role of tight junction dysfunction in pathology.

Both gain-of-function and loss-of-function mutations in tight junction genes can be introduced to cell or animal models by a variety of genetic tools. The expansion of such a toolbox is at the forefront of genetic research.

Finally, I hope this book will suit the needs of researchers to analyze the tight junction, provide a step-by-step guide for researchers from beginning to sophisticated levels, and serve as a companion to conceptual books on tight junction.

Acknowledgments

I am deeply grateful to the support of many brilliant mentors, colleagues, and fellows over the years, in particular Dr. Daniel Goodenough, who introduced me to the field of tight junction biology, nourished my career development, and motivated me in countless scientific setbacks. Production of this book is not possible without the help of the highly professional editorial team in Elsevier, particularly my project managers, Andre Wolff, Susan Ikeda, and Swapna Srinivasan. Finally, I am sincerely indebted to the National Institute of Diabetes and Digestive and Kidney Diseases, National Science Foundation, Department of Defense, and American Heart Association for their continuous support of my thinking, writing, and experimenting for over 10 years.

Chapter 1

Introduction

Chapter 1.1

Cell junction

1.1.1 Junction category

Mammalian cells possess four intercellular junction systems formed by characteristic transmembrane molecules and submembranous protein plaques (Table 1.1.1). The intercellular junction systems are as follows:

- Tight junction (*zonula occludens*) is a multiprotein complex whose general function is to seal the paracellular space to prevent leakage of solutes and water.
- Adherens junction (*zonula adherens*) is a multiprotein architecture whose general role is cell adhesion.
- Desmosome (*macula adherens*) provides a strong type of cell adhesion and is found in stratified epithelium.
- Gap junction (*macula communicans*) creates an intercellular channel allowing direct molecular exchange from cell to cell.

1.1.2 Tight junction

Tight junction (TJ) is the most apical member of intercellular junction systems. When viewed with transmission electron microscopy, TJ appears as a zone in which adjacent plasma membranes are fused (Fig. 1.1.1) (Farquhar & Palade, 1963). Freeze-fracture replica electron microscopy reveals that TJ is made of a continuous network of "fibrils" or "strands" on the protoplasmic (P) fracture face and complementary empty grooves on the exoplasmic (E) fracture face (Fig. 1.1.2) (Goodenough & Revel, 1970). The proteins making the TJ fibrils or strands are claudins. Claudins are tetraspan membrane proteins consisting of a family with 28 members ranging in molecular mass from 20 to 28 kDa. Claudins have four transmembrane domains, two

A Laboratory Guide to the Tight Junction. DOI: https://doi.org/10.1016/B978-0-12-818647-3.00001-5

TABLE 1.1.1 Major types of cell junction.

	Cellular localization	Integral protein	Plaque protein	Associated filament
Tight junction	Epithelial cell Endothelial cell	Claudins Occludin	ZOs	Microfilament (actin)
Adherens junction	Epithelial cell Endothelial cell Cardiomyocyte Mesenchymal and neural cell	Cadherins Nectins	Catenins	Microfilament (actin)
Desmosome	Stratified epithelial cell	Desmogleins Desmocollins	Desmoplakin plakoglobin	Intermediate filament (cytokeratin, vimentin, desmin)
Gap junction	Epithelial cell Endothelial cell	Connexins	ZOs	–

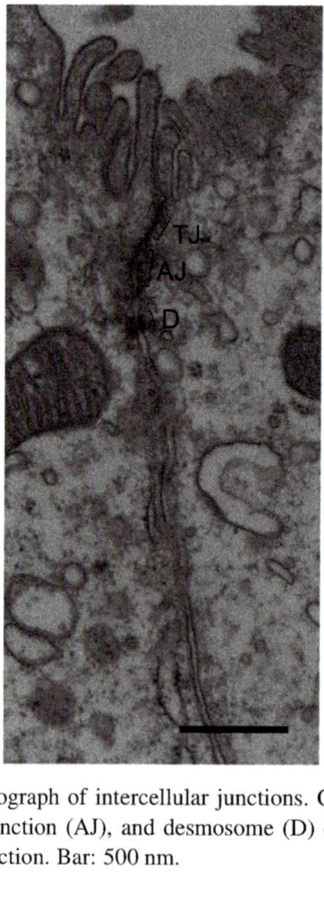

FIGURE 1.1.1 Electron micrograph of intercellular junctions. Characteristic subapical triad of tight junction (TJ), adherens junction (AJ), and desmosome (D) of mouse kidney proximal epithelial cells in a longitudinal section. Bar: 500 nm.

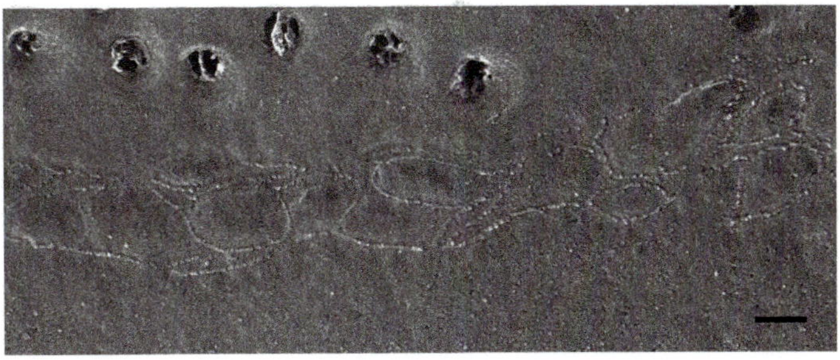

FIGURE 1.1.2 Freeze-fracture replica of tight junction. Tight junctions in mouse kidney distal epithelial cells appear as anastomosing strands on the protoplasmic (P) fracture face. Bar: 100 nm.

extracellular loops (ECLs: ECL1 and ECL2), amino- and carboxyl-terminal cytoplasmic domains, and a short cytoplasmic turn. The charges in ECL1 regulate paracellular ion selectivity by electrostatic interactions. The carboxyl-terminal domain of claudin contains a PDZ (postsynaptic density 95/discs large/ZO-1) binding motif (YV) that is critical for interaction with the TJ plaque protein ZO-1 (Hou, Rajagopal, & Yu, 2013).

1.1.3 Adherens junction

Adherens junction appears under transmission electron microscopy as plasma membrane apposition separated by an intercellular cleft of 10–20 nm in width (Fig. 1.1.1). The cleft is occupied by rod-shaped molecules known as cadherins (Takeichi, 1990). The classic cadherin family comprises approximately 20 members that share a common domain organization. The cadherin protein features an amino-terminal extracellular domain or ectodomain that is followed by a transmembrane domain and a carboxyl-terminal intracellular domain. The ectodomain contains five secondary structural repeats, termed extracellular cadherin (EC) domains and numbered from EC1 to EC5. Each EC domain is made of approximately 110 amino acids and resembles the immunoglobulin domain found in antibodies (Shapiro & Weis, 2009). The intracellular domain binds to a class of cytoplasmic proteins, known as catenins. The catenins, in turn, interact with F-actin and other cytoskeletal proteins (Rimm, Koslov, Kebriaei, Cianci, & Morrow, 1995).

1.1.4 Desmosome

Desmosome is in many ways similar to adherens junction (Fig. 1.1.1). Desmosome is composed of the desmosome-intermediate filament complex, which is a network of desmosomal cadherin proteins, plaque proteins, and intermediate filaments (Garrod, 1993). Desmosomal cadherins include desmogleins and desmocollins. The extracellular domains of desmogleins and desmocollins mediate cell adhesion, whereas the cytoplasmic tails of these cadherins interact with the desmosomal plaque proteins such as plakoglobin and desmoplakin (Kowalczyk et al., 1994). The interaction between desmoplakin and intermediate filaments tether desmosomes to the cytoskeletal network. Desmogleins and desmocollins both contain four extracellular cadherin (EC1 to EC4) domains and a fifth domain termed the extracellular anchor (EA). The EA domain is not present in classic cadherin proteins.

1.1.5 Gap junction

Gap junction (GJ) is seen by transmission electron microscopy as an area of plasma membrane apposition separated by a 2–3 nm "gap" of intercellular space (Fig. 1.1.3). Freeze-fractured gap junction shows a lattice of densely packed particles on the P face and complementary pits on the E face. These particles are made of the connexin proteins (Fig. 1.1.3) (Goodenough & Paul, 2009). Connexins are tetraspan membrane proteins consisting of a family of 21 members ranging in molecular mass from 23 to 62 kDa.

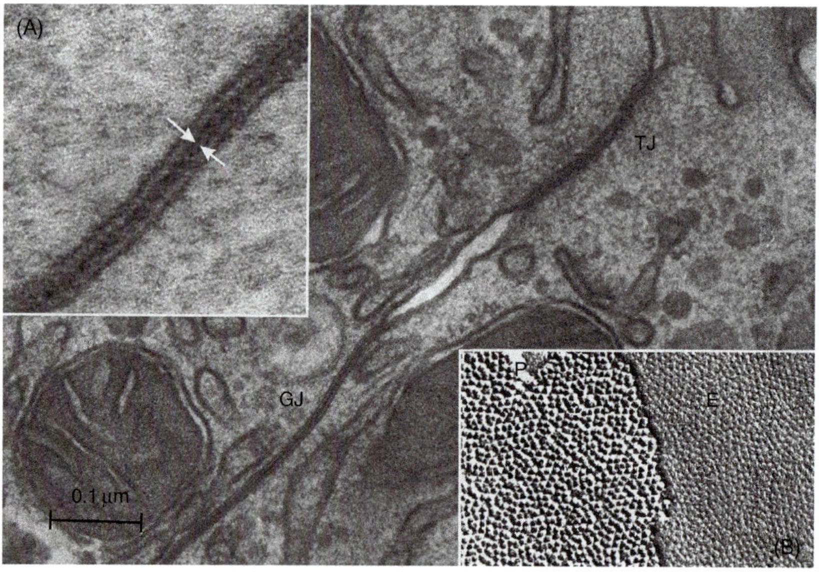

FIGURE 1.1.3 Electron microscopy of gap junction. Gap junction (GJ) in the mouse hepatocyte is viewed by transmission electron microscopy. (Inset A) A high magnification view of the gap junction revealing the 2–3 nm "gap" (white arrows) separating the plasma membranes. (Inset B) A freeze-fracture replica of a gap junction showing the characteristic particles on the protoplasmic (P) fracture face and complementary pits on the ectoplasmic (E) fracture face. Bar: 100 nm. *Reproduced with permission from Goodenough, D. A., & Paul, D. L. (2009). Gap junctions.* Cold Spring Harbor Perspectives in Biology, 1, *a002576.*

Connexins contain four transmembrane domains, two extracellular loops (ECLs: ECL1 and ECL2), unstructured N and C cytoplasmic termini, and a cytoplasmic turn. Six connexin molecules assemble into a hemichannel, known as connexon, in the plasma membrane (Musil & Goodenough, 1993). Two connexons, by head-to-head docking, form a gap junction. Connexin binds to the PDZ domain in ZO-1 via its C-terminal domain (Giepmans & Moolenaar, 1998). ZO-1 is preferentially associated with the periphery of gap junction plaques and it may anchor gap junctions to filamentous actin cytoskeleton (Hunter, Barker, Zhu, & Gourdie, 2005).

References

Farquhar, M. G., & Palade, G. E. (1963). Junctional complexes in various epithelia. *The Journal of Cell Biology*, *17*, 375–412.

Garrod, D. R. (1993). Desmosomes and hemidesmosomes. *Current Opinion in Cell Biology*, *5*, 30–40.

Giepmans, B. N., & Moolenaar, W. H. (1998). The gap junction protein connexin43 interacts with the second PDZ domain of the zona occludens-1 protein. *Current Biology: CB*, *8*, 931–934.

Goodenough, D. A., & Paul, D. L. (2009). Gap junctions. *Cold Spring Harbor Perspectives in Biology*, *1*, a002576.

Goodenough, D. A., & Revel, J. P. (1970). A fine structural analysis of intercellular junctions in the mouse liver. *The Journal of Cell Biology*, *45*, 272–290.

Hou, J., Rajagopal, M., & Yu, A. S. (2013). Claudins and the kidney. *Annual Review of Physiology*, *75*, 479–501.

Hunter, A. W., Barker, R. J., Zhu, C., & Gourdie, R. G. (2005). Zonula occludens-1 alters connexin43 gap junction size and organization by influencing channel accretion. *Molecular Biology of the Cell*, *16*, 5686–5698.

Kowalczyk, A. P., Stappenbeck, T. S., Parry, D. A., Palka, H. L., Virata, M. L., Bornslaeger, E. A., … Green, K. J. (1994). Structure and function of desmosomal transmembrane core and plaque molecules. *Biophysical Chemistry*, *50*, 97–112.

Musil, L. S., & Goodenough, D. A. (1993). Multisubunit assembly of an integral plasma membrane channel protein, gap junction connexin43, occurs after exit from the ER. *Cell*, *74*, 1065–1077.

Rimm, D. L., Koslov, E. R., Kebriaei, P., Cianci, C. D., & Morrow, J. S. (1995). Alpha 1(E)-catenin is an actin-binding and -bundling protein mediating the attachment of F-actin to the membrane adhesion complex. *Proceedings of the National Academy of Sciences of the United States of America*, *92*, 8813–8817.

Shapiro, L., & Weis, W. I. (2009). Structure and biochemistry of cadherins and catenins. *Cold Spring Harbor Perspectives in Biology*, *1*, a003053.

Takeichi, M. (1990). Cadherins: A molecular family important in selective cell-cell adhesion. *Annual Review of Biochemistry*, *59*, 237–252.

Chapter 1.2

Cell adhesion

1.2.1 Cadherin interaction

Cadherins mediate Ca^{++}-dependent cell adhesion at adherens junction and desmosome (Meng & Takeichi, 2009). The cadherin ectodomain contains five repeating secondary structures, termed extracellular cadherin (EC) domain. The crystal structures of cadherin EC domains have been resolved (Fig. 1.2.1A) (Boggon et al., 2002). Each EC domain is composed of seven β-strands, numbered A–G and arranged as two antiparallel β-sheets. The connections between successive EC domains are rigidified by the Ca^{++} ions to form a strong curvature. Single-molecule fluorescence microscopy reveals that cadherins bind in *trans* via dimerization of their EC1 domains (Zhang, Sivasankar, Nelson, & Chu, 2009). The prevailing model of EC1 dimerization is based upon the domain swapping theory, which states that the amino-terminal β-strand (the A strand) in the EC1 domain swaps so that the A strand of one protomer replaces the A strand of the other protomer (Fig. 1.2.1B) (Chen, Posy, Ben-Shaul, Shapiro, & Honig, 2005). The *cis* cadherin interaction involves a complementary fit between the convex surface in the EC2 domain of one molecule and the concave surface in the EC1 domain of another molecule (Fig. 1.2.1C).

1.2.2 Claudin interaction

Claudins mediate Ca^{++}-independent cell adhesion at tight junction (Kubota et al., 1999). The crystal structure of claudin offers by far the most important hint of how claudins interact in *cis* and *trans* (Fig. 1.2.2) (Suzuki et al., 2014). The transmembrane domains in claudin form typical left-handed α-helices, and large portions of the two extracellular loops (ECLs) form a β-sheet structure. The β-sheet structure extends from the membrane surface and comprises five β-strands (β1–β5), four contributed by ECL1 and one by ECL2. A conserved hydrophobic residue (Met68) from the ECL1 domain in one protomer fits into the hydrophobic pocket formed by the conserved residues (Phe146, Phe147, and Leu158) in the third transmembrane domain and the ECL2 domain of the adjacent protomer. Such tandem *cis* interactions align the claudin protomers into a linear polymer. *Trans* claudin interaction depends upon the β-sheet structure, thereby involving both extracellular loops. A few loci in ECL1 and ECL2 have been demonstrated to play important roles in *trans* claudin interaction (Daugherty, Ward, Smith, Ritzenthaler, & Koval, 2007; Piontek et al., 2008).

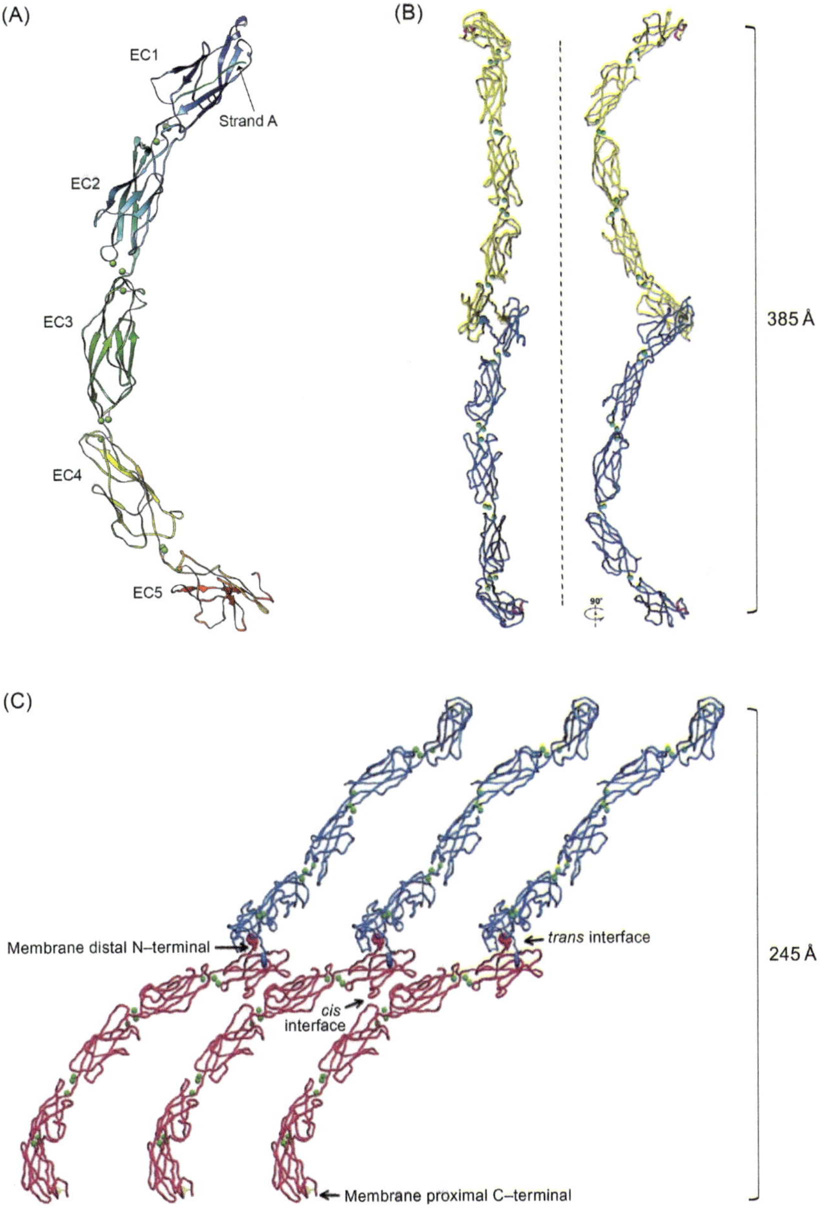

FIGURE 1.2.1 Cadherin structure. (A) Crystal structure of *Xenopus* C-cadherin ectodomain in ribbon representation. The green spheres depict the bound Ca^{++} ions. (B) Experimentally derived backbone worm trace of two C-cadherin ectodomains connected through *trans* dimer interaction. The molecules are oriented as if emanating from opposing cell surfaces. (C) C-cadherin ectodomains, joined by both *cis* and *trans* interfaces, are arrayed as if emanating from juxtaposed cell surfaces. Molecules from either putative cell surface are shown in blue or pink. *Reproduced with permission from Boggon, T. J., Murray, J., Chappuis-Flament, S., Wong, E., Gumbiner, B. M., & Shapiro, L. (2002). C-cadherin ectodomain structure and implications for cell adhesion mechanisms.* Science (New York, NY), 296, *1308–1313.*

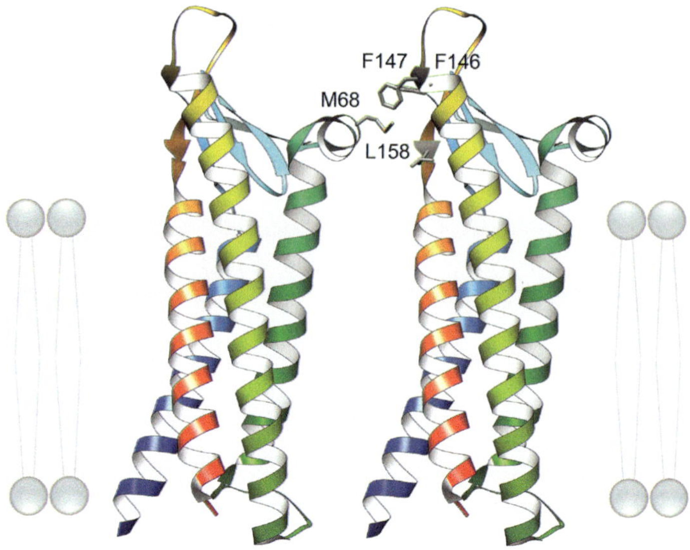

FIGURE 1.2.2 *Cis* claudin interaction. Ribbon representation of two linearly aligned claudin-15 protomers in the crystal when viewed in parallel to the membrane. Critical residues for the lateral interaction are labeled and shown in stick representation. The depicted structure is based upon the X-ray analysis by Suzuki et al. (2014).

1.2.3 Connexin interaction

Connexins form an intercellular channel to permit the exchange of ions and small molecules between cells. The crystal structure of two membrane-spanning connexons forming a gap junction channel is now available (Fig. 1.2.3) (Maeda et al., 2009). The connexin protomer forms a typical four-helix bundle, in which any pair of adjacent helices is antiparallel. The protomers in each hexameric connexon are related by a sixfold symmetry axis perpendicular to the membrane plane. Most of the prominent *cis* protomer interactions are in the extracellular portion of the transmembrane domains (TM1–4). The *trans* protomer interactions between the hexameric connexons are located in the extracellular half of transmembrane helices TM2 and TM4 and in the extracellular loops. Most of the residues involved in *cis* and *trans* protomer interactions are conserved within the connexin family and mutations of these residues are associated with diseases, which indicate that connexin oligomerization is required for gap junction function (Laird, 2006).

1.2.4 Cadherin compatibility and tissue morphogenesis

Classical cadherin ectodomains are remarkably similar. Sequences of the EC1 domains of E- and N-cadherins are 57% identical, and only a few

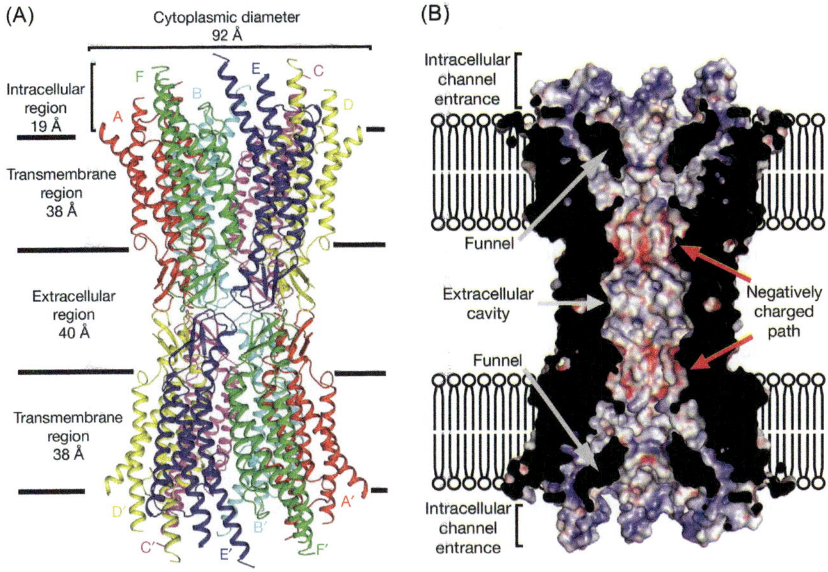

FIGURE 1.2.3 Structure of gap junction channel. (A) The protomers in two Cx26 hemichannels, which are related by a twofold axis, are shown in ribbon representation. (B) Vertical cross section through the Cx26 channel shows the electrostatic surface potential inside the channel. The channel features a wide cytoplasmic opening, which is restricted by the funnel structure, a negatively charged path and an extracellular cavity in the middle. The displayed potentials range from −40 (*red*) to +40 (*blue*) kT/e. *Reproduced with permission from Maeda, S., Nakagawa, S., Suga, M., Yamashita, E., Oshima, A., Fujiyoshi, Y., & Tsukihara, T. (2009). Structure of the connexin 26 gap junction channel at 3.5 A resolution.* Nature, 458, *597–602.*

residues within the strand-swap interface differ. Yet cells expressing E- or N-cadherin can be sorted based upon homotypic interactions in a coculture experiment (Tamura, Shan, Hendrickson, Colman, & Shapiro, 1998). It seems certain that cell sorting specificity resides within the EC1 domain (Nose, Tsuji, & Takeichi, 1990). A theory posits that multiple *trans* interactions amplify small affinity differences at the molecular level into large differences in intercellular adhesive strength (Chen et al., 2005). Differential cell adhesion plays a vital role in tissue morphogenesis. One striking example comes from the morphogenesis of neural crest cells (Hatta, Takagi, Fujisawa, & Takeichi, 1987). A series of cadherin switches occur during this process. Neural crest precursors first downregulate E-cadherin to form the border between neural and nonneural ectoderm. During neurulation, these precursors are incorporated into the neural fold and the neural tube when they start to express N-cadherin. Eventually, they downregulate cadherin expression in order to delaminate from the neural tube and become migratory.

1.2.5 Claudin compatibility and tissue barrier

Claudin compatibility can be addressed on both *cis* and *trans* levels. For example, claudin-16 and claudin-19 interact in *cis* but not in *trans* (Hou et al., 2008). When claudin-16 and claudin-19 are introduced into the same epithelial cells, tight junction permeability displays a synergy that is more than the combination of individual claudin effects in singly expressed cells (Hou et al., 2008). Mutations disrupting claudin-16 and claudin-19 *cis* interaction reduce the tight junction permeability by increasing the tight junction ultrastructural complexity, that is, the number of tight junction strands (Gong et al., 2015). Coculture experiments indicate that no heterotypic channel can be made by claudin-16 and claudin-19 due to the lack of *trans* interaction (Gong et al., 2015). According to the crystal structure of claudin, the fourth β-strand in ECL1 faces the extracellular space and contains charged amino acids that act as a selectivity filter to dictate the paracellular ion permeability. Because the charges in the selectivity filter have to be symmetrically aligned, homotypic channels are favored by such physical nature.

1.2.6 Connexin compatibility and electric synapse

A connexon may dock with an identical connexon to form a homotypic channel or with a connexon containing different connexins to form a heterotypic channel, mindful that only some combinations are permitted (White, Bruzzone, Wolfram, Paul, & Goodenough, 1994). The gap junction channels composed of different connexons may have different physiological properties, such as single-channel conductance, permeability profile to tracers, and biologically relevant signal transduction. Such structural asymmetry is important to establish the rectifying neuronal synapse, in which action potential is transmitted orthodromically but not antidromically. For example, Cx45/Cx43 heterotypic junctions allow fast rectification to influence dendrodendritic interactions in the central nervous system or to modulate reentry circuits in the myocardium (Bukauskas, Angele, Verselis, & Bennett, 2002). Because neither connexin displays homotypic voltage gating, the rectification observed cannot be predicted from the properties of the individual channels. A model proposes that the asymmetry of the heterotypic channel results in a separation of positive and negative charges across the two junctional membranes and that the rectification of ionic currents occurs within the channel rather than resulting from voltage-induced connexin conformational changes.

References

Boggon, T. J., Murray, J., Chappuis-Flament, S., Wong, E., Gumbiner, B. M., & Shapiro, L. (2002). C-cadherin ectodomain structure and implications for cell adhesion mechanisms. *Science (New York, NY)*, *296*, 1308–1313.

Bukauskas, F. F., Angele, A. B., Verselis, V. K., & Bennett, M. V. (2002). Coupling asymmetry of heterotypic connexin 45/ connexin 43-EGFP gap junctions: Properties of fast and slow gating mechanisms. *Proceedings of the National Academy of Sciences of the United States of America, 99*, 7113–7118.

Chen, C. P., Posy, S., Ben-Shaul, A., Shapiro, L., & Honig, B. H. (2005). Specificity of cell-cell adhesion by classical cadherins: Critical role for low-affinity dimerization through beta-strand swapping. *Proceedings of the National Academy of Sciences of the United States of America, 102*, 8531–8536.

Daugherty, B. L., Ward, C., Smith, T., Ritzenthaler, J. D., & Koval, M. (2007). Regulation of heterotypic claudin compatibility. *The Journal of Biological Chemistry, 282*, 30005–30013.

Gong, Y., Renigunta, V., Zhou, Y., Sunq, A., Wang, J., Yang, J., ... Hou, J. (2015). Biochemical and biophysical analyses of tight junction permeability made of claudin-16 and claudin-19 dimerization. *Molecular Biology of the Cell.*

Hatta, K., Takagi, S., Fujisawa, H., & Takeichi, M. (1987). Spatial and temporal expression pattern of N-cadherin cell adhesion molecules correlated with morphogenetic processes of chicken embryos. *Developmental Biology, 120*, 215–227.

Hou, J., Renigunta, A., Konrad, M., Gomes, A. S., Schneeberger, E. E., Paul, D. L., ... Goodenough, D. A. (2008). Claudin-16 and claudin-19 interact and form a cation-selective tight junction complex. *The Journal of Clinical Investigation, 118*, 619–628.

Kubota, K., Furuse, M., Sasaki, H., Sonoda, N., Fujita, K., Nagafuchi, A., & Tsukita, S. (1999). Ca(2 +)-independent cell-adhesion activity of claudins, a family of integral membrane proteins localized at tight junctions. *Current Biology: CB, 9*, 1035–1038.

Laird, D. W. (2006). Life cycle of connexins in health and disease. *The Biochemical Journal, 394*, 527–543.

Maeda, S., Nakagawa, S., Suga, M., Yamashita, E., Oshima, A., Fujiyoshi, Y., & Tsukihara, T. (2009). Structure of the connexin 26 gap junction channel at 3.5 A resolution. *Nature, 458*, 597–602.

Meng, W., & Takeichi, M. (2009). Adherens junction: Molecular architecture and regulation. *Cold Spring Harbor Perspectives in Biology, 1*, a002899.

Nose, A., Tsuji, K., & Takeichi, M. (1990). Localization of specificity determining sites in cadherin cell adhesion molecules. *Cell, 61*, 147–155.

Piontek, J., Winkler, L., Wolburg, H., Muller, S. L., Zuleger, N., Piehl, C., ... Blasig, I. E. (2008). Formation of tight junction: Determinants of homophilic interaction between classic claudins. *FASEB Journal: Official Publication of the Federation of American Societies for Experimental Biology, 22*, 146–158.

Suzuki, H., Nishizawa, T., Tani, K., Yamazaki, Y., Tamura, A., Ishitani, R., ... Fujiyoshi, Y. (2014). Crystal structure of a claudin provides insight into the architecture of tight junctions. *Science (New York, NY), 344*, 304–307.

Tamura, K., Shan, W. S., Hendrickson, W. A., Colman, D. R., & Shapiro, L. (1998). Structure-function analysis of cell adhesion by neural (N-) cadherin. *Neuron, 20*, 1153–1163.

White, T. W., Bruzzone, R., Wolfram, S., Paul, D. L., & Goodenough, D. A. (1994). Selective interactions among the multiple connexin proteins expressed in the vertebrate lens: The second extracellular domain is a determinant of compatibility between connexins. *The Journal of Cell Biology, 125*, 879–892.

Zhang, Y., Sivasankar, S., Nelson, W. J., & Chu, S. (2009). Resolving cadherin interactions and binding cooperativity at the single-molecule level. *Proceedings of the National Academy of Sciences of the United States of America, 106*, 109–114.

Chapter 1.3

Paracellular channel

1.3.1 Ion channel in tight junction

Paracellular ion transport through the tight junction is passive, driven by electrochemical gradients, and demonstrates channel-like properties such as electric conductance, charge, and size selectivity similar to those found in membrane ion channels. The paracellular channel denotes a new type of ion channel located in the tight junction and serving to join the two exterior compartments. The paracellular channel is made from the claudin proteins. Each claudin protein has its own unique property of ion permeability (Table 1.3.1). Notably, the measurement of claudin permeability depends upon the background of endogenous claudins present in the tight junction. Because the transport property of tight junction is determined by the combination of all claudin proteins expressed in a cell, studying any individual claudin will need to take into account the permeabilities of the remaining claudins.

1.3.2 Electric conductance of paracellular channel

The conductance of paracellular cation channels, for example, claudin-2, −10b, −15, −16, and −21 is at $\sim 10\,\text{mS/cm}^2$ (Hou, Paul, & Goodenough, 2005; Tanaka et al., 2016; Van Itallie et al., 2006; Van Itallie, Fanning, & Anderson, 2003; Yu et al., 2009). The conductance of paracellular anion channels, for example, claudin-4 and −17 is by one magnitude lower and at $\sim 1\,\text{mS/cm}^2$ (Hou, Renigunta, Yang, & Waldegger, 2010; Krug et al., 2012). Brownian dynamics simulations predict the maximal single-channel conductance for claudin to be around 100 pS (Yu et al., 2009). The overall paracellular conductance (G_p) is determined by the single claudin channel conductance (G_s), the open probability of claudin channel (P_o), the claudin molecular density along each tight junction (TJ) strand (N), and the number of TJ strands (n), according to the following equation (Claude, 1978).

$$G_p = G_s \times P_o^n \times N \tag{1.3.1}$$

Assuming that cation-selective and anion-selective claudins share similar single-channel conductance, open probability, and on-strand molecular density levels, the TJ strand number will be the single most likely factor that differs between cation-selective and anion-selective paracellular pathways.

TABLE 1.3.1 Claudin permeabilities evaluated in epithelial cells.

Species	Permeability	Cell background	Reference
Claudin-1	Cation and anion barrier	MDCK-II	Inai, Kobayashi, and Shibata (1999)
Claudin-2	Cation channel	MDCK-I, -II, -C7, LLC-PK1, Caco2	Amasheh et al. (2002), Fujita et al. (2008), Hou, Gomes, Paul, and Goodenough (2006), Van Itallie et al. (2003), Yu et al. (2009)
	Divalent cation channel	MDCK-I, Caco2	Fujita et al. (2008) and Yu et al. (2009)
Claudin-3	Cation and anion barrier	MDCK-II	Milatz et al. (2010)
Claudin-4	Cation barrier	MDCK-II	Van Itallie, Rahner, and Anderson (2001)
	Anion channel	LLC-PK1, M-1, mIMCD3	Hou et al. (2006, 2010)
Claudin-5	Cation barrier	MDCK-II	Wen, Watry, Marcondes, and Fox (2004)
Claudin-6	Anion barrier	MDCK-II	Sas, Hu, Moe, and Baum (2008)
Claudin-7	Cation channel and anion barrier	LLC-PK1	Alexandre, Lu, and Chen (2005)
	Anion channel	LLC-PK1	Hou et al. (2006)
Claudin-8	Cation barrier	MDCK-II	Yu, Enck, Lencer, and Schneeberger (2003)
	Anion channel	M-1, mIMCD3	Hou et al. (2010)
Claudin-9	Anion barrier	MDCK-II	Sas et al. (2008)
Claudin-10a	Anion channel	MDCK-II	Van Itallie et al. (2006)
Claudin-10b	Cation channel	LLC-PK1	Van Itallie et al. (2006)
Claudin-11	Cation barrier	MDCK-II	Van Itallie et al. (2003)
Claudin-12	Divalent cation channel	Caco2	Fujita et al. (2008)

(Continued)

TABLE 1.3.1 (Continued)

Species	Permeability	Cell background	Reference
Claudin-14	Cation barrier	MDCK-II	Ben-Yosef et al. (2003)
Claudin-15	Cation channel	LLC-PK1	Van Itallie et al. (2003)
Claudin-16	Cation channel	LLC-PK1	Hou et al. (2005)
	Divalent cation channel	MDCK-C7	Kausalya et al. (2006)
Claudin-17	Anion channel	MDCK-C7, LLC-PK1	Krug et al. (2012)
Claudin-18	Cation barrier	MDCK-II	Jovov et al. (2007)
Claudin-19	Cation barrier	MDCK-II	Angelow, El-Husseini, Kanzawa, and Yu (2007)
	Anion barrier	LLC-PK1	Hou et al. (2008)
Claudin-21	Cation channel	MDCK-I	Tanaka et al. (2016)

Caco2, Human colon carcinoma cell line 2; *LLC-PK1,* Lilly Laboratories cell-porcine kidney 1; *MDCK-I, -II, or -C7,* Madin-Darby canine kidney type I, type II, or type C7 cell; *M-1,* mouse cortical collecting duct type 1 cell; *mIMCD3,* mouse inner medullary collecting duct type 3 cell.

1.3.3 Ion selectivity of paracellular channel

Domain swapping experiments have demonstrated the pivotal role of the first extracellular loop (ECL1) of claudin in dictating the ion selectivity of paracellular channel. When the first, but not the second extracellular loop in claudin-2, a cation-selective claudin, is replaced with the corresponding domain in claudin-4, an anion-selective claudin, the paracellular channel made of claudin-2 becomes anion selective, and vice versa (Fig. 1.3.1) (Colegio, Van Itallie, Rahner, & Anderson, 2003). The charged amino acid residues in the ECL1 of claudin may create a favorable electrostatic environment to facilitate the permeation of ions. According to the crystal structure of claudin, the fourth β-strand in ECL1 faces the extracellular space and contains the charged amino acids that can act as a selectivity filter to confer ion selectivity to the paracellular channel (Suzuki et al., 2014). Sequence comparison of the putative selectivity filter in claudins reveals a conserved locus, that is, D65 in claudin-2 and K65 in claudin-4 (Fig. 1.3.2). The electric

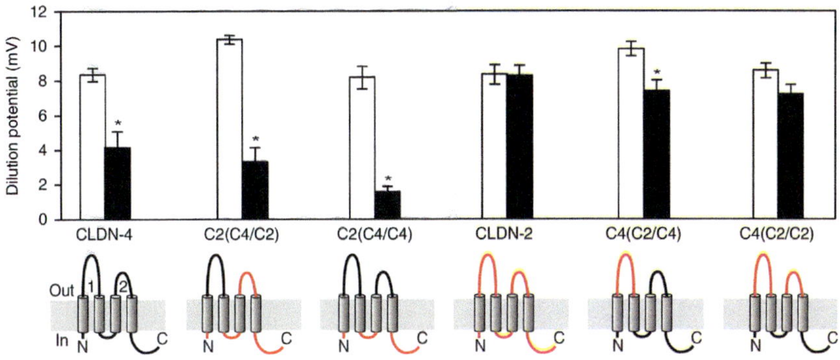

FIGURE 1.3.1 Extracellular domain in claudin determines paracellular ion selectivity. Lower panel: domain swapping of claudin extracellular loops (*black segments*, claudin-4; *red segments*, claudin-2). Upper panel: dilution potentials compared between the MDCK-II cells expressing (*filled bars*) and not expressing (*open bars*) the chimeric claudin protein. More positive dilution potential indicates higher cation selectivity. *, $P < .05$ compared to nonexpressing cells. *Reproduced with permission from Hou, J. (2018).* The paracellular channel – biology, physiology and disease. *Academic Press.*

```
                                                    Fourth β-strand
Claudin-1       27 LPQWKIYSYAG--DNIVTAQAIYEGLWMSCVSQS-TGQIQCKVFDSLLN-LNS 75
Claudin-2       27 LPNWRTSSYVG--ASIVTAVGFSKGLWMECATHS-TGITQCDIYSTLLG-LPA 75
Claudin-3       26 LPMWRVSAFIG--SSIITAQITWEGLWMNCVVQS-TGQMQCKMYDSLLA-LPQ 74
Claudin-4       27 LPMWRVTAFIG--SNIVTAQTSWEGLWMNCVVQS-TGQMQCKMYDSMLA-LPQ 75
Claudin-5       27 LPMWQVTAFLD--HNIVTAQTTWKGLWMSCVVQS-TGHMQCKVYESVLA-LSA 75
Claudin-6       27 LPMWKVTAFIG--NSIVVAQMVWEGLWMSCVVQS-TGQMQCKVYDSLLA-LPQ 75
Claudin-7       27 IPQWQMSSYAG--DNIITAQAMYKGLWMECVTQS-TGMMSCKMYDSVLA-LPG 75
Claudin-8       27 MPQWRVSAFIE--SNIVVFENRWEGLWMNCMRHA-NIRMQCKVYDSLLA-LSP 75
Claudin-9       27 LPLWKVTAFIG--NSIVVAQVVWEGLWMSCVVQS-TGQMQCKVYDSLLA-LPQ 75
Claudin-10a     25 SNEWKVTTR-A--SSVITATWVYQGLWMNCAGNA-LGSFHCRPHFTIFK-VEG 72
Claudin-10b     27 TDYWKVSTI-D--GTVITTATYFANLWKICVTDS-TGVANCKEFPSMLA-LDG 74
Claudin-11      27 TNDWVVTCSYTIPTCRKMDELGSKGLWADCVM-A-TGLYHCKPLVDILI-LPG 76
Claudin-12      30 LPNWRKLRLITF-NRNEKNLTIYTGLWVKCARY--DGSSDCLMYDRTWY-LSV 78
Claudin-13      27 LPVWRVT-FPD--DETDPDATIWEGLWHICQVRE-NRWIQCTLYDTRIL-VAQ 74
Claudin-14      27 LPHWRRTAHVG--TNILTAVSYLKGLWMECVWHS-TGIYQCQIYRSLLA-LPR 75
Claudin-15      26 NSYWRVSTV-H--GNVITTNTIFENLWYSCATDS-LGVSNCWDFPSMLA-LSG 73
Claudin-16      26 TDCWMVNAD-----DSLEVSTKCRGLWWECVTNAFDGIRTCDEYDSIYAEHPL 73
Claudin-17      27 LPQWRVSAFIG--SNIIIFERIWEGLWMNCIQQA-MVTLQCKFYNSILA-LPP 75
Claudin-18      27 MDMWSTQDL-Y--DNPVTAVFQYEGLWRSCVQQS-SGFTECRPYFTILG-LPA 74
Claudin-19      27 LPQWKQSSYAG--DAIITAVGLYEGLWMSCASQS-TGQVQCKLYDSLLA-LDG 75
Claudin-20      27 LPNWKVNAYAG--PNIVTAVVQVQGLWVDCTWYS-TGMFSCTLKYSILS-LPV 75
Claudin-21      30 LPQWKTLTLDL--N---EMETWVSGLWEACVNQE-EAGTVCKAFESFLS-LPQ 75
                      *                    .**  *          *
```

FIGURE 1.3.2 Alignment of the first extracellular loop domain in claudin. The mouse claudin sequences are shown here. Negatively charged acidic residues are labeled in red; positively charged basic residues in blue. The amino acid loci making the putative selectivity filter, corresponding to the fourth β-strand in claudin-15 crystal structure, are highlighted in yellow. The claudin-21 gene is also known as the claudin-25 pseudogene in the mouse genome. *Reproduced with permission from Hou, J. (2018).* The paracellular channel – biology, physiology and disease. *Academic Press.*

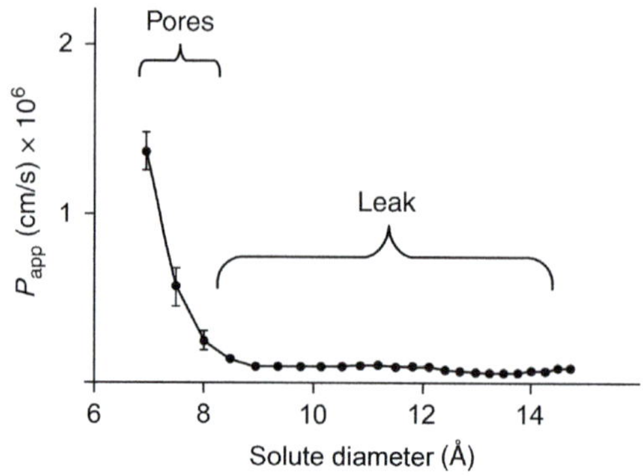

FIGURE 1.3.3 The two-pathway model of paracellular size selectivity. Idealized data reveal the permeability of uncharged polyethylene glycol molecules of increasing sizes across an epithelium. The pathway for molecules of <8 Å in diameter is formed by the claudin channels that are size selective. The pathway for molecules >8 Å in diameter shows no size selectivity but is dependent upon TJ dynamics or through tricellular junction. *Reproduced with permission from Hou, J. (2018).* The paracellular channel – biology, physiology and disease. *Academic Press.*

charge, carried by the amino acid on this locus, is believed to stabilize the permeant ions by electrostatic interaction, according to Eisenman's theory.

1.3.4 Size selectivity of paracellular channel

The size selectivity of paracellular channel is biphasic, consisting of a high-capacity, size-restrictive pathway and a low-capacity, size-independent pathway (Fig. 1.3.3) (Watson, Rowland, & Warhurst, 2001). The high-capacity, size-restrictive pathway behaves as a system of small pores with diameter of <8 Å. Expression of claudin-2 selectively increases the small pore density in tight junction but has no effect on the permeability of tracers that are larger than 8 Å in diameter (Van Itallie et al., 2008). The permeability of uncharged solutes appears dissociated from the permeability of electrolytes in this type of pore, which suggests that charged molecules experience additional electrostatic interactions with the claudin proteins making the pore (Van Itallie et al., 2008). The low-capacity, size-independent pathway allows permeation of molecules with diameter of >8 Å. It could represent a novel pathway through tricellular tight junction (*vide infra*) or transient breaks in bicellular tight junction as a part of the dynamic behaviors of claudin interaction.

1.3.5 Paracellular water channel

Whether paracellular channels permeate water is highly controversial. An optical measurement has found near zero water permeability across the tight junction, or more precisely, the bicellular tight junction

(Kovbasnjuk, Leader, Weinstein, & Spring, 1998). Tricellular tight junction is a new concept and may contain the long-sought paracellular water channel. Regular bicellular tight junctions cannot seal tricellular corners. In fact, the tricellular tight junction consists of two new classes of proteins—tricellulin and angulin. An angulin knockout mouse model develops the phenotypes similar to the hereditary human disease—*diabetes insipidus*, which include polydipsia, polyuria, and renal concentrating defect (Gong et al., 2017). The renal tubular epithelium that is impermeable to water in the wildtype mouse becomes highly permeable to water in the angulin knockout mouse. Molecular analyses reveal normal gene expression profiles and subcellular localization patterns for the claudin proteins making the bicellular tight junction in the knockout mouse kidney. The paracellular permeabilities to Na^+ and Cl^- are not altered in the knockout mouse kidney tubules, which suggests that water permeation is separated from ion permeation in the tight junction (Gong et al., 2017). In the theory, tricellular tight junction can permeate molecules with sizes up to 10 nm in diameter. Water is likely among many uncharged molecules that pass through the tricellular tight junction instead of the bicellular tight junction.

References

Alexandre, M. D., Lu, Q., & Chen, Y. H. (2005). Overexpression of claudin-7 decreases the paracellular Cl- conductance and increases the paracellular Na + conductance in LLC-PK1 cells. *Journal of Cell Science*, *118*, 2683–2693.

Amasheh, S., Meiri, N., Gitter, A. H., Schoneberg, T., Mankertz, J., Schulzke, J. D., & Fromm, M. (2002). Claudin-2 expression induces cation-selective channels in tight junctions of epithelial cells. *Journal of Cell Science*, *115*, 4969–4976.

Angelow, S., El-Husseini, R., Kanzawa, S. A., & Yu, A. S. (2007). Renal localization and function of the tight junction protein, claudin-19. *American Journal of Physiology Renal Physiology*, *293*, F166–F177.

Ben-Yosef, T., Belyantseva, I. A., Saunders, T. L., Hughes, E. D., Kawamoto, K., Van Itallie, C. M., . . . Wilcox, E. R. (2003). Claudin 14 knockout mice, a model for autosomal recessive deafness DFNB29, are deaf due to cochlear hair cell degeneration. *Human Molecular Genetics*, *12*, 2049–2061.

Claude, P. (1978). Morphological factors influencing transepithelial permeability: A model for the resistance of the zonula occludens. *The Journal of Membrane Biology*, *39*, 219–232.

Colegio, O. R., Van Itallie, C., Rahner, C., & Anderson, J. M. (2003). Claudin extracellular domains determine paracellular charge selectivity and resistance but not tight junction fibril architecture. *American Journal of Physiology Cell Physiology*, *284*, C1346–C1354.

Fujita, H., Sugimoto, K., Inatomi, S., Maeda, T., Osanai, M., Uchiyama, Y., . . . Yokozaki, H. (2008). Tight junction proteins claudin-2 and -12 are critical for vitamin D-dependent Ca2 + absorption between enterocytes. *Molecular Biology of the Cell*, *19*, 1912–1921.

Gong, Y., Himmerkus, N., Sunq, A., Milatz, S., Merkel, C., Bleich, M., & Hou, J. (2017). ILDR1 is important for paracellular water transport and urine concentration mechanism.

Proceedings of the National Academy of Sciences of the United States of America, 114, 5271–5276.

Hou, J., Gomes, A. S., Paul, D. L., & Goodenough, D. A. (2006). Study of claudin function by RNA interference. *The Journal of Biological Chemistry, 281*, 36117–36123.

Hou, J., Paul, D. L., & Goodenough, D. A. (2005). Paracellin-1 and the modulation of ion selectivity of tight junctions. *Journal of Cell Science, 118*, 5109–5118.

Hou, J., Renigunta, A., Konrad, M., Gomes, A. S., Schneeberger, E. E., Paul, D. L., … Goodenough, D. A. (2008). Claudin-16 and claudin-19 interact and form a cation-selective tight junction complex. *The Journal of Clinical Investigation, 118*, 619–628.

Hou, J., Renigunta, A., Yang, J., & Waldegger, S. (2010). Claudin-4 forms paracellular chloride channel in the kidney and requires claudin-8 for tight junction localization. *Proceedings of the National Academy of Sciences of the United States of America, 107*, 18010–18015.

Inai, T., Kobayashi, J., & Shibata, Y. (1999). Claudin-1 contributes to the epithelial barrier function in MDCK cells. *European Journal of Cell Biology, 78*, 849–855.

Jovov, B., Van Itallie, C. M., Shaheen, N. J., Carson, J. L., Gambling, T. M., Anderson, J. M., & Orlando, R. C. (2007). Claudin-18: A dominant tight junction protein in Barrett's esophagus and likely contributor to its acid resistance. *American Journal of Physiology Gastrointestinal and Liver Physiology, 293*, G1106–G1113.

Kausalya, P. J., Amasheh, S., Gunzel, D., Wurps, H., Muller, D., Fromm, M., & Hunziker, W. (2006). Disease-associated mutations affect intracellular traffic and paracellular Mg2 + transport function of Claudin-16. *The Journal of Clinical Investigation, 116*, 878–891.

Kovbasnjuk, O., Leader, J. P., Weinstein, A. M., & Spring, K. R. (1998). Water does not flow across the tight junctions of MDCK cell epithelium. *Proceedings of the National Academy of Sciences of the United States of America, 95*, 6526–6530.

Krug, S. M., Gunzel, D., Conrad, M. P., Rosenthal, R., Fromm, A., Amasheh, S., … Fromm, M. (2012). Claudin-17 forms tight junction channels with distinct anion selectivity. *Cellular and Molecular Life Sciences: CMLS, 69*, 2765–2778.

Milatz, S., Krug, S. M., Rosenthal, R., Gunzel, D., Muller, D., Schulzke, J. D., … Fromm, M. (2010). Claudin-3 acts as a sealing component of the tight junction for ions of either charge and uncharged solutes. *Biochimica et Biophysica Acta, 1798*, 2048–2057.

Sas, D., Hu, M., Moe, O. W., & Baum, M. (2008). Effect of claudins 6 and 9 on paracellular permeability in MDCK II cells. *American Journal of Physiology Regulatory, Integrative and Comparative Physiology, 295*, R1713–R1719.

Suzuki, H., Nishizawa, T., Tani, K., Yamazaki, Y., Tamura, A., Ishitani, R., … Fujiyoshi, Y. (2014). Crystal structure of a claudin provides insight into the architecture of tight junctions. *Science (New York, NY), 344*, 304–307.

Tanaka, H., Yamamoto, Y., Kashihara, H., Yamazaki, Y., Tani, K., Fujiyoshi, Y., … Tsukita, S. (2016). Claudin-21 has a paracellular channel role at tight junctions. *Molecular and Cellular Biology, 36*, 954–964.

Van Itallie, C., Rahner, C., & Anderson, J. M. (2001). Regulated expression of claudin-4 decreases paracellular conductance through a selective decrease in sodium permeability. *The Journal of Clinical Investigation, 107*, 1319–1327.

Van Itallie, C. M., Fanning, A. S., & Anderson, J. M. (2003). Reversal of charge selectivity in cation or anion-selective epithelial lines by expression of different claudins. *American Journal of Physiology Renal Physiology, 285*, F1078–F1084.

Van Itallie, C. M., Holmes, J., Bridges, A., Gookin, J. L., Coccaro, M. R., Proctor, W., … Anderson, J. M. (2008). The density of small tight junction pores varies among cell types and is increased by expression of claudin-2. *Journal of Cell Science, 121*, 298–305.

Van Itallie, C. M., Rogan, S., Yu, A., Vidal, L. S., Holmes, J., & Anderson, J. M. (2006). Two splice variants of claudin-10 in the kidney create paracellular pores with different ion selectivities. *American Journal of Physiology Renal Physiology*, *291*, F1288–F1299.

Watson, C. J., Rowland, M., & Warhurst, G. (2001). Functional modeling of tight junctions in intestinal cell monolayers using polyethylene glycol oligomers. *American Journal of Physiology Cell Physiology*, *281*, C388–C397.

Wen, H., Watry, D. D., Marcondes, M. C., & Fox, H. S. (2004). Selective decrease in paracellular conductance of tight junctions: Role of the first extracellular domain of claudin-5. *Molecular and Cellular Biology*, *24*, 8408–8417.

Yu, A. S., Cheng, M. H., Angelow, S., Gunzel, D., Kanzawa, S. A., Schneeberger, E. E., ... Coalson, R. D. (2009). Molecular basis for cation selectivity in claudin-2-based paracellular pores: Identification of an electrostatic interaction site. *The Journal of General Physiology*, *133*, 111–127.

Yu, A. S., Enck, A. H., Lencer, W. I., & Schneeberger, E. E. (2003). Claudin-8 expression in Madin-Darby canine kidney cells augments the paracellular barrier to cation permeation. *The Journal of Biological Chemistry*, *278*, 17350–17359.

Chapter 1.4

Perijunctional cytoskeleton

1.4.1 Actin polymerization

Eukaryotic cells contain three main kinds of cytoskeletal filaments: microfilaments, intermediate filaments, and microtubules. Actin is a family of globular proteins that form the microfilaments. It can be present as either a free monomer known as G-actin (globular) or as part of a linear polymer known as F-actin (filamentous). Actin proteins polymerize to drive the extension of the plasma membrane at the cell leading edge (Pollard & Borisy, 2003). The directionality of the pushing force produced by actin polymerization originates from the structural polarity of actin polymer, in which the barbed end or "plus end" polymerizes faster than the pointed end or "minus end" (Fig. 1.4.1) (Woodrum, Rich, & Pollard, 1975). Inside a cell, actin polymerization is oriented invariably with the barbed end facing the plasma membrane. Depolymerization occurs at the pointed end to release actin monomers for protein recycling.

1.4.2 Actin reorganization

Actin reorganization drives cell junction formation. Before cells make contact, actin filaments form the concentric ring, also known as the "circumferential actin cable," and a dense meshwork between the ring and the plasma membrane (Fig. 1.4.2A) (Gloushankova et al., 1998). As cell contacts are established, cadherin puncta are connected to the actin ring via radial actin bundles (Fig. 1.4.2B) (Adams, Chen, Smith, & Nelson, 1998). Subsequently, radial actin bundles are replaced by parallel contractile bundles underneath the region of contact, which is referred to as the "perijunctional actin belt" (Fig. 1.4.2C) (Hirokawa, Keller, Chasan, & Mooseker, 1983). The mechanisms underlying such actin reorganization include branched actin polymerization and myosin-mediated tension. Polymerization of branched actin networks at the edges of a cell junction generates a pushing force necessary to create cell membrane apposition (Fig. 1.4.3A). The Arp2/3 complex plays a vital role in branched actin polymerization (Michael & Yap, 2013). As cell junction matures, the branched actin networks give way to thick actin bundles known as actin arcs and myosin is enriched on the actin arcs to allow contraction at the edges of the junction. The actin bundles along the junctional zone may break and retract toward the edges, in a process termed as "actin bundle snapping" (Krendel & Bonder, 1999). Actin bundle snapping could be a consequence of increased tension generated by actin arcs.

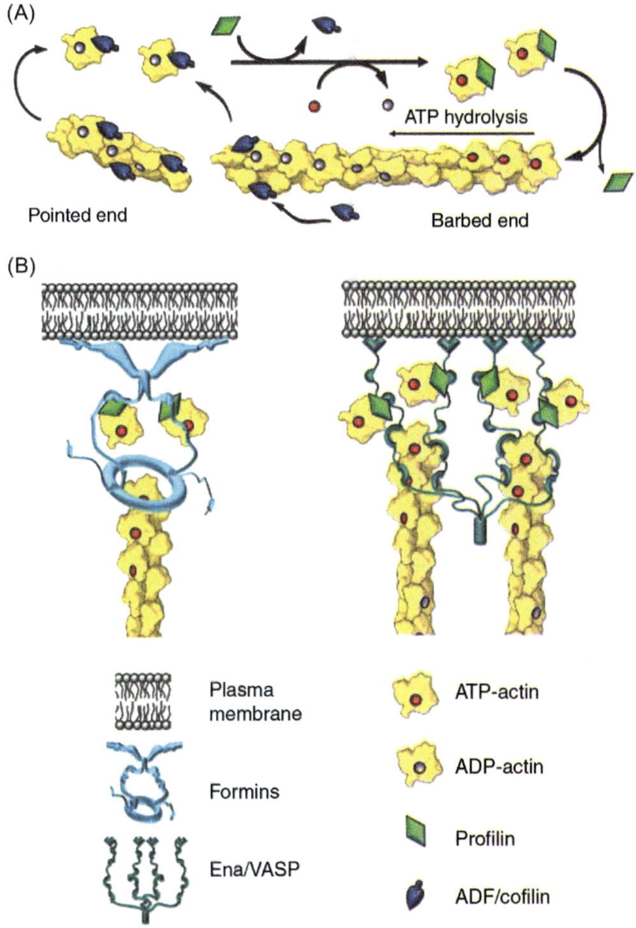

FIGURE 1.4.1 Actin polymerization. (A) Polymerization of the actin filament preferentially occurs at the barbed end from ATP–actin–profilin complexes. After incorporation of an actin monomer into the filament, profilin dissociates, and polymerization-triggered ATP hydrolysis and subsequent release of inorganic phosphate from actin subunits make filaments more susceptible to depolymerization and increase their affinity for actin-depolymerizing factor (ADF)/cofilin. ADF/cofilin severs actin filaments and promotes their depolymerization. Released actin subunits then bind to profilin to produce new ATP–actin–profilin complexes. (B) Formin and Ena/VASP (vasodilator-stimulated phosphoprotein) proteins interact with the barbed ends of actin filaments, promote their elongation by recruiting ATP–actin–profilin complexes, and protect the barbed ends from capping. They also anchor the barbed ends to the plasma membrane. *Reproduced with permission from Svitkina, T. (2018). The actin cytoskeleton and actin-based motility.* Cold Spring Harbor Perspectives in Biology, 10, *a018267.*

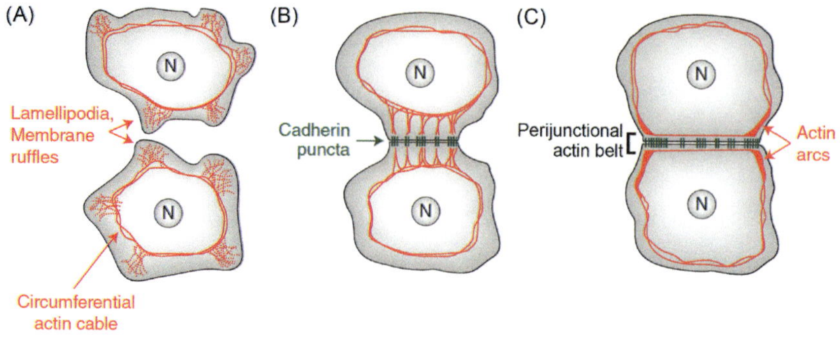

FIGURE 1.4.2 Actin reorganization during cell junction formation. (A) Before epithelial cells make contacts, actin filaments form the circumferential actin cable and dense meshworks between the ring and the membrane protrusions (lamellipodia and membrane ruffles). (B) Cadherin puncta form at the tips of these protrusions and are connected to the circumferential actin cable via radial actin bundles. (C) As cell junctions expand and mature, actin remodeling along the junction results in the formation of perijunctional actin belt and thick bundles known as actin arcs concentrated on the junctional edges. *Reproduced with permission from Cavey, M., & Lecuit, T. (2009). Molecular bases of cell-cell junctions stability and dynamics.* Cold Spring Harbor Perspectives in Biology, 1, *a002998.*

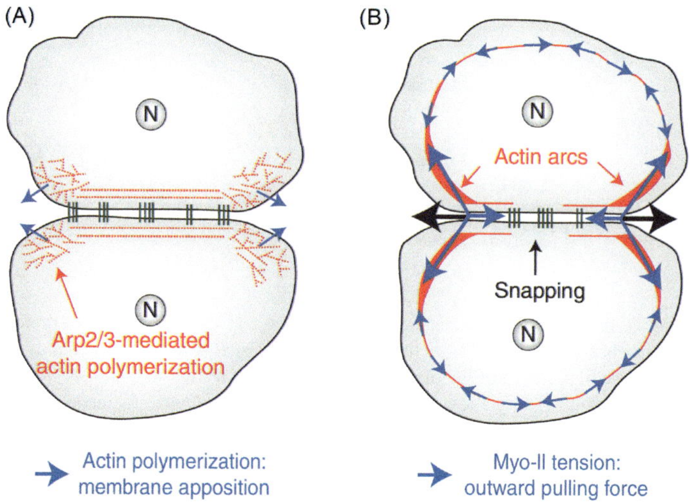

FIGURE 1.4.3 Forces driving actin reorganization during cell junction formation. (A) Polymerization of branched actin networks at the edges of a cell junction generates a pushing force (*blue arrows*) to bring cell membranes into close proximity, which are then ligated by homophilic cadherin dimers. (B) The sum of myosin-mediated tension applied at the junctional edges is initially null. The snapping of actin bundles produces a net outward pulling force (*black arrows*), which drives cell junction expansion. *Reproduced with permission from Cavey, M., & Lecuit, T. (2009). Molecular bases of cell-cell junctions stability and dynamics.* Cold Spring Harbor Perspectives in Biology, 1, *a002998.*

Retraction of these bundles after snapping results in a net outward force pulling the edges to drive cell junction expansion (Fig. 1.4.3B).

1.4.3 Contractile apparatus

Actomyosin constitutes the principal contractile apparatus in nonmuscle cells. Contractility is generated by the process of myosin II sliding actin filaments via their physical interaction once myosin II is activated by phosphorylation of its regulatory light chain (RLC) (Kasza & Zallen, 2011). Myosin II activation entails the assembly of myosin into antiparallel minifilaments so that each contains 10–30 individual myosin molecules. Such antiparallel arrangement allows myosin to slide F-actin inward, thereby generating a contractile force (Fig. 1.4.4). In the best-characterized model, phosphorylation of the RLC promotes myosin assembly and binding to F-actin (Sellers, 1991). RLC phosphorylation is, in turn, determined by the balance between myosin light-chain kinase (MLCK) and myosin light-chain phosphatase (MLCP). MLCK responds to changes in intracellular Ca^{2+} levels to increase the contractility (Sellers, 1991). MLCP decreases the contractility and its own activity is inhibited

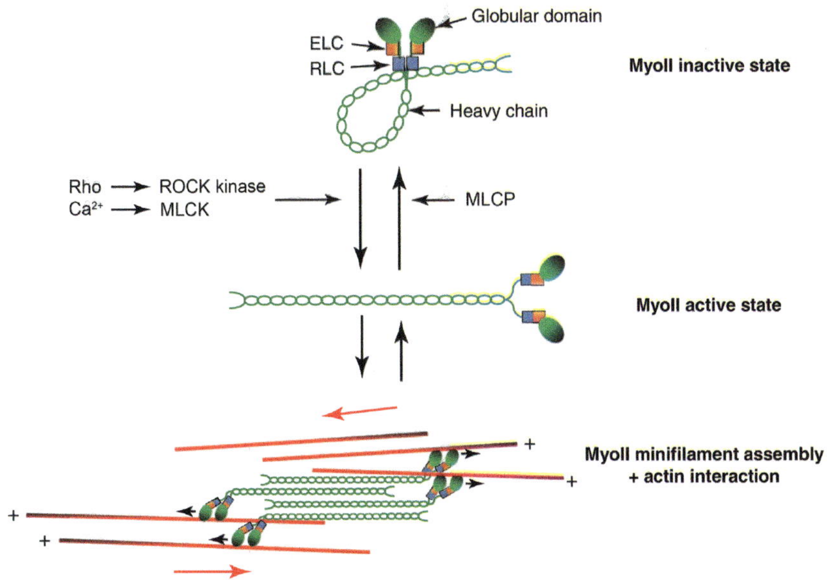

FIGURE 1.4.4 Actomyosin assembly. Myosin II (MyoII) changes its conformation after phosphorylation of its RLC by kinases. Activated MyoII forms antiparallel minifilaments to slide actin filaments inward, thereby generating a contractile force. The red arrows represent the directions of sliding actin filaments. *ELC*, Essential light chain; *RLC*, regulatory light chain. *Reproduced with permission from Levayer, R., & Lecuit, T. (2012). Biomechanical regulation of contractility: spatial control and dynamics.* Trends in Cell Biology, 22, *61–81.*

by the Rho kinase (ROCK) that mediates the effects of the small GTPase, Rho, a major regulator of the contractile apparatus (Kimura et al., 1996).

1.4.4 Mechanosensitive signal transduction

The symmetry of cell junction implies that the contractile force exerted on one cell will be experienced by its neighboring cell. Direct evidence for cadherin as a force sensor comes from experiments using recombinant cadherin ligands to probe the interaction strength with cellular cadherins (Ladoux et al., 2010). Homophilic cadherin interaction is necessary for mechanosensing to occur. The mechanosensitive signals are then conveyed by a complex network of intracellular pathways to regulate the contractile apparatus. Among the pathways, Rho GTPase and α-catenin directly participate in cadherin mechanotransduction. The classic model posits that homophilic interaction triggers cadherin binding to α-catenin, which, in turn, recruits the actin-binding protein vinculin into the cell junction (Watabe-Uchida et al., 1998). Activation of the Rho signaling is vital for cadherin to recruit myosin II to the site of cell adhesion. The abundance level of junctional myosin II and the tension level of actomyosin belt are both reduced when the Rho signaling is blocked (Ratheesh & Yap, 2012). ROCK serves as a second messenger of Rho signaling. When ROCK is inhibited, the myosin recruitment to cell junction is interrupted (Shewan et al., 2005).

1.4.5 Contractility and paracellular permeability

The concept that actomyosin contraction regulates tight junction integrity was originally proposed to explain the effects of cytochalasin treatment, which severs actin filaments, on paracellular permeability across epithelial cell monolayers (Madara, Barenberg, & Carlson, 1986). MLCK, which induces the contraction of actomyosin ring through myosin II regulatory light-chain phosphorylation, has emerged as a key regulator of paracellular permeability. Short, or smooth muscle, MLCK is not expressed in epithelium. Long MLCK, derived from the same gene as short MLCK, uses an upstream promoter to produce different 5′-transcriptional and translational start sites and additional amino-terminal sequences. Two long MLCK isoforms, MLCK1, or full-length long MLCK, and MLCK2, which lacks a single exon within the unique, long MLCK sequence, are expressed by epithelial cells (Kamm & Stull, 2001). Among the MLCK isoforms, MLCK1 is concentrated at the perijunctional actomyosin ring and MLCK1 knockdown reduces the paracelluar permeability in cultured enterocytes (Clayburgh et al., 2004). MLCK can not only induce the reorganization of perijunctional actin belt, but also regulate the dynamic behavior of tight junction proteins. In tight junctions of colonic epithelia, MLCK inhibition stabilizes the ZO-1 protein by reducing its molecular mobility (Yu et al., 2010).

References

Adams, C. L., Chen, Y. T., Smith, S. J., & Nelson, W. J. (1998). Mechanisms of epithelial cell-cell adhesion and cell compaction revealed by high-resolution tracking of E-cadherin-green fluorescent protein. *The Journal of Cell Biology*, *142*, 1105–1119.

Clayburgh, D. R., Rosen, S., Witkowski, E. D., Wang, F., Blair, S., Dudek, S., ... Turner, J. R. (2004). A differentiation-dependent splice variant of myosin light chain kinase, MLCK1, regulates epithelial tight junction permeability. *The Journal of Biological Chemistry*, *279*, 55506–55513.

Gloushankova, N. A., Krendel, M. F., Alieva, N. O., Bonder, E. M., Feder, H. H., Vasiliev, J. M., & Gelfand, I. M. (1998). Dynamics of contacts between lamellae of fibroblasts: Essential role of the actin cytoskeleton. *Proceedings of the National Academy of Sciences of the United States of America*, *95*, 4362–4367.

Hirokawa, N., Keller, T. C., 3rd, Chasan, R., & Mooseker, M. S. (1983). Mechanism of brush border contractility studied by the quick-freeze, deep-etch method. *The Journal of Cell Biology*, *96*, 1325–1336.

Kamm, K. E., & Stull, J. T. (2001). Dedicated myosin light chain kinases with diverse cellular functions. *The Journal of Biological Chemistry*, *276*, 4527–4530.

Kasza, K. E., & Zallen, J. A. (2011). Dynamics and regulation of contractile actin-myosin networks in morphogenesis. *Current Opinion in Cell Biology*, *23*, 30–38.

Kimura, K., Ito, M., Amano, M., Chihara, K., Fukata, Y., Nakafuku, M., ... Okawa, K. (1996). Regulation of myosin phosphatase by Rho and Rho-associated kinase (Rho-kinase). *Science (New York, NY)*, *273*, 245–248.

Krendel, M. F., & Bonder, E. M. (1999). Analysis of actin filament bundle dynamics during contact formation in live epithelial cells. *Cell Motility and the Cytoskeleton*, *43*, 296–309.

Ladoux, B., Anon, E., Lambert, M., Rabodzey, A., Hersen, P., Buguin, A., ... Mege, R. M. (2010). Strength dependence of cadherin-mediated adhesions. *Biophysical Journal*, *98*, 534–542.

Madara, J. L., Barenberg, D., & Carlson, S. (1986). Effects of cytochalasin D on occluding junctions of intestinal absorptive cells: Further evidence that the cytoskeleton may influence paracellular permeability and junctional charge selectivity. *The Journal of Cell Biology*, *102*, 2125–2136.

Michael, M., & Yap, A. S. (2013). The regulation and functional impact of actin assembly at cadherin cell-cell adhesions. *Seminars in Cell and Developmental Biology*, *24*, 298–307.

Pollard, T. D., & Borisy, G. G. (2003). Cellular motility driven by assembly and disassembly of actin filaments. *Cell*, *112*, 453–465.

Ratheesh, A., & Yap, A. S. (2012). A bigger picture: Classical cadherins and the dynamic actin cytoskeleton. *Nature reviews Molecular Cell Biology*, *13*, 673–679.

Sellers, J. R. (1991). Regulation of cytoplasmic and smooth muscle myosin. *Current Opinion in Cell Biology*, *3*, 98–104.

Shewan, A. M., Maddugoda, M., Kraemer, A., Stehbens, S. J., Verma, S., Kovacs, E. M., & Yap, A. S. (2005). Myosin 2 is a key Rho kinase target necessary for the local concentration of E-cadherin at cell-cell contacts. *Molecular Biology of the Cell*, *16*, 4531–4542.

Watabe-Uchida, M., Uchida, N., Imamura, Y., Nagafuchi, A., Fujimoto, K., Uemura, T., ... Takeichi, M. (1998). alpha-Catenin-vinculin interaction functions to organize the apical junctional complex in epithelial cells. *The Journal of Cell Biology*, *142*, 847–857.

Woodrum, D. T., Rich, S. A., & Pollard, T. D. (1975). Evidence for biased bidirectional polymerization of actin filaments using heavy meromyosin prepared by an improved method. *The Journal of Cell Biology*, *67*, 231–237.

Yu, D., Marchiando, A. M., Weber, C. R., Raleigh, D. R., Wang, Y., Shen, L., & Turner, J. R. (2010). MLCK-dependent exchange and actin binding region-dependent anchoring of ZO-1 regulate tight junction barrier function. *Proceedings of the National Academy of Sciences of the United States of America*, *107*, 8237–8241.

Chapter 1.5

Junction signaling

1.5.1 Catenin signaling

The signaling pathway of cell junction that draws immediate attention involves β-catenin (Fig. 1.5.1). β-Catenin binds to the cytoplasmic carboxyl-terminus of cadherin via its central domain (Huber & Weis, 2001). Concomitantly, β-catenin binds to α-catenin via its amino-terminal domain (Pokutta & Weis, 2000). The cadherin–β-catenin complex may sequester α-catenin in adherens junction, which, when released, is capable of lowering Arp2/3 activity, reducing actin branching and cellular protrusion and facilitating contact inhibition (Yamada, Pokutta, Drees, Weis, & Nelson, 2005). β-Catenin also plays a major role in the canonical Wnt signaling pathway (*vide infra*). Cadherin can bind to catenins apart from β-catenin. Among them, plakoglobin (γ-catenin) has been found in the nucleus to activate gene transcription (Garcia-Gras et al., 2006). The p120-catenin subfamily includes p120, and ARVCF-, δ-, and p0071-catenins. Several p120 subfamily members have been shown to either directly or indirectly modulate the small GTPases such as Rac1 and RhoA (Anastasiadis, 2007).

1.5.2 Small GTPases

In epithelial cells, Rac1 and Cdc42, which are members of the Rho family GTPases, localize to the sites of adherens junctions (Kuroda et al., 1998). On the other hand, RhoA, another member of the Rho family GTPases, remains in the cytosol (Takaishi, Sasaki, Kotani, Nishioka, & Takai, 1997). Homophilic cadherin interactions can induce rapid activation of Rac1 and Cdc42 via PI-3 kinase and p120-catenin pathways (Fig. 1.5.1) (Noren, Liu, Burridge, & Kreft, 2000; Pece, Chiariello, Murga, & Gutkind, 1999). RhoA is also activated during the initiation and expansion of epithelial cell junctions (Yamada & Nelson, 2007). The Rho family GTPases can reciprocally regulate cadherin-mediated cell adhesion to provide a mechanism of feedback control. Specifically, Rac1 and Cdc42 regulate E-cadherin activity through direct interaction with the cadherin–catenin complex (Fukata et al., 1999). RhoA affects E-cadherin-mediated adhesion by remodeling the perijunctional cytoskeleton (see Chapter 1.4: Perijunctional cytoskeleton).

1.5.3 Receptor tyrosine kinases

Cadherins and Ig-family adhesion receptors, such as nectins and CAMs, can regulate the receptor tyrosine kinase (RTK) pathways including the pathways

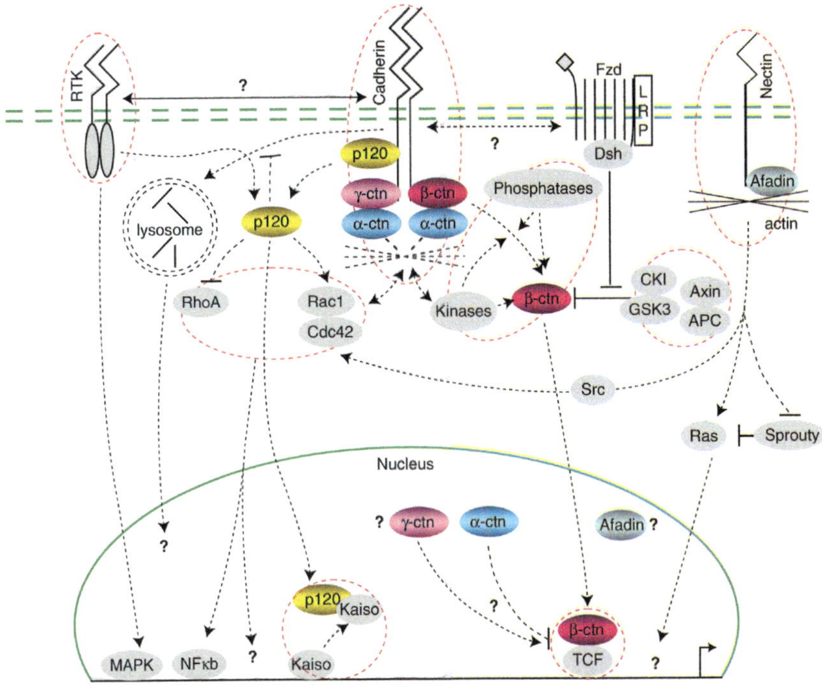

FIGURE 1.5.1 Adherens junction signaling to the nucleus. Proteins bound to the cytoplasmic domains of cadherin include various catenins, with β-catenin further associating with α-catenin and α-catenin with actin microfilaments. The p120-catenin subfamily modulates cadherin lateral clustering, cadherin endocytosis, and activities of Rho family GTPases. β-Catenin can act as a transcriptional coactivator in conjunction with LEF/TCF, whereas p120-catenin relieves Kaiso-mediated gene repression. Signaling from the nectin/afadin complex to the nucleus may occur directly through afadin's interaction with transcriptional factors or indirectly through Ras or Rho. Other pathways known to interact with cadherin include the receptor tyrosine kinase (RTK) pathway and the canonical Wnt pathway comprising the frizzled/LRP/β-catenin complex. *Reproduced with permission from McCrea, P. D., Gu, D., & Balda, M. S. (2009). Junctional music that the nucleus hears: cell-cell contact signaling and the modulation of gene activity.* Cold Spring Harbor Perspectives in Biology, 1, *a002923.*

activated by various growth factors, such as EGF and FGF (Fig. 1.5.1). E-cadherin is found in complex with EGF, FGF, and cErbB2 receptors (Bryant, Wylie, & Stow, 2005; Ochiai et al., 1994; Qian, Karpova, Sheppard, McNally, & Lowy, 2004). In breast cancer cells, E-cadherin adhesion reduces cell proliferation promoted by the EGF ligand (Qian et al., 2004). In endothelial cells, VE-cadherin interacts with VEGF receptor and reduces VEGF receptor activity (Grazia Lampugnani et al., 2003). The cadherin-RTK interaction is believed to dampen RTK signaling by modulation of RTK-ligand interaction, RTK desensitization, and RTK endocytosis.

1.5.4 Canonical Wnt signaling

The protein β-catenin plays a critical role in canonical Wnt signaling. Canonical Wnt signaling includes the following steps. In the absence of Wnt ligands, cytoplasmic β-catenin is recruited into the destruction complex, where it interacts with APC and axin, and is phosphorylated by axin-bound casein kinase 1α (CK1α) and GSK3β. Phosphorylated β-catenin is then targeted for proteasome dependent degradation. In the presence of Wnt ligands, LRP5-LRP6 surface receptors are phosphorylated by GSK3β, and Dishevelled is recruited to the plasma membrane, where it interacts with Frizzled receptors and disassembles the destruction complex. β-Catenin, now released from the destruction complex, translocates to the nucleus to activate gene transcription in conjunction with LEF1 and TCF transcription factors (Fig. 1.5.2) (Klaus & Birchmeier, 2008). Cadherin-mediated cell adhesion competes with Wnt signaling for the same pool of β-catenin proteins. Conceivably, overexpression of cadherins can inhibit canonical Wnt signaling (Heasman et al., 1994). One theory posits that cadherin triggers the relocation of β-catenin from the nucleus to the plasma membrane, which prohibits LEF1-mediated gene transcription (Stockinger, Eger, Wolf, Beug, & Foisner, 2001). β-Catenin appears able to alter its structural conformation to suit different signaling needs. Wnt signaling generates a monomeric, intramolecularly folded form of β-catenin that binds to LEF1/TCF but not cadherins, whereas cadherin-bound β-catenin forms a dimer that can further interact with α-catenin (Gottardi & Gumbiner, 2004).

1.5.5 Hippo signaling

The Hippo pathway in *Drosophila* comprises the Hpo kinase (MST1/2 in mammals), which together with the Sav adaptor protein (SAV1 in mammals), phosphorylates and activates a complex of the Wts kinase (LATS1/2 in mammals) and its cofactor Mats (MOBKL1A/B in mammals). When Hpo or MST1/2 is active, Wts or LATS1/2 binds and phosphorylates the transcriptional coactivator Yki in flies or YAP/TAZ in mammals, causing its inactivation by nuclear exclusion and subsequent proteasomal degradation. When Hpo/MST1/2 is not active, Yki/YAP/TAZ accumulates in the nucleus where it binds to TEAD and other transcription factors and drive the expression of target genes that promote cell proliferation and resist cell apoptosis (Pan, 2007). Cell junctions are major hubs where components of the Hippo pathway are assembled and regulated (Fig. 1.5.3) (Yu, Zhao, & Guan, 2015). Several upstream regulators of the Hippo pathway are localized to tight junction, or integral parts of tight junction, which include the Crb and aPKC polarity complex, the Angiomotin (Amot) family proteins, and the Zonula occludens (ZO) proteins. Among them, ZO-2 directly binds to YAP via its PDZ domain and knockdown of ZO-2 abrogates the nuclear localization of

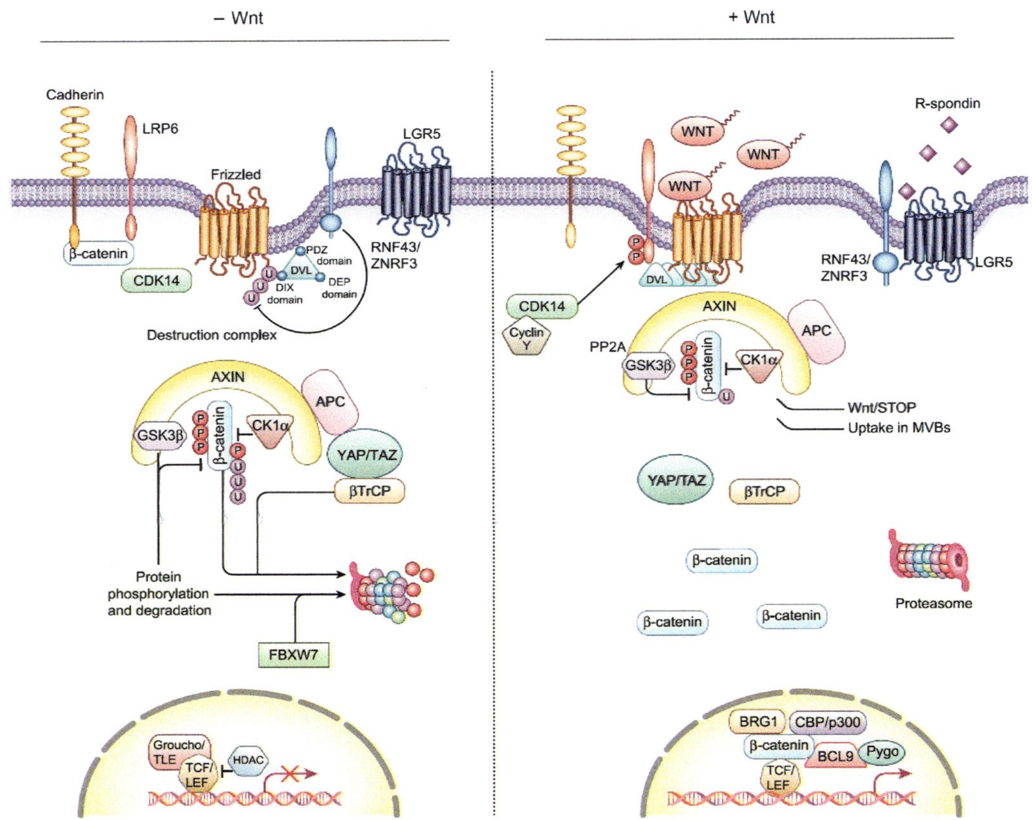

FIGURE 1.5.2 Canonical Wnt pathway. In the absence of a Wnt ligand, β-catenin is phosphorylated in the destruction complex made of axin and APC by the kinases CK1α and GSK3β, which results in the ubiquitination and degradation of β-catenin in the proteasome. In the presence of a Wnt ligand, a Fzd/LRP5/6 complex is formed and bound by Dishevelled, leading to the phosphorylation of LRP5/6 receptors by GSK3β. Phosphorylated LRP5/6 recruits axin to the plasma membrane, which leads to the disassembly of destruction complex and the releasing of β-catenin. β-Catenin is then translocated to the nucleus, a process involving BCL9-2. β-Catenin interacts with TCF/LEF, and the recruitment of cofactors such as BCL9, Pygo, and CBP, as well as the replacement of Groucho, leads to the activation of target genes. *Reproduced with permission from Zhan, T., Rindtorff, N., & Boutros, M. (2017). Wnt signaling in cancer.* Oncogene *36, 1461–1473.*

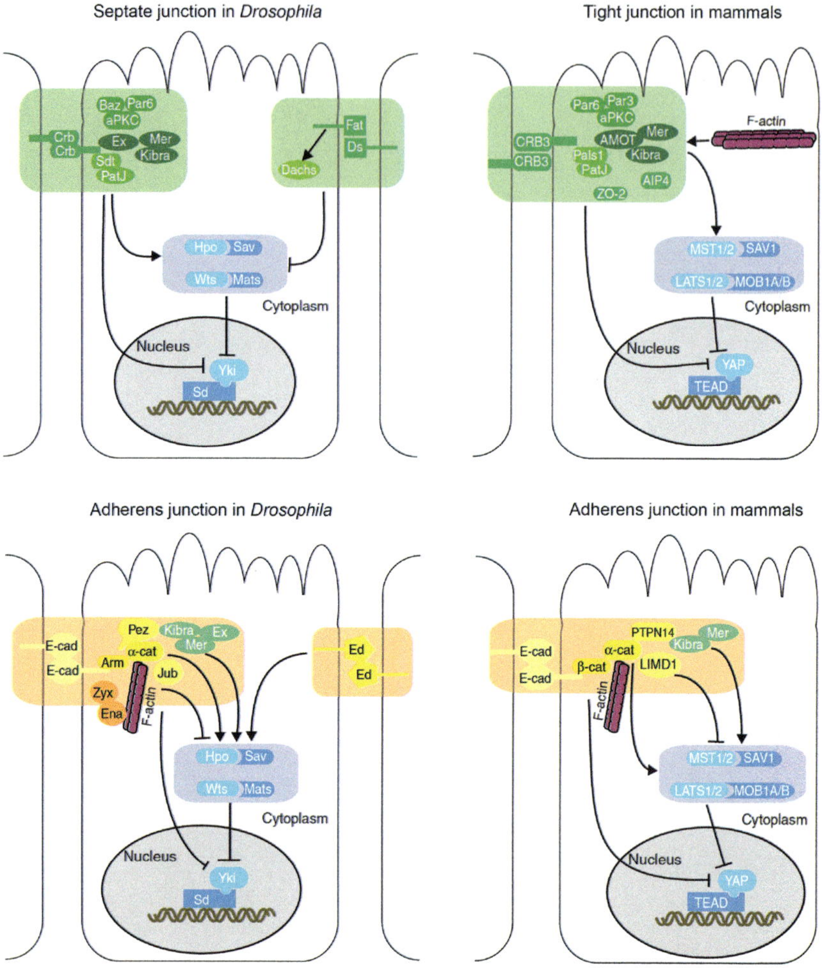

FIGURE 1.5.3 Molecular mechanisms connecting cell junctions to the Hippo pathway. In *Drosophila* septate junction, Crumbs (Crb) and atypical protein kinase C (aPKC) promote Hippo signaling via Mer, Ex, and kidney and brain expressed protein (KIBRA), while Fat and Dachs inhibit the Wts kinase. In mammalian tight junction, the Crb complex acts through Angiomotin (AMOT), which suppresses YAP activity by direct binding and activates LATS2 by recruitment. ZO-2 binds to YAP and sequesters it in the cytoplasm. In *Drosophila* adherens junction, E-cad/α-cat/β-cat complex, Echinoid (Ed), and Pez activate Hippo signaling, while Ajuba (Jub) and Zyxin (Zyx) inhibit it. In mammalian adherens junction, E-cad/α-cat/β-cat complex and PTPN14 promote Hippo signaling activity, while the Ajuba protein LIMD1 inhibits the activity of LATS1/2 kinases. α-Cat and PTPN14 sequester and inhibit YAP by direct protein binding. *Reproduced with permission from Karaman, R., & Halder, G. (2018).* Cell junctions in Hippo signaling. Cold Spring Harbor Perspectives in Biology, 10, *a028753.*

YAP in epithelial cells (Oka et al., 2010). Proteins making the adherens junction can also participate in the Hippo signaling, which include α-catenin, Merlin (NF2), and the LIM-domain containing proteins Ajuba and Zyxin (Gumbiner & Kim, 2014). Different from tight junctions, adherens junctions do not simply provide inert sites for the assembly of Hippo pathway, but they can transduce the signals of mechanical stress in an epithelial tissue to the Hippo pathway (Benham-Pyle, Pruitt, & Nelson, 2015). Loss of adherens junction components, such as E-cadherin, α-catenin, or β-catenin, induces nuclear accumulation and activation of YAP in epithelial and endothelial cells (Kim, Koh, Chen, & Gumbiner, 2011). Mechanistically, α-catenin interacts with phosphorylated YAP and prevents YAP from entering the nucleus to turn on growth-related genes (Schlegelmilch et al., 2011).

References

Anastasiadis, P. Z. (2007). p120-ctn: A nexus for contextual signaling via Rho GTPases. *Biochim Biophys Acta, 1773*, 34–46.

Benham-Pyle, B. W., Pruitt, B. L., & Nelson, W. J. (2015). Cell adhesion. Mechanical strain induces E-cadherin-dependent Yap1 and beta-catenin activation to drive cell cycle entry. *Science (New York, NY), 348*, 1024–1027.

Bryant, D. M., Wylie, F. G., & Stow, J. L. (2005). Regulation of endocytosis, nuclear translocation, and signaling of fibroblast growth factor receptor 1 by E-cadherin. *Molecular Biology of the Cell, 16*, 14–23.

Fukata, M., Kuroda, S., Nakagawa, M., Kawajiri, A., Itoh, N., Shoji, I., ... Kikuchi, A. (1999). Cdc42 and Rac1 regulate the interaction of IQGAP1 with beta-catenin. *The Journal of Biological Chemistry, 274*, 26044–26050.

Garcia-Gras, E., Lombardi, R., Giocondo, M. J., Willerson, J. T., Schneider, M. D., Khoury, D. S., & Marian, A. J. (2006). Suppression of canonical Wnt/beta-catenin signaling by nuclear plakoglobin recapitulates phenotype of arrhythmogenic right ventricular cardiomyopathy. *Journal of Clinical Investigation, 116*, 2012–2021.

Gottardi, C. J., & Gumbiner, B. M. (2004). Distinct molecular forms of beta-catenin are targeted to adhesive or transcriptional complexes. *The Journal of Cell Biology, 167*, 339–349.

Grazia Lampugnani, M., Zanetti, A., Corada, M., Takahashi, T., Balconi, G., Breviario, F., ... Daniel, T. O. (2003). Contact inhibition of VEGF-induced proliferation requires vascular endothelial cadherin, beta-catenin, and the phosphatase DEP-1/CD148. *The Journal of Cell Biology, 161*, 793–804.

Gumbiner, B. M., & Kim, N. G. (2014). The Hippo-YAP signaling pathway and contact inhibition of growth. *Journal of Cell Science, 127*, 709–717.

Heasman, J., Crawford, A., Goldstone, K., Garner-Hamrick, P., Gumbiner, B., McCrea, P., ... Wylie, C. (1994). Overexpression of cadherins and underexpression of beta-catenin inhibit dorsal mesoderm induction in early *Xenopus* embryos. *Cell, 79*, 791–803.

Huber, A. H., & Weis, W. I. (2001). The structure of the beta-catenin/E-cadherin complex and the molecular basis of diverse ligand recognition by beta-catenin. *Cell, 105*, 391–402.

Kim, N. G., Koh, E., Chen, X., & Gumbiner, B. M. (2011). E-cadherin mediates contact inhibition of proliferation through Hippo signaling-pathway components. *Proceedings of the National Academy of Sciences of the United States of America, 108*, 11930–11935.

Klaus, A., & Birchmeier, W. (2008). Wnt signalling and its impact on development and cancer. *Nature Reviews Cancer, 8*, 387–398.

Kuroda, S., Fukata, M., Nakagawa, M., Fujii, K., Nakamura, T., Ookubo, T., ... Tani, H. (1998). Role of IQGAP1, a target of the small GTPases Cdc42 and Rac1, in regulation of E-cadherin-mediated cell-cell adhesion. *Science (New York, NY), 281*, 832–835.

Noren, N. K., Liu, B. P., Burridge, K., & Kreft, B. (2000). p120 catenin regulates the actin cytoskeleton via Rho family GTPases. *The Journal of Cell Biology, 150*, 567–580.

Ochiai, A., Akimoto, S., Kanai, Y., Shibata, T., Oyama, T., & Hirohashi, S. (1994). c-erbB-2 gene product associates with catenins in human cancer cells. *Biochemical and Biophysical Research Communications, 205*, 73–78.

Oka, T., Remue, E., Meerschaert, K., Vanloo, B., Boucherie, C., Gfeller, D., ... Gettemans, J. (2010). Functional complexes between YAP2 and ZO-2 are PDZ domain-dependent, and regulate YAP2 nuclear localization and signalling. *Biochemical Journal, 432*, 461–472.

Pan, D. (2007). Hippo signaling in organ size control. *Genes and Development, 21*, 886–897.

Pece, S., Chiariello, M., Murga, C., & Gutkind, J. S. (1999). Activation of the protein kinase Akt/PKB by the formation of E-cadherin-mediated cell-cell junctions. Evidence for the association of phosphatidylinositol 3-kinase with the E-cadherin adhesion complex. *The Journal of Biological Chemistry, 274*, 19347–19351.

Pokutta, S., & Weis, W. I. (2000). Structure of the dimerization and beta-catenin-binding region of alpha-catenin. *Molecular Cell, 5*, 533–543.

Qian, X., Karpova, T., Sheppard, A. M., McNally, J., & Lowy, D. R. (2004). E-cadherin-mediated adhesion inhibits ligand-dependent activation of diverse receptor tyrosine kinases. *The EMBO Journal, 23*, 1739–1748.

Schlegelmilch, K., Mohseni, M., Kirak, O., Pruszak, J., Rodriguez, J. R., Zhou, D., ... Brummelkamp, T. R. (2011). Yap1 acts downstream of alpha-catenin to control epidermal proliferation. *Cell, 144*, 782–795.

Stockinger, A., Eger, A., Wolf, J., Beug, H., & Foisner, R. (2001). E-cadherin regulates cell growth by modulating proliferation-dependent beta-catenin transcriptional activity. *The Journal of Cell Biology, 154*, 1185–1196.

Takaishi, K., Sasaki, T., Kotani, H., Nishioka, H., & Takai, Y. (1997). Regulation of cell-cell adhesion by rac and rho small G proteins in MDCK cells. *The Journal of Cell Biology, 139*, 1047–1059.

Yamada, S., & Nelson, W. J. (2007). Localized zones of Rho and Rac activities drive initiation and expansion of epithelial cell-cell adhesion. *The Journal of Cell Biology, 178*, 517–527.

Yamada, S., Pokutta, S., Drees, F., Weis, W. I., & Nelson, W. J. (2005). Deconstructing the cadherin-catenin-actin complex. *Cell, 123*, 889–901.

Yu, F. X., Zhao, B., & Guan, K. L. (2015). Hippo pathway in organ size control, tissue homeostasis, and cancer. *Cell, 163*, 811–828.

Chapter 2

Biochemical approaches for tight junction

Chapter 2.1

Biochemistry of tight junction

2.1.1 Biochemical organization of tight junction

Tight junction (TJ) is a subcellular architecture consisting of both proteins and lipids. TJ appears as a continuous network of "fibrils" or "strands" on freeze fractured membrane replicas under electron microscopy (Fig. 2.1.1). These "fibrils" or "strands" are arranged as linear parallel arrays of intramembrane particles of ~10 nm in diameter (Goodenough & Revel, 1970). It has been proven that these intramembrane particles consist in proteins (vide infra). Less is known about the lipidic nature of TJ. Unlike other areas in the plasma membrane, TJ contains abundant detergent-insoluble glycolipid microdomains (Nusrat et al., 2000).

2.1.2 Tight junction enriched protein fraction

Because TJ is a morphologically and biochemically distinct region of the plasma membrane, it can be in the theory isolated by subcellular fractionation techniques. A bile canaliculus-enriched membrane fraction from the hepatocyte plasma membrane contains a large amount of TJs (Neville, 1960). Biochemical protocols designed to dissolve nonjunctional proteins will in the theory yield a preparation highly enriched in junctional proteins. The first attempt to isolate TJ enriched protein fraction was made by dissolving the bile canaliculus-enriched membrane fraction with an anionic detergent—sodium deoxycholate (DOC) (Stevenson & Goodenough, 1984). The resultant TJ enriched protein fraction, that is, the DOC-insoluble fraction,

A Laboratory Guide to the Tight Junction. DOI: https://doi.org/10.1016/B978-0-12-818647-3.00002-7

FIGURE 2.1.1 Tight junction ultrastructure. Freeze-fracture replica electron micrograph reveals the TJ strands in porcine LLC-PK1 cells. The TJ strands consist of linear parallel arrays of intramembrane particles (*white arrowhead*). Bar: 100 nm. *Reproduced with permission from Gong, Y., Renigunta, V., Zhou, Y., Sunq, A., Wang, J., Yang, J., Renigunta, A., Baker, L. A., & Hou, J. (2015). Biochemical and biophysical analyses of tight junction permeability made of claudin-16 and claudin-19 dimerization.* Molecular Biology of the Cell, 26, *4333–4346.*

was denatured in 1% SDS with boiling and then used to immunize laboratory rats to generate monoclonal antibodies against putative TJ proteins. One antibody recognized a 220-kDa protein in the TJ, later referred to as ZO-1 (Stevenson, Siliciano, Mooseker, & Goodenough, 1986). A nonionic detergent—NP-40 was also used to solubilize the proteins that are not stabilized by strong intercellular interactions in the TJ membrane (Tsukita & Tsukita, 1989). Monoclonal antibodies generated against the NP-40-insoluble fraction led to the discovery of occludin (Furuse et al., 1993).

2.1.3 Tight junction integral protein fraction

The TJ enriched protein fraction contains ZO-1, vinculin and α-actinin, which are the peripheral proteins playing a role in TJ and cytoskeleton interaction (Stevenson et al., 1986; Tsukita & Tsukita, 1989). A more stringent biochemical protocol utilized a denaturant—guanidine-HCl to remove these peripheral proteins, leaving behind the core proteins engaged in intercellular interactions, that is, the TJ integral protein fraction (Furuse, Fujita, Hiiragi, Fujimoto, & Tsukita, 1998). There are only nine proteins in the TJ integral protein fraction, including occludin and claudin (Fig. 2.1.2). Transfection experiments in mouse L fibroblasts that normally lack the TJ have revealed that ectopically expressed claudin proteins can reconstitute the TJ-like

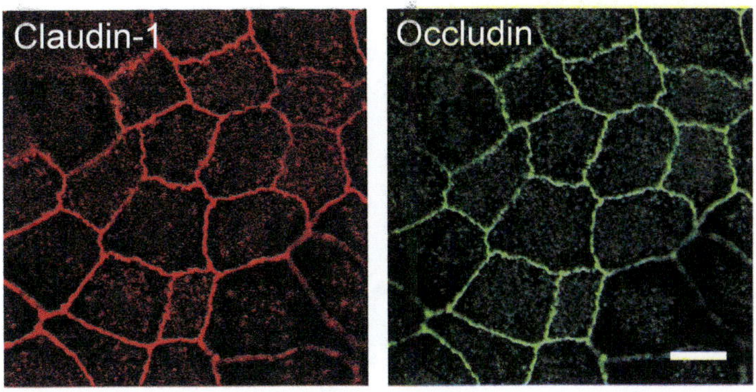

FIGURE 2.1.2 Tight junction localization of occludin and claudin. The TJ proteins—claudin-1 and occludin—were immunostained in the Madin-Darby canine kidney (MDCK) cells to demonstrate their TJ localization. Bar: 10 μm. *Reproduced with permission from Furuse, M., Fujita, K., Hiiragi, T., Fujimoto, K., & Tsukita, S. (1998). Claudin-1 and -2: Novel integral membrane proteins localizing at tight junctions with no sequence similarity to occludin.* The Journal of Cell Biology *141, 1539–1550.*

"fibrils" or "strands" in the cell junctional membrane (Furuse, Sasaki, Fujimoto, & Tsukita, 1998). Although occludin itself cannot reconstitute such TJ-like structures, when introduced into the claudin-expressing L fibroblasts, occludin became incorporated into the claudin-based TJ strands (Furuse, Sasaki et al., 1998).

2.1.4 Molecular structure of claudin protein

Claudins are tetraspan proteins consisting of a family with at least 27 members ranging in molecular mass from 20 to 28 kDa (Hou, 2018). The crystal structures of three claudin proteins have been resolved: claudin-4 (PDB: 5B2G), claudin-15 (PDB: 4P79), and claudin-19 (PDB: 3X29). The transmembrane domains (TM1-4) of claudin-15 form typical left-handed α-helixes, and large portions of the two extracellular loops (ECL1 and ECL2) form a β-sheet structure (Fig. 2.1.3). The β-sheet domain extends from the membrane surface and comprises five β strands (β1–β5), four contributed by ECL1 and one by ECL2. The β-sheet structure is stabilized by a disulfide bond between the cysteine residues at β3 and β4 (Cys52 and Cys62, respectively), which are conserved among all members of the claudin family as part of the consensus motif (-GLWCC). All of the charged residues in ECL1 extend away from the β-sheet surface. Among them, two negatively charged residues, Asp55 and Asp64, are located to the distal edge of the β-sheet domain. Mutagenic studies indicate that these two loci dictate the ion selectivity of the paracellular channel made by claudin-15 (Colegio, Van Itallie, McCrea, Rahner, & Anderson, 2002).

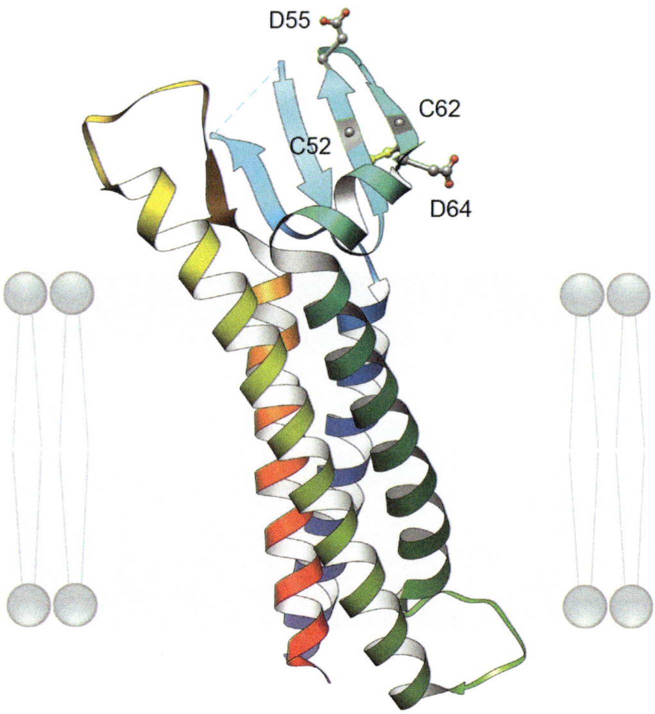

FIGURE 2.1.3 Molecular structure of claudin. Crystal structure of monomeric claudin-15 in ribbon representation when viewed from the side of the membrane. The color changes gradually from the N-terminus (*blue*) to the C-terminus (*red*). A disulfide bond (*yellow*) is made between the cysteine residues C52 and C62. Two important negatively charged residues, D55 and D64, are also labeled. *Reproduced with permission from Hou, J. (2018).* The paracellular channel – biology, physiology and disease. *Academic Press.*

2.1.5 Models of claudin interaction

Conceptually, claudins can interact during TJ formation in a number of ways (Fig. 2.1.4). In Model A, interactions are only permitted with the same type of claudin, both within the plasma membrane of one cell (homomeric, *cis*) and between the plasma membrane of adjacent cells (homotypic, *trans*). In Model B, only homomeric interaction is permitted within a single plasma membrane but mixed (heterotypic, *trans*) interactions are allowed between cells. Model C demonstrates the complementary possibility, where mixed (heteromeric, *cis*) interactions are permitted within a cell but only homotypic interactions permitted between cells. Finally, Model D illustrates a situation where both heteromeric and heterotypic interactions are allowed. The *cis* and *trans* interactions have been observed experimentally for several claudins using yeast two-hybrid assay, coimmunoprecipitation assay, and fluorescence resonance energy transfer assay (Daugherty, Ward, Smith, Ritzenthaler, &

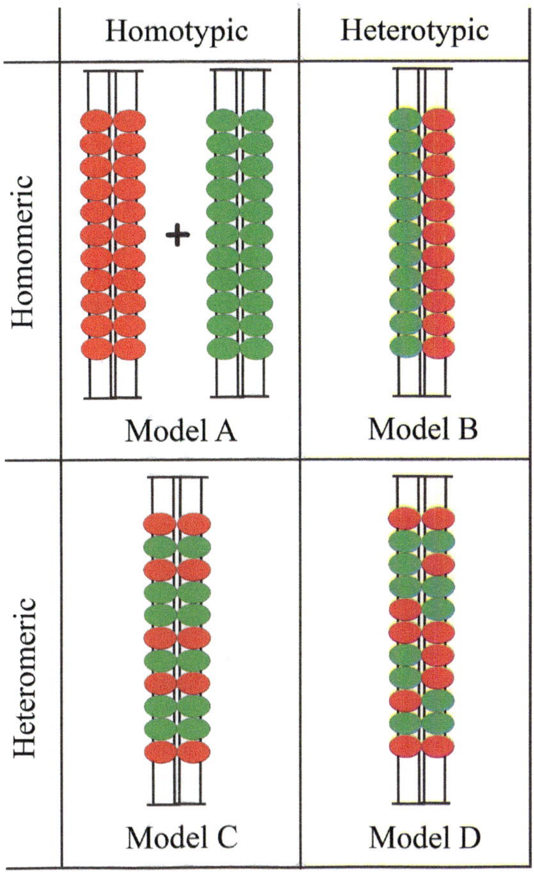

FIGURE 2.1.4 Models of claudin interaction. Four possible models of claudin interaction in each paired TJ strand are illustrated. *Reproduced with permission from Hou, J. (2018).* The paracellular channel – biology, physiology and disease. *Academic Press.*

Koval, 2007; Hou et al., 2008; Rossa et al., 2014). Mutagenic studies have identified multiple amino acid loci in the transmembrane domains and the extracellular loops of claudins required for their *cis* and *trans* interactions (Angelow & Yu, 2009; Daugherty et al., 2007; Gong et al., 2015; Piontek et al., 2008).

2.1.6 Claudin interaction with tight junction plaque proteins

The TJ plaque contains a large number of adaptor proteins that form a protein network via multiple protein interaction motifs to anchor the TJ onto the cytoskeleton. An important group of TJ plaque proteins is characterized by the presence of the so-called PDZ domain. The PDZ domain mediates protein interactions with other PDZ-containing proteins, and with specific sequences, most often ending in Val, at the carboxyl-terminal ends of membrane proteins (Fanning & Anderson, 1996). Intriguingly, all known claudins, with the exception of claudin-12, have a Val at their carboxyl termini. ZO-1 is the first discovered, most abundant, PDZ-containing protein in the TJ plaque. ZO-1 interacts with claudins (Itoh et al., 1999). When the mouse mammary gland tumor Eph4 cells were depleted with ZO-1 and ZO-2 expression, no TJ can form despite normal claudin gene expression (Umeda et al., 2006). Ectopically expressed ZO-1 in ZO protein depleted Eph4 cells recruited claudins to the junctional areas where they were polymerized into TJ strands (Umeda et al., 2006).

2.1.7 Tight junction anchorage onto cytoskeleton

TJ is associated with a highly ordered cytoskeletal structure, referred to as the perijunctional actomyosin ring, which comprises the antiparallel actin filaments and the conventional myosin, myosin II (Drenckhahn & Dermietzel, 1988). This structure has demonstrated contractile activities both in vitro and in vivo (Mooseker et al., 1982). The pharmacologic agents that disrupt the actin filaments in the perijunctional actomyosin ring, such as cytochalasins, can alter the TJ assembly and permeability (Madara, Barenberg, & Carlson, 1986; Meza, Ibarra, Sabanero, Martinez-Palomo, & Cereijido, 1980). In contrast, the microtubule-disrupting agents have shown little effect on TJ structure or function (Gonzalez-Mariscal, Chavez de Ramirez, & Cereijido, 1985). The TJ anchorage onto the perijunctional

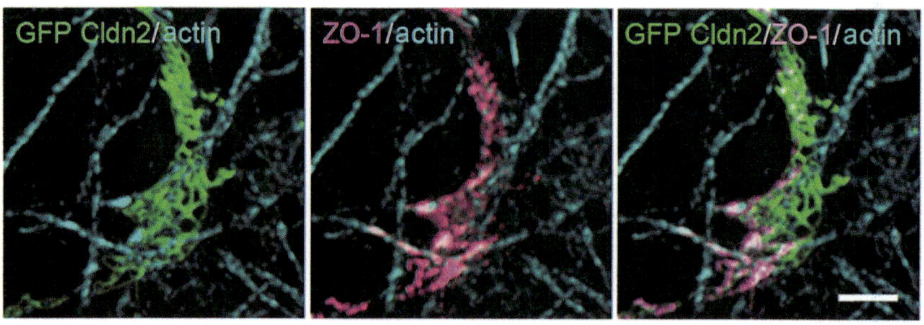

FIGURE 2.1.5 Localization of claudin-2, ZO-1, and actin in Rat-1 fibroblast cells. Superresolution optical micrographs taken with structured illumination microscopy (SIM) reveal that claudin-2 forms a network of strands that are associated with ZO-1. ZO-1 is sandwiched between the claudin strands and the underlying actin filaments. Bar: 1.5 μm. *Reproduced with permission from Van Itallie, C. M., Tietgens, A. J., & Anderson, J. M. (2017). Visualizing the dynamic coupling of claudin strands to the actin cytoskeleton through ZO-1.* Molecular Biology of the Cell, 28, *524–534.*

actomyosin ring is mediated by the TJ plaque proteins including the PDZ-containing proteins such as ZO-1; whereas, the amino-terminal domains in ZO-1 interact with the TJ integral proteins, for example, claudins, the carboxyl-terminal domains in ZO-1 interact with the actin filaments (Itoh, Nagafuchi, Moroi, & Tsukita, 1997). Superresolution optical techniques have allowed visualizing the dynamic coupling of claudin-based TJ strands to the actin cytoskeleton via ZO-1 protein binding (Fig. 2.1.5) (Van Itallie, Tietgens, & Anderson, 2017).

References

Angelow, S., & Yu, A. S. (2009). Structure-function studies of claudin extracellular domains by cysteine-scanning mutagenesis. *The Journal of Biological Chemistry*, *284*, 29205–29217.

Colegio, O. R., Van Itallie, C. M., McCrea, H. J., Rahner, C., & Anderson, J. M. (2002). Claudins create charge-selective channels in the paracellular pathway between epithelial cells. *American Journal of Physiology Cell Physiology*, *283*, C142–147.

Daugherty, B. L., Ward, C., Smith, T., Ritzenthaler, J. D., & Koval, M. (2007). Regulation of heterotypic claudin compatibility. *The Journal of Biological Chemistry*, *282*, 30005–30013.

Drenckhahn, D., & Dermietzel, R. (1988). Organization of the actin filament cytoskeleton in the intestinal brush border: A quantitative and qualitative immunoelectron microscope study. *The Journal of Cell Biology*, *107*, 1037–1048.

Fanning, A. S., & Anderson, J. M. (1996). Protein-protein interactions: PDZ domain networks. *Current Biology: CB*, *6*, 1385–1388.

Furuse, M., Fujita, K., Hiiragi, T., Fujimoto, K., & Tsukita, S. (1998). Claudin-1 and -2: Novel integral membrane proteins localizing at tight junctions with no sequence similarity to occludin. *The Journal of Cell Biology*, *141*, 1539–1550.

Furuse, M., Hirase, T., Itoh, M., Nagafuchi, A., Yonemura, S., Tsukita, S., & Tsukita, S. (1993). Occludin: A novel integral membrane protein localizing at tight junctions. *The Journal of Cell Biology*, *123*, 1777–1788.

Furuse, M., Sasaki, H., Fujimoto, K., & Tsukita, S. (1998). A single gene product, claudin-1 or -2, reconstitutes tight junction strands and recruits occludin in fibroblasts. *The Journal of Cell Biology*, *143*, 391–401.

Gong, Y., Renigunta, V., Zhou, Y., Sunq, A., Wang, J., Yang, J., … Hou, J. (2015). Biochemical and biophysical analyses of tight junction permeability made of claudin-16 and claudin-19 dimerization. *Molecular Biology of the Cell*.

Gonzalez-Mariscal, L., Chavez de Ramirez, B., & Cereijido, M. (1985). Tight junction formation in cultured epithelial cells (MDCK). *The Journal of Membrane Biology*, *86*, 113–125.

Goodenough, D. A., & Revel, J. P. (1970). A fine structural analysis of intercellular junctions in the mouse liver. *The Journal of Cell Biology*, *45*, 272–290.

Hou, J. (2018). *The paracellular channel – Biology, physiology and disease*. Academic Press.

Hou, J., Renigunta, A., Konrad, M., Gomes, A. S., Schneeberger, E. E., Paul, D. L., … Goodenough, D. A. (2008). Claudin-16 and claudin-19 interact and form a cation-selective tight junction complex. *The Journal of Clinical Investigation*, *118*, 619–628.

Itoh, M., Furuse, M., Morita, K., Kubota, K., Saitou, M., & Tsukita, S. (1999). Direct binding of three tight junction-associated MAGUKs, ZO-1, ZO-2, and ZO-3, with the COOH termini of claudins. *The Journal of Cell Biology*, *147*, 1351–1363.

Itoh, M., Nagafuchi, A., Moroi, S., & Tsukita, S. (1997). Involvement of ZO-1 in cadherin-based cell adhesion through its direct binding to alpha catenin and actin filaments. *The Journal of Cell Biology*, *138*, 181–192.

Madara, J. L., Barenberg, D., & Carlson, S. (1986). Effects of cytochalasin D on occluding junctions of intestinal absorptive cells: Further evidence that the cytoskeleton may influence paracellular permeability and junctional charge selectivity. *The Journal of Cell Biology*, *102*, 2125–2136.

Meza, I., Ibarra, G., Sabanero, M., Martinez-Palomo, A., & Cereijido, M. (1980). Occluding junctions and cytoskeletal components in a cultured transporting epithelium. *The Journal of Cell Biology*, *87*, 746–754.

Mooseker, M. S., Bonder, E. M., Grimwade, B. G., Howe, C. L., Keller, T. C., 3rd, Wasserman, R. H., & Wharton, K. A. (1982). Regulation of contractility, cytoskeletal structure, and filament assembly in the brush border of intestinal epithelial cells. *Cold Spring Harbor Symposia on Quantitative Biology*, *46*(Pt 2), 855–870.

Neville, D. M., Jr. (1960). The isolation of a cell membrane fraction from rat liver. *The Journal of Biophysical and Biochemical Cytology*, *8*, 413–422.

Nusrat, A., Parkos, C. A., Verkade, P., Foley, C. S., Liang, T. W., Innis-Whitehouse, W., ... Madara, J. L. (2000). Tight junctions are membrane microdomains. *Journal of Cell Science*, *113*(Pt 10), 1771–1781.

Piontek, J., Winkler, L., Wolburg, H., Muller, S. L., Zuleger, N., Piehl, C., ... Blasig, I. E. (2008). Formation of tight junction: Determinants of homophilic interaction between classic claudins. *Faseb Journal*, *22*, 146–158.

Rossa, J., Protze, J., Kern, C., Piontek, A., Gunzel, D., Krause, G., & Piontek, J. (2014). Molecular and structural transmembrane determinants critical for embedding claudin-5 into tight junctions reveal a distinct four-helix bundle arrangement. *The Biochemical Journal*, *464*, 49–60.

Stevenson, B. R., & Goodenough, D. A. (1984). Zonulae occludentes in junctional complex-enriched fractions from mouse liver: Preliminary morphological and biochemical characterization. *The Journal of Cell Biology*, *98*, 1209–1221.

Stevenson, B. R., Siliciano, J. D., Mooseker, M. S., & Goodenough, D. A. (1986). Identification of ZO-1: A high molecular weight polypeptide associated with the tight junction (zonula occludens) in a variety of epithelia. *The Journal of Cell Biology*, *103*, 755–766.

Tsukita, S., & Tsukita, S. (1989). Isolation of cell-to-cell adherens junctions from rat liver. *The Journal of Cell Biology*, *108*, 31–41.

Umeda, K., Ikenouchi, J., Katahira-Tayama, S., Furuse, K., Sasaki, H., Nakayama, M., ... Tsukita, S. (2006). ZO-1 and ZO-2 independently determine where claudins are polymerized in tight-junction strand formation. *Cell*, *126*, 741–754.

Van Itallie, C. M., Tietgens, A. J., & Anderson, J. M. (2017). Visualizing the dynamic coupling of claudin strands to the actin cytoskeleton through ZO-1. *Molecular Biology of the Cell*, *28*, 524–534.

Chapter 2.2

Tight junction isolation by subcellular fractionation

2.2.1 Background knowledge

2.2.1.1 Subcellular fractionation

Subcellular fractionation is a biological technique to separate different cellular organelles while preserving their individual functions. Fractionation is achieved by differential centrifugation, that is, the sequential increase in gravitational force resulting in the sequential separation of cellular organelles according to their density. TJ is a morphologically and functionally differentiated subcellular architecture composed of a unique set of proteins and lipids. In the theory, TJ is well suited for purification by subcellular fractionation. Mammalian liver is an advantageous starting material for TJ isolation because of the large amount of TJs present in the bile canaliculus membrane (Goodenough & Revel, 1970). Unlike many other organs, the liver is made up of a predominant type of cell—the hepatocyte, which reduces the complexity in the biochemically distinct subtypes of TJs contributed by different types of cells in an organ. The use of liver tissue is not without any disadvantage. First, the liver contains a large amount of extracellular matrix, especially in older animals. Second, the liver contains a large number of red blood cells, which need to be lysed prior to the subcellular fractionation. Finally, the cell junction in the bile canaliculus is a mixture of tight junction and adherens junction (AJ) (Tsukita & Tsukita, 1989). Subcellular fractionation is less likely to separate the TJ from the AJ based upon the rate of sedimentation.

2.2.1.2 Detergent and denaturant extraction

In the theory, the TJ architecture is stabilized by multiple protein interactions both within the plasma membrane of a cell and between the plasma membranes of neighboring cells. A successful TJ isolation protocol may exploit this feature of TJ by selectively removing the nonjunctional proteins that are not part of the TJ interactome. Detergents can dissolve most nonjunctional proteins, leaving behind an insoluble fraction enriched with TJs (Fig. 2.2.1) (Furuse et al., 1993; Stevenson & Goodenough, 1984). Denaturants such as guanidine and urea break protein interactions in the TJ. When used at right concentrations, denaturants can strip the associated cytoskeletal proteins from the TJ (Furuse, Fujita, Hiiragi, Fujimoto, & Tsukita, 1998).

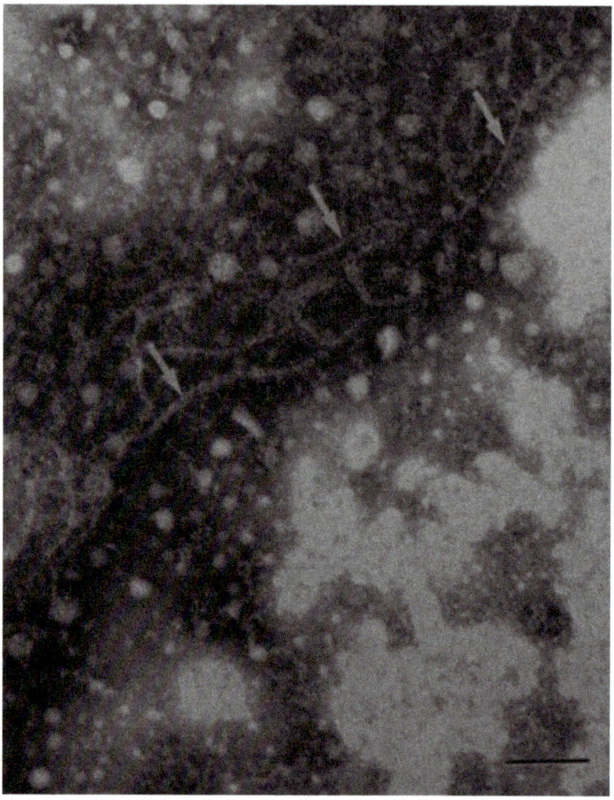

FIGURE 2.2.1 Tight junction in detergent extracted bile canaliculus membrane. The bile canaliculus fraction from the mouse liver was treated with 0.5% DOC, negatively stained with 1% Phosphotungstic acid and visualized by electron microscopy. TJs can be seen as stain-excluding fibrils embedded in the amorphous cytoplasmic dense layer. Bar: 100 nm. *Reproduced with permission from Stevenson, B. R., & Goodenough, D. A. (1984). Zonulae occludentes in junctional complex-enriched fractions from mouse liver: Preliminary morphological and biochemical characterization.* The Journal of Cell Biology, 98, *1209–1221.*

2.2.2 Materials and reagents

2.2.2.1 Animals

Strain of mouse: C57BL/6; age of mouse: 8-week old; gender of mouse: male.

2.2.2.2 Homogenizer

Dounce tissue homogenizer: 10 mL size.

2.2.2.3 Centrifuge

Beckman Ultracentrifuge Optima L-100XP.
Beckman SW40 Ti swinging-bucket rotor.

2.2.2.4 Buffers

Bicarbonate buffer (1 mM)

1 mM	$NaHCO_3$
1 mM	EGTA
1 mM	Phenylmethanesulfonyl fluoride (PMSF)

Bring to 1 L with dH_2O, stir to dissolve, and adjust pH to 8.0.
Filter, sterilize, and store @4°C.
25% Sucrose in bicarbonate buffer

25 g	Sucrose

Bring to 100 mL with bicarbonate buffer (1 mM) and stir to dissolve.
Filter, sterilize, and store @4°C.
41% Sucrose in bicarbonate buffer

41 g	Sucrose

Bring to 100 mL with bicarbonate buffer (1 mM) and stir to dissolve.
Filter, sterilize, and store @4°C.
50% Sucrose in bicarbonate buffer

50 g	Sucrose

Bring to 100 mL with bicarbonate buffer (1 mM) and stir to dissolve.
Filter, sterilize, and store @4°C
67% Sucrose in bicarbonate buffer

67 g	Sucrose

Bring to 100 mL with bicarbonate buffer (1 mM) and stir to dissolve.
Filter, sterilize, and store @4°C.
Extraction buffer with no detergent

100 mM	KCl
1 mM	EGTA
10 mM	HEPES

Bring to 1 L with dH_2O, stir to dissolve and adjust pH to 7.5.
Filter, sterilize, and store @4°C.
Extraction buffer with 1% sodium deoxycholate

100 mM	KCl
1 mM	EGTA
10 mM	HEPES
1% (w/v)	Sodium deoxycholate

Bring to 1 L with dH_2O, stir to dissolve, and adjust pH to 7.5.
Filter, sterilize, and store @4°C.
Extraction buffer with 6 M guanidine-HCl

100 mM	KCl
1 mM	EGTA
10 mM	HEPES
6 M	Guanidine-HCl

Bring to 1 L with dH_2O, stir to dissolve, and adjust pH to 7.5.
Filter, sterilize, and store @4°C.

2.2.3 Experimental procedure

1. Sacrifice the animals by cervical dislocation and dissect the livers immediately.
2. Homogenize 5 g of liver tissues with a 10-mL Dounce tissue homogenizer in 2 mL of ice-cold bicarbonate buffer.
3. Add the homogenate to 8 mL of bicarbonate buffer and stir at 4°C for 30 minutes.
4. Centrifuge the homogenate at $5000 \times g$ for 10 minutes at 4°C.
5. Discard the supernatant and carefully aspirate the white fatty material floating on the surface.
6. Gently resuspend the soft upper pellet (*liver membrane fraction*) with 1 mL of ice-cold bicarbonate buffer and transfer to a new tube. Discard the white hard lower pellet (*liver nuclear fraction*).
7. Dilute the suspension with bicarbonate buffer to 10 mL and centrifuge at $10{,}000 \times g$ for 10 minutes at 4°C.
8. Discard the supernatant and resuspend the pellet with 0.5 mL of ice-cold bicarbonate buffer.
9. Add 1.5 mL of 67% sucrose in bicarbonate buffer dropwise to the suspension. Mix by inverting 10 times.
10. Load the sample onto a 25%–41%–50% sucrose step gradient and centrifuge at $76{,}000 \times g$ for 2 hours at 4°C with Beckman SW40 Ti swinging-bucket rotor.
11. Collect the 25–41 interface and dilute it with bicarbonate buffer to 10 mL and centrifuge at $20{,}000 \times g$ for 10 minutes at 4°C.
12. Discard the supernatant and resuspend the pellet with 1 mL of ice-cold extraction buffer with no detergent (*bile canaliculus fraction*).
13. Add 1 mL of extraction buffer with 1% sodium deoxycholate (DOC) dropwise to the bile canaliculus fraction (the final concentration of DOC = 0.5%). Mix by stirring for 30 minutes at 4°C.
14. Centrifuge the detergent extracted bile canaliculus fraction at $50{,}000 \times g$ for 30 minutes at 4°C.

15. Discard the supernatant and resuspend the pellet with 1 mL of ice-cold extraction buffer with no detergent (*TJ enriched protein fraction*).
16. Add 2 mL of extraction buffer with 6 M guanidine-HCl dropwise to the TJ enriched protein fraction (the final concentration of guanidine = 4 M). Mix by stirring for 30 minutes at 4°C.
17. Centrifuge the guanidine extracted fraction at 50,000 × *g* for 30 minutes at 4°C.
18. Discard the supernatant and resuspend the pellet with 1 mL of ice-cold extraction buffer with no detergent (*TJ integral protein fraction*).

2.2.4 Data analysis

The protein composition in each purified fraction, that is, liver membrane fraction, bile canaliculus fraction, TJ enriched protein fraction, and TJ integral protein fraction, can be analyzed by Western blot using antibodies against the protein markers of cytoskeleton, plasma membrane, TJ, and AJ (Fig. 2.2.2). The stepwise enrichment of TJ integral proteins such as occludin and claudins would indicate a success of the biochemical protocol. The concentration of detergent and denaturant in the protocol can be adjusted experimentally to increase or decrease the stringency of the protein extraction condition. The more stringent an extraction condition is, the less protein interaction remains in the purified sample.

2.2.5 Troubleshooting

2.2.5.1 Contamination of intracellular membrane

Centrifugation separates membranous structures principally on the basis of their buoyant density. The density of a membranous structure depends on the protein composition of the membrane. In the theory, the density of the plasma membrane is higher than that of the intracellular membrane, which includes the mitochondrion, the lysosome, the endosome, the Golgi apparatus, and the endoplasmic reticulum. In practice, however, the achievement of a pure fraction of plasma membrane is a barely accessible goal. Each type of intracellular organelle is composed of several populations of structures that vary in size, density, and other properties on which separation relies. The complexity of tissues and cells, together with the structural alterations induced by homogenization and fractionation, further increases the biochemical heterogeneity in the sample material. Using antibodies or chemical binders such as lectins to the glycosylated proteins in the plasma membrane can increase the purity of the isolated membrane fraction (Lee et al., 2008).

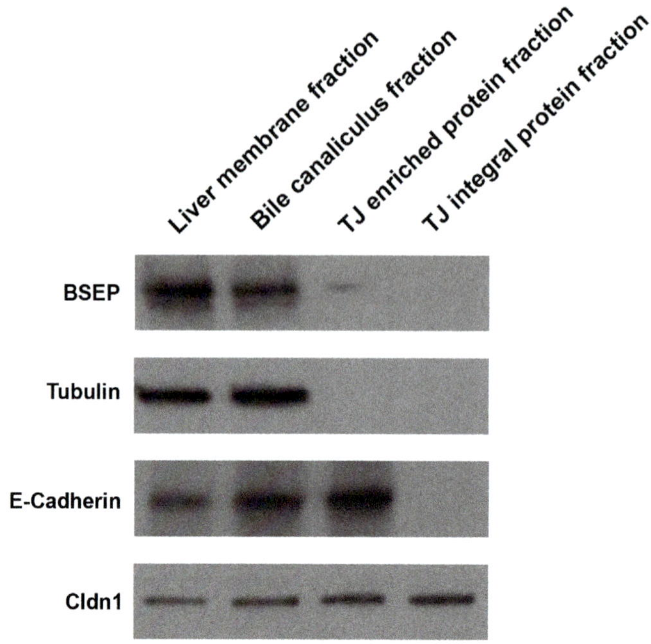

FIGURE 2.2.2 Stepwise enrichment of TJ protein from liver cell membrane. An aliquot of 20 μL from liver membrane fraction, bile canaliculus fraction, TJ enriched protein fraction, and TJ integral protein fraction was fractionated by SDS-PAGE and blotted with antibodies against BSEP (plasma membrane marker), tubulin (cytoskeletal marker), E-Cadherin (AJ marker), and Cldn1 (TJ marker). Note that only TJ integral proteins such as Cldn1 can withstand guanidine extraction.

2.2.5.2 Contamination of lipid-raft microdomains

The lipid rafts (also known as lipid microdomains) are discrete plasma membrane microdomains enriched in cholesterol, glycosphingolipids, and glycosylphosphatidylinositol (GPI)-anchored proteins, and insoluble at low concentrations of nonionic detergents. The TJ resembles the lipid raft in several ways, including high cholesterol content, posttranslational palmitoylation, protein clustering, and detergent insolubility (Francis et al., 1999; Shigetomi, Ono, Inai, & Ikenouchi, 2018; Van Itallie, Gambling, Carson, & Anderson, 2005). From a biochemical point of view, the TJ might be considered as a subclass of lipid-raft membrane microdomains. In the theory, the TJ can withstand stronger detergents than the lipid raft. However, the use of a stringent solubilization condition may break some protein interactions in the TJ. Immunopurification based upon antibodies against TJ-specific proteins might provide a more feasible tool to separate the TJ from the lipid raft.

2.2.5.3 Contamination of adherens junction

Because the TJ and the AJ both mediate cell adhesion, an important part of the biochemical protocol is to ensure that the TJ can be separated from the AJ. Calcium chelation breaks cadherin interactions and destabilizes the AJ (Meng & Takeichi, 2009). The use of calcium chelator such as EDTA or EGTA is, therefore, essential for the purity of isolated TJs.

2.2.5.4 Extrapolation to other organ systems

The biochemical protocol for TJ isolation can be applied to other organ systems, such as the lungs, the kidneys, and the intestines. Compared to the liver, the cell populations in these organs are more heterogeneous. An initial step of cell separation is often needed to enrich the epithelial cells that are abundant with TJs. In vitro cultured epithelial cells are a good source of starting material for TJ isolation. Genetic manipulation and clonal selection can be easily carried out in cell cultures, which ensures that the purified TJs are biochemically identical. Various peptide tags can be introduced to the TJ protein population, which facilitates the immunoprecipitation of TJ from the plasma membrane (Tang, 2006).

2.2.6 Concluding remarks

Subcellular fractionation can provide a crude preparation of tight junctions from various organ and cell models, which allows further biochemical and biophysical analyses. The purified tight junctions contain hundreds to thousands of proteins, many of which have not been cloned. The interactome within the purified tight junction is well preserved by subcellular fractionation, which offers a valuable opportunity to study protein interactions important for tight junction formation. Finally, the purified tight junction can be subjected to various staining assays for light microscopy and electron microscopy. Compared to intact cells or tissues, a purified biochemical fraction is a simpler protein–lipid environment, which might create a unique platform for single-molecule super-resolution imaging studies.

References

Francis, S. A., Kelly, J. M., McCormack, J., Rogers, R. A., Lai, J., Schneeberger, E. E., & Lynch, R. D. (1999). Rapid reduction of MDCK cell cholesterol by methyl-beta-cyclodextrin alters steady state transepithelial electrical resistance. *European Journal of Cell Biology*, *78*, 473–484.

Furuse, M., Fujita, K., Hiiragi, T., Fujimoto, K., & Tsukita, S. (1998). Claudin-1 and -2: Novel integral membrane proteins localizing at tight junctions with no sequence similarity to occludin. *The Journal of Cell Biology*, *141*, 1539–1550.

Furuse, M., Hirase, T., Itoh, M., Nagafuchi, A., Yonemura, S., Tsukita, S., & Tsukita, S. (1993). Occludin: A novel integral membrane protein localizing at tight junctions. *The Journal of Cell Biology*, *123*, 1777–1788.

Goodenough, D. A., & Revel, J. P. (1970). A fine structural analysis of intercellular junctions in the mouse liver. *The Journal of Cell Biology*, *45*, 272–290.

Lee, Y. C., Block, G., Chen, H., Folch-Puy, E., Foronjy, R., Jalili, R., ... Lindemose, S., et al. (2008). One-step isolation of plasma membrane proteins using magnetic beads with immobilized concanavalin A. *Protein Expression and Purification*, *62*, 223–229.

Meng, W., & Takeichi, M. (2009). Adherens junction: Molecular architecture and regulation. *Cold Spring Harbor Perspectives in Biology*, *1*, a002899.

Shigetomi, K., Ono, Y., Inai, T., & Ikenouchi, J. (2018). Adherens junctions influence tight junction formation via changes in membrane lipid composition. *The Journal of Cell Biology*, *217*, 2373–2381.

Stevenson, B. R., & Goodenough, D. A. (1984). Zonulae occludentes in junctional complex-enriched fractions from mouse liver: Preliminary morphological and biochemical characterization. *The Journal of Cell Biology*, *98*, 1209–1221.

Tang, V. W. (2006). Proteomic and bioinformatic analysis of epithelial tight junction reveals an unexpected cluster of synaptic molecules. *Biology Direct*, *1*, 37.

Tsukita, S., & Tsukita, S. (1989). Isolation of cell-to-cell adherens junctions from rat liver. *The Journal of Cell Biology*, *108*, 31–41.

Van Itallie, C. M., Gambling, T. M., Carson, J. L., & Anderson, J. M. (2005). Palmitoylation of claudins is required for efficient tight-junction localization. *Journal of Cell Science*, *118*, 1427–1436.

Chapter 2.3

Immunoprecipitation of *cis* and *trans* claudin interactions

2.3.1 Background knowledge

2.3.1.1 Coimmunoprecipitation

Many of the protein interactions that exist in the cell membrane can be preserved when a cell is lysed under nondenaturing conditions. The immunoprecipitation approach takes advantage of this fact to detect physiologically relevant protein interactions with antibodies. For example, if protein A interacts with protein B, then an antibody against protein A may coprecipitate protein B, and vice versa. Detection of an interaction by this method requires that the protein A–B complex remains intact through a series of lysing and washing steps. Therefore low-affinity or transient interactions might not be observed with this method. Moreover, this approach is only applicable to proteins that can be solubilized from the cell membrane. Large insoluble protein structures are difficult to detect by this approach.

2.3.1.2 Preservation of protein interaction

The TJ in fact is a detergent-resistant microdomain in the cell membrane (Tsukita, Furuse, & Itoh, 2001). Solubilizing the TJ integral proteins such as claudins requires a stringent condition, that is, the use of denaturants such as guanidine or urea (Furuse, Fujita, Hiiragi, Fujimoto, & Tsukita, 1998). These denaturants act by breaking the protein interactions among claudins in the TJ. Therefore, native TJs, from either in vitro cell culture or ex vivo tissue dissection, after being solubilized by denaturants, are not a practical model for studying protein interactions. The insect Sf9 and Sf21 cells, the human embryonic HEK293 cells, or the mouse L fibroblast cells, on the other hand, lack endogenous claudin proteins because they do not form cell junctions (Furuse, Sasaki, & Tsukita, 1999; Gong et al., 2015; Hou et al., 2008). Ectopically expressed claudins can be solubilized from the plasma membrane in these cell models with relatively milder conditions, for example, the use of nonionic detergents such as Triton X-100 to preserve interactions.

2.3.1.3 *Cis* versus *trans* interaction

Claudins can interact in *cis* or *trans* mode (Section 2.1.5). To study *cis* interactions, a pair of claudin proteins are coexpressed in the model cells such as

Sf9, Sf21, HEK293, or L cells. Because these cells make no TJ, the interaction in the plasma membrane is predominantly in *cis*. *Trans* interactions, in the theory, are more difficult to detect in such cell models. Two claudin proteins have to be separately introduced to the cells, which are then cocultured at high density to promote cell–cell contact.

2.3.2 Materials and reagents

2.3.2.1 Cell model for ectopic gene expression

HEK293 cells; available from ATCC.

2.3.2.2 Plasmids

pcDNA3.1 mammalian expression vector (pcDNA3.1-Cldn16; pcDNA3.1-Cldn19); available from Invitrogen.

pcDNA3.1-His mammalian expression vector (pcDNA3.1-His-Cldn4); available from Invitrogen.

pEGFP-C1 mammalian expression vector (pEGFP-Cldn4); available from Clontech.

2.3.2.3 Cell culture medium

Complete medium

500 mL	Dulbecco's Modified Eagle's medium (DMEM) (high glucose/ + sodium pyruvate/ + glutamine); available from Gibco
50 mL	Fetal bovine serum (FBS); available from Gibco
5 mL	Penicillin/streptomycin; available from Gibco

2.3.2.4 Buffers

$CaCl_2$ (2.5 M)

27.7 g	$CaCl_2$

Bring to 100 mL with dH_2O, and stir to dissolve.

Filter, sterilize, and store @4°C.

2 × HBS buffer

140 mM	NaCl
1.5 mM	Na_2HPO_4
50 mM	HEPES

Bring to 100 mL with dH_2O, stir to dissolve, and adjust pH to 7.0.

Filter, sterilize, and store @4°C.

Bicarbonate buffer (1 mM)

1 mM	$NaHCO_3$
1 mM	EGTA
1 mM	Phenylmethanesulfonyl fluoride (PMSF)

Bring to 1 L with dH_2O, stir to dissolve, and adjust pH to 8.0.
Filter, sterilize, and store @4°C.
Extraction buffer with no detergent

100 mM	KCl
1 mM	EGTA
10 mM	HEPES

Bring to 1 L with dH_2O, stir to dissolve, and adjust pH to 7.5.
Filter, sterilize, and store @4°C.
Extraction buffer with 1% Triton X-100

100 mM	KCl
1 mM	EGTA
10 mM	HEPES
1% (v/v)	Triton X-100

Bring to 1 L with dH_2O, stir to dissolve, and adjust pH to 7.5.
Filter, sterilize, and store @4°C.
Extraction buffer with 2% Triton X-100

100 mM	KCl
1 mM	EGTA
10 mM	HEPES
2% (v/v)	Triton X-100

Bring to 1 L with dH_2O, stir to dissolve, and adjust pH to 7.5.
Filter, sterilize, and store @4°C.

2.3.3 Experimental procedure

2.3.3.1 Immunoprecipitation of *cis* claudin interactions

Day 1

1. Grow HEK293 cells to confluence in one T225 flask.
2. Trypsinize the T225 flask, centrifuge the cells at 250 × *g* for 5 minutes at room temperature, and resuspend cells in 5 mL of DMEM complete medium.
3. Prepare four 150 mm culture dishes and add 20 mL of DMEM complete medium to each dish.
4. Add 1 mL resuspended cells to each dish.
5. Return the cells to incubator and culture for overnight (12–16 hours) at 37°C.

Day 2

1. Take the dishes from the incubator and change fresh medium.
   (*Because HEK293 cells are fragile, tilt the flask when removing the medium and adding fresh medium so that the cell monolayer is not disturbed.*)
2. Mix the following DNAs (made w/ Endo-free Qiagen midi-Kits) in a 15 mL centrifuge tube.

20 μg	pcDNA3.1-Cldn16
20 μg	pcDNA3.1-Cldn19
x μL	Nuclease-free H_2O
1.9 mL	Total volume

3. Add 100 μL of 2.5 M $CaCl_2$ to DNA mixture. Vortex to mix.
4. In a separate 15 mL centrifuge tube, add 1 mL of 2 × HBS.
5. Add DNA mixture (Step 2) to 2 × HBS (Step 3) dropwise with a Pasteur pipette. Flick the HBS tube while adding DNA mixture. When finishing adding, vortex gently (using low speed) for 20 seconds.
   (*This is sufficient to transfect one 150 mm dish.*)
6. Let the DNA-HBS mixture stand for 20 minutes at room temperature.
7. Add 2 mL of DNA-HBS mixture dropwise to a 150 mm flask.
   (*When adding the mixture, hold the pipette stable while swirling the flask slowly so that the drops can be distributed evenly in different areas of the flask. After adding the mixture, continue swirling the flask for 20 seconds so that the DNA mixture can fully dissolve into the medium.*)
8. Return the flasks into the incubator for 12–16 hours.

Day 3

1. Take the dishes of transfected HEK293 cells out of the incubator and change fresh medium. Return the flasks into the incubator for 48 hours.

Day 5

1. Aspirate medium from the dishes. Wash cells once with 10 mL of ice-cold 1 × PBS.
2. Add 10 mL of ice-cold bicarbonate buffer to each dish. Brush the cells off the dish and transfer the cells to a 50 mL centrifuge tube.
3. Centrifuge the cells at 250 × *g* for 5 minutes at room temperature. Homogenize 5 g of cells with a 10-mL Dounce tissue homogenizer in 2 mL of ice-cold bicarbonate buffer.
4. Add the homogenate to 8 mL of bicarbonate buffer and stir at 4°C for 30 minutes.
5. Centrifuge the homogenate at 5,000 × *g* for 10 minutes at 4°C.
6. Discard the supernatant and resuspend the soft upper pellet (*the membrane fraction*) with 5 mL of ice-cold bicarbonate buffer and transfer to a new tube. Discard the hard lower pellet (*the nuclear fraction*).

7. Centrifuge the membrane lysate at 5,000 × *g* for 10 minutes at 4°C.
8. Discard the supernatant and resuspend the pellet with 1 mL of ice-cold extraction buffer with no detergent.
9. Add 1 mL of extraction buffer with 2% Triton X-100 dropwise to the membrane lysate (the final concentration of Triton X-100 = 1%). Mix by stirring for 30 minutes at 4°C.
10. Centrifuge the detergent extracted membrane lysate at 50,000 × *g* for 30 minutes at 4°C.
11. Collect the supernatant.
12. Remove nonspecific binding to protein A/G from the membrane lysate.

500 µL	Detergent extracted membrane lysate
100 µL	50% (v/v) protein A/G
5 µL	BSA (100 mg/mL)

13. Rotate at 4°C for 1 hour.
14. Centrifuge the lysate-protein A/G mixture at 50,000 × *g* for 5 minutes at 4°C.
15. Collect the supernatant.
16. Perform immunoprecipitation with antibodies.

150 µL	Cleared membrane lysate
30 µL	50% (v/v) protein A/G
1.5 µL	BSA (100 mg/mL)
15 µL	Antibody (anti-Cldn16, anti-Cldn19, or irrelevant IgG)—5 µg

17. Rotate at 4°C for 3 hours.
18. Centrifuge the lysate-protein A/G mixture at 50,000 × *g* for 5 minutes at 4°C.
19. Wash with 500 µL of extraction buffer with 1% Triton X-100 for three times.
20. Centrifuge the lysate-protein A/G mixture at 50,000 × *g* for 5 minutes at 4°C.
21. Discard the supernatant. Resuspend the pellet with 100 µL Laemmli 2 × buffer and run Western blot with anti-Cldn16 or anti-Cldn19 antibody (Fig. 2.3.1).

2.3.3.2 Immunoprecipitation of *trans* claudin interactions

Day 1

6. Grow HEK293 cells to confluence in one T225 flask.
7. Trypsinize the cells from the T225 flask, centrifuge the cells at 250 × *g* for 5 minutes at room temperature, and resuspend cells in 5 mL of DMEM complete medium.
8. Prepare four 150 mm culture dishes and add 20 mL of DMEM complete medium to each dish.

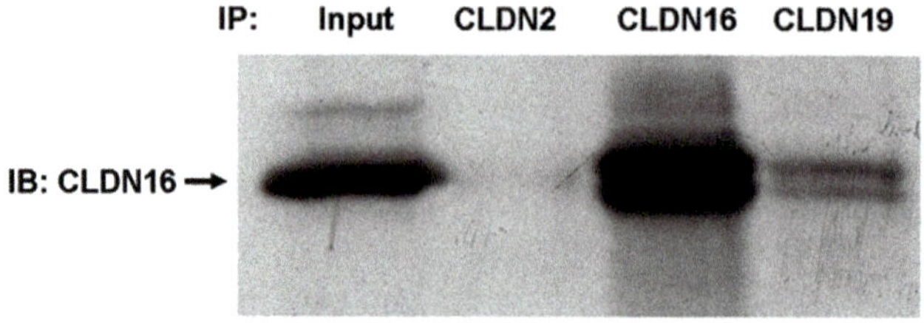

FIGURE 2.3.1 Immunoprecipitation of *cis* claudin interaction. HEK293 cells were cotransfected with claudin-16 and claudin-19 proteins, lysed with a buffer containing 1% Triton X-100, and analyzed by coimmunoprecipitation with antibodies against claudin-16 and claudin-19. Antibodies used for coimmunoprecipitation are shown above the lanes; antibody for blot visualization is shown at left. *Reproduced with permission from Hou, J., Renigunta, A., Konrad, M., Gomes, A. S., Schneeberger, E. E., Paul, D. L., Waldegger, S., & Goodenough, D. A. (2008). Claudin-16 and claudin-19 interact and form a cation-selective tight junction complex.* The Journal of Clinical Investigation, 118, *619–628.*

9. Add 1 mL of resuspended cells to each dish.
10. Return the cells to incubator and culture for overnight (12–16 hours) at 37°C.

Day 2

9. Take the dishes from the incubator and change fresh medium.
   (*Because HEK293 cells are fragile, tilt the flask when removing the medium and adding fresh medium so that the cell monolayer is not disturbed.*)
10. Mix the following DNAs (made w/ Endo-free Qiagen midi-Kits) in a 15 mL centrifuge tube.

20 μg	pcDNA3.1-His-Cldn4 or pEGFP-Cldn4
x μL	Nuclease-free H_2O
1.9 mL	Total volume

11. Add 100 μL of 2.5 M $CaCl_2$ to the DNA mixture. Vortex to mix.
12. In a separate 15 mL centrifuge tube, add 1 mL of 2 × HBS.
13. Add DNA mixture (Step 2) to 2 × HBS (Step 3) dropwise with a Pasteur pipette. Flick the HBS tube while adding the DNA mixture. When finishing adding, vortex gently (using low speed) for 20 seconds.
   (*This is sufficient to transfect one 150-mm dish.*)
14. Let the DNA-HBS mixture stand for 20 minutes at room temperature.
15. Add 2 mL DNA-HBS mixture dropwise to a 150 mm flask.
   (*When adding the mixture, hold the pipette stable while swirling the flask slowly so that the drops can be distributed evenly in different areas of the flask. After adding the mixture, continue swirling the flask for 20 seconds so that the DNA mixture can fully dissolve into the medium.*)
16. Return the flasks into the incubator for 12–16 hours.

Day 3

2. Take the dishes of transfected HEK293 cells out of the incubator and change fresh medium. Return the flasks into the incubator for 48 hours.

Day 4

1. Trypsinize the cells from the 150 mm dish, centrifuge the cells at 250 × *g* for 5 minutes at room temperature, and resuspend cells in 5 mL of DMEM complete medium.
2. Mix 5 mL of His-Cldn4 expression cells with 5 mL of EGFP-Cldn4 expression and load to a 100 mm dish.
   (*Coculture of cells at high density will maximize trans claudin interactions.*)
3. Return the flasks into the incubator for 12–16 hours.

Day 5

22. Aspirate medium from the dishes. Wash cells once with 10 mL ice-cold 1 × PBS.
23. Add 10 mL of ice-cold bicarbonate buffer to each dish. Brush the cells off the dish and transfer the cells to a 50 mL centrifuge tube.
24. Centrifuge the cells at 250 × *g* for 5 minutes at room temperature. Homogenize 5 g of cells with a 10-mL Dounce tissue homogenizer in 2 mL of ice-cold bicarbonate buffer.
25. Add the homogenate to 8 mL of bicarbonate buffer and stir at 4°C for 30 minutes.
26. Centrifuge the homogenate at 5,000 × *g* for 10 minutes at 4°C.
27. Discard the supernatant and resuspend the soft upper pellet (*the membrane fraction*) with 5 mL of ice-cold bicarbonate buffer and transfer to a new tube. Discard the hard lower pellet (*the nuclear fraction*).
28. Centrifuge the membrane lysate at 5,000 × *g* for 10 minutes at 4°C.
29. Discard the supernatant and resuspend the pellet with 1 mL of ice-cold extraction buffer with no detergent.
30. Add 1 mL of extraction buffer with 2% Triton X-100 dropwise to the membrane lysate (the final concentration of Triton X-100 = 1%). Mix by stirring for 30 minutes at 4°C.
31. Centrifuge the detergent extracted membrane lysate at 50,000 × *g* for 30 minutes at 4°C.
32. Collect the supernatant.
33. Remove nonspecific binding to protein A/G from the membrane lysate.

500 μL	Detergent extracted membrane lysate
100 μL	50% (v/v) protein A/G
5 μL	BSA (100 mg/mL)

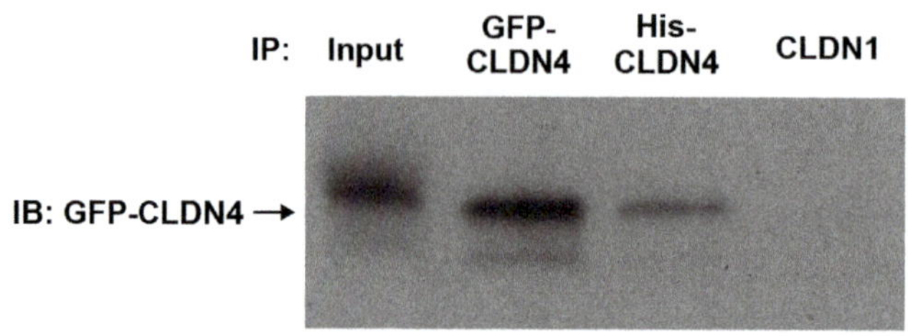

FIGURE 2.3.2 Immunoprecipitation of *trans* claudin interaction. Mouse L fibroblast cells were independently transfected with GFP-cldn4 and His-cldn4 proteins, cocultured, lysed with a buffer containing 1% Triton X-100, and analyzed by coimmunoprecipitation with antibodies against GFP and His tag. Antibodies used for coimmunoprecipitation are shown above the lanes; antibody for blot visualization is shown at left. *Reproduced with permission from Gong, Y., Yu, M., Yang, J., Gonzales, E., Perez, R., Hou, M., Tripathi, P., Hering-Smith, K. S., Hamm, L. L., & Hou, J. (2014). The Cap1-claudin-4 regulatory pathway is important for renal chloride reabsorption and blood pressure regulation.* Proceedings of the National Academy of Sciences of the United States of America, 111, *E3766–E3774.*

34. Rotate at 4°C for 1 hour.
35. Centrifuge the lysate-protein A/G mixture at 50,000 × *g* for 5 minutes at 4°C.
36. Collect the supernatant.
37. Perform immunoprecipitation with antibodies.

150 μL	Cleared membrane lysate
30 μL	50% (v/v) protein A/G
1.5 μL	BSA (100 mg/mL)
15 μL	Antibody (anti-His, anti-GFP or irrelevant IgG)—5 μg

38. Rotate at 4°C for 3 hours.
39. Centrifuge the lysate-protein A/G mixture at 50,000 × *g* for 5 minutes at 4°C.
40. Wash with 500 μL of extraction buffer with 1% Triton X-100 for three times.
41. Centrifuge the lysate-protein A/G mixture at 50,000 × *g* for 5 minutes at 4°C.
42. Discard the supernatant. Resuspend the pellet with 100 μL Laemmli 2 × buffer and run Western blot with anti-His or anti-GFP antibody (Fig. 2.3.2).

2.3.4 Data analysis

If the purpose of the experiment is to test whether two claudin proteins interact, the presence of claudin interactions in the immunoprecipitated protein complex can be detected by Western blot using selective antibodies against each claudin species. If the purpose is to identify novel binding partners for claudins, the immunoprecipitated proteins are usually fractionated by

SDS-PAGE, recovered after silver staining, and analyzed by mass spectrometry to reveal the identity of the proteins.

2.3.5 Troubleshooting

2.3.5.1 Lysis condition

When lysing the cells, it is important to consider that the solubilization conditions do not disrupt the protein interactions that exist in the membrane. In general, a higher salt concentration (200–1000 mM of NaCl) and the presence of an ionic detergent (e.g., 0.1%–1% SDS or sodium deoxycholate) are more disruptive than lower salt concentration (100–150 mM of NaCl) and nonionic detergent (e.g., 0.1%–1% NP-40 or Triton X-100). Mechanical processes such as sonication and reducing agents such as mercaptoethanol also tend to disrupt the protein interactions.

2.3.5.2 Choice of antibody

The antibody used for immunoprecipitation should be well characterized. There are several ways to define the suitability of an antibody to immunoprecipitating purpose. (1) The antibody can immunoprecipitate the protein against which it was raised from crude cell lysates. (2) The antibody has no binding affinity to a cell line that lacks the target protein (the KO cell line). (3) The immunoprecipitated protein can be recognized by different antibodies directed against different epitopes on the target protein. Conversely, some antibodies recognize the epitopes that might be obscured by particular protein interactions. Thus not all antibodies should be expected to immunoprecipitate the target protein.

2.3.5.3 Nonspecific protein interactions

False-positive results arise from the presence of proteins in the washed immunoprecipitates that are not normally associated with the target protein in the cell membrane. There are several approaches to reduce false-positive results due to nonspecific protein interactions. (1) Increase the ionic strength in the washing buffer because higher concentrations of salt decrease the protein binding affinity. In the theory, a genuine protein interaction has a higher affinity than nonspecific protein interactions. (2) Decrease the amount of primary antibody in the immunoprecipitation so that the total number of proteins that precipitate nonspecifically can be reduced. The concentration of the primary antibody should be titrated to a point where the signal obtained from specific protein interactions is maximized relative to nonspecific protein interactions. (3) Preclear the cell lysate with a control antibody that is closed matched to the specific antibody in host species and immunoglobulin isotype (Harlow & Lane, 1988). For a mouse monoclonal antibody, the proper control is another monoclonal antibody of the same isotype; for a rabbit serum, it is the preimmune serum from the same rabbit; and for a rabbit polyclonal antibody, it is an irrelevant rabbit polyclonal antibody.

2.3.6 Concluding remarks

Immunoprecipitation is generally regarded as the most rigorous demonstration of physiological interactions between two proteins. Compared to other approaches such as yeast two-hybrid assay, in vitro pull-down assay, and atomic force microscopy, immunoprecipitation permits interrogating the protein interactions that exist within intact cell niche. Immunoprecipitation can also be used to identify novel proteins that interact with a known target. As TJ is a high-order membrane architecture consisting in over 400 proteins, the delineation of individual interactions from the TJ interactome will be a high priority for TJ research.

References

Furuse, M., Fujita, K., Hiiragi, T., Fujimoto, K., & Tsukita, S. (1998). Claudin-1 and -2: Novel integral membrane proteins localizing at tight junctions with no sequence similarity to occludin. *The Journal of Cell Biology*, *141*, 1539–1550.

Furuse, M., Sasaki, H., & Tsukita, S. (1999). Manner of interaction of heterogeneous claudin species within and between tight junction strands. *The Journal of Cell Biology*, *147*, 891–903.

Gong, Y., Renigunta, V., Zhou, Y., Sunq, A., Wang, J., Yang, J., ... Hou, J. (2015). Biochemical and biophysical analyses of tight junction permeability made of claudin-16 and claudin-19 dimerization. *Molecular Biology of the Cell*.

Harlow, E., & Lane, D. (1988). *A laboratory manual* (579). New York: Cold Spring Harbor Laboratory.

Hou, J., Renigunta, A., Konrad, M., Gomes, A. S., Schneeberger, E. E., Paul, D. L., ... Goodenough, D. A. (2008). Claudin-16 and claudin-19 interact and form a cation-selective tight junction complex. *The Journal of Clinical Investigation*, *118*, 619–628.

Tsukita, S., Furuse, M., & Itoh, M. (2001). Multifunctional strands in tight junctions. *Nature reviews Molecular Cell Biology*, *2*, 285–293.

Chapter 2.4

Isolation of claudin oligomer by chemical cross-linking

2.4.1 Background knowledge

2.4.1.1 Chemical cross-linkers

Cross-linking is the process of chemically joining two or more molecules by a covalent bond. Cross-linkers contain reactive ends to specific functional groups (e.g., primary amine, sulfhydryl, etc.) on proteins or peptides. The availability of these chemical groups in proteins and peptides make them targets for study using cross-linking methods. The popular cross-linkers and their target groups in proteins are summarized in Table 2.4.1. Among the target groups, primary amines ($-NH_2$) are the most commonly used. Primary amines exist at the N-terminus of each polypeptide chain (also known as the α-amine) and in the side chain of lysine (Lys, K) residues (also known as the ε-amine). Because of their positive charge at physiologic conditions, primary amines are usually outward facing (i.e., on the outer surface of protein), making them accessible for conjugation without denaturing the protein structure.

2.4.1.2 Ectopic claudin expression model

It is now accepted that claudins polymerize to form the basic TJ architecture (Tsukita, Furuse, & Itoh, 2001). The polymerization process takes place via *cis* and *trans* claudin interactions (Section 2.1.5). In mammals, there are 27 claudin protein molecules. A claudin protein may interact with the same type of claudin or a different type of claudin. Because the epithelium or the endothelium from either in vitro culture or ex vivo extraction often expresses multiple claudin proteins, the study of selective claudin polymerization is difficult in such a model. The insect Sf9 and Sf21 cells, the human embryonic HEK293 cells, or the mouse L fibroblast cells, on the other hand, lack endogenous claudin proteins because they do not form cell junctions (Furuse, Sasaki, & Tsukita, 1999; Gong et al., 2015; Hou et al., 2008). When ectopically expressed, a selected pair of claudins can be analyzed for oligomeric properties they spontaneously form. Notably, as there is no TJ made by these cells, the oligomer predominantly exists in *cis*.

2.4.2 Materials and reagents

2.4.2.1 Cell model for ectopic gene expression

HEK293 cells; available from ATCC.

TABLE 2.4.1 Popular cross-linker reactive groups and target groups.

Reactive class	Target group	Reactive group
Amine-reactive	$-NH_2$	NHS ester
		Aldehyde
		Imidoester pentafluorophenyl ester Hydroxymethyl phosphine
		Sulfonyl chloride
		Fluorobenzene
		Epoxide
Carboxyl-reactive	$-COOH$	Carbodiimide
Sulfhydryl-reactive	$-SH$	Maleimide
		Haloacetyl
		Pyridyldisulfide
		Thiosulfonate
		Vinylsulfone
Hydroxyl-reactive	$-OH$	Isocyanate
Photoreactive	Random	Diazirine
		Aryl azide

2.4.2.2 Plasmids

pcDNA3.1 mammalian expression vector (pcDNA3.1-Cldn16; pcDNA3.1-Cldn19); available from Invitrogen.

2.4.2.3 Cell culture medium

Complete medium

500 mL	Dulbecco's Modified Eagle's medium (DMEM) (high glucose/ + sodium pyruvate/ + glutamine); available from Gibco
50 mL	Fetal bovine serum (FBS); available from Gibco
5 mL	Penicillin/streptomycin; available from Gibco

2.4.2.4 Buffers

$CaCl_2$ (2.5 M)

27.7 g $CaCl_2$

Bring to 100 mL with dH_2O and stir to dissolve.
Filter, sterilize, and store @4°C.
2 × HBS buffer

140 mM	NaCl
1.5 mM	Na_2HPO_4
50 mM	HEPES

Bring to 100 mL with dH_2O, stir to dissolve, and adjust pH to 7.0.
Filter, sterilize, and store @4°C.
Bicarbonate buffer (1 mM)

1 mM	$NaHCO_3$
1 mM	EGTA
1 mM	Phenylmethanesulfonyl fluoride (PMSF)

Bring to 1 L with dH_2O, stir to dissolve, and adjust pH to 8.0.
Filter, sterilize, and store @4°C.
Extraction buffer with no detergent

100 mM	KCl
1 mM	EGTA
10 mM	HEPES

Bring to 1 L with dH_2O, stir to dissolve, and adjust pH to 7.5.
Filter, sterilize, and store @4°C.
Extraction buffer with 1% Triton X-100

100 mM	KCl
1 mM	EGTA
10 mM	HEPES
1% (v/v)	Triton X-100

Bring to 1 L with dH_2O, stir to dissolve, and adjust pH to 7.5.
Filter, sterilize, and store @4°C.
Extraction buffer with 2% Triton X-100

100 mM	KCl
1 mM	EGTA
10 mM	HEPES
2% (v/v)	Triton X-100

Bring to 1 L with dH_2O, stir to dissolve, and adjust pH to 7.5.
Filter, sterilize, and store @4°C.
5% Sucrose in extraction buffer

5 g	Sucrose

Bring to 100 mL with extraction buffer and stir to dissolve.
Filter, sterilize, and store @4°C.
20% Sucrose in extraction buffer

20 g	Sucrose

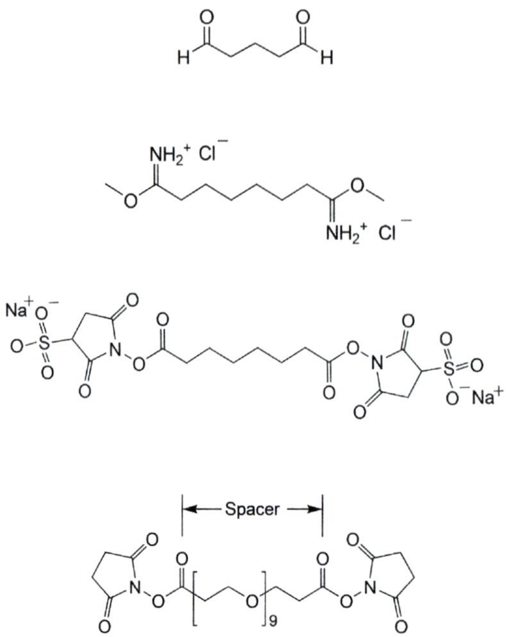

Glutaraldehyde, Spacer arm, 5.0 Å

DMS, Spacer arm, 11.0 Å

BS3, Spacer arm, 11.4 Å

BS(PEG)$_9$, Spacer arm, 35.8 Å

FIGURE 2.4.1 Cross-linkers and the lengths of their spacer arms.

Bring to 100 mL with extraction buffer and stir to dissolve.
Filter, sterilize, and store @4°C.
Glycine buffer (2.5 M)

18.8 g Glycine

Bring to 100 mL with dH_2O, stir to dissolve, and adjust pH to 7.5.
Filter, sterilize, and store @4°C.

2.4.2.5 Chemical cross-linkers

Glutaraldehyde solution: 50% in H_2O; Electron Microscopy Sciences.

BS(PEG)$_9$, (PEGylated bis(sulfosuccinimidyl)suberate): stock 100 mM in DMSO; Thermo Fisher Scientific.

BS3, bis(sulfosuccinimidyl) suberate: stock 5 mg/mL in H_2O; Thermo Fisher Scientific.

DMS, dimethyl suberimidate: stock 100 mM in H_2O; Thermo Fisher Scientific (Fig. 2.4.1).

2.4.2.6 Equipment

Beckman Ultracentrifuge Optima L-100XP.

Beckman SW40 Ti swinging-bucket rotor.
Thermo Fisher Scientific Sucrose gradient maker.

2.4.3 Experimental procedure

2.4.3.1 HEK293 cell transfection

Day 1

11. Grow HEK293 cells to confluence in one T225 flask.
12. Trypsinize the cells from the T225 flask, centrifuge the cells at $250 \times g$ for 5 minutes at room temperature, and resuspend the cells in 5 mL of DMEM complete medium.
13. Prepare four 150 mm culture dishes, and add 20 mL of DMEM complete medium to each dish.
14. Add 1 mL of resuspended cells to each dish.
15. Return the cells to the incubator and culture for overnight (12–16 hours) at 37°C.

Day 2

17. Take the dishes from the incubator and change fresh medium.
    (Because HEK293 cells are fragile, tilt the flask when removing the medium and adding fresh medium so that the cell monolayer is not disturbed.)
18. Mix the following DNAs (made w/ Endo-free Qiagen midi-Kits) in a 15 mL centrifuge tube.

20 μg	pcDNA3.1-Cldn16
20 μg	pcDNA3.1-Cldn19
x μL	Nuclease-free H_2O
1.9 mL	Total volume

19. Add 100 μL of 2.5 M $CaCl_2$ to DNA mixture. Vortex to mix.
20. In a separate 15 mL centrifuge tube, add 1 mL of $2 \times$ HBS.
21. Add DNA mixture (Step 2) to $2 \times$ HBS (Step 3) dropwise with a Pasteur pipette. Flick the HBS tube while adding DNA mixture. When finishing adding, vortex gently (using low speed) for 20 seconds.
    (This is sufficient to transfect one 150-mm dish.)
22. Let the DNA-HBS mixture stand for 20 minutes at room temperature.
23. Add 2 mL of DNA-HBS mixture dropwise to a 150-mm flask.
    (When adding the mixture, hold the pipette stable while swirling the flask slowly so that the drops can be distributed evenly in different areas of the flask. After adding the mixture, continue swirling the flask for 20 seconds so that the DNA mixture can fully dissolve into the medium.)
24. Return the flasks into the incubator for 12–16 hours.

Day 3

3. Take the dishes of transfected HEK293 cells out of the incubator and change fresh medium. Return the flasks into the incubator for 48 hours.

2.4.3.2 Solubilization of claudin protein

Day 5

43. Aspirate medium from the dishes. Wash cells once with 10 mL of ice-cold 1 × PBS.
44. Add 10 mL of ice-cold bicarbonate buffer to each dish. Brush the cells off the dish and transfer the cells to a 50 mL centrifuge tube.
45. Centrifuge the cells at 250 × *g* for 5 minutes at room temperature. Homogenize 5 g of cells with a 10-mL Dounce tissue homogenizer in 2 mL of ice-cold bicarbonate buffer.
46. Add the homogenate to 8 mL of bicarbonate buffer and stir at 4°C for 30 minutes.
47. Centrifuge the homogenate at 5,000 × *g* for 10 minutes at 4°C.
48. Discard the supernatant and resuspend the soft upper pellet (*the membrane fraction*) with 5 mL of ice-cold bicarbonate buffer and transfer to a new tube. Discard the hard lower pellet (*the nuclear fraction*).
49. Centrifuge the membrane lysate at 5,000 × *g* for 10 minutes at 4°C.
50. Discard the supernatant and resuspend the pellet with 1 mL of ice-cold extraction buffer with no detergent.
51. Add 1 mL of extraction buffer with 2% Triton X-100 dropwise to the membrane lysate (the final concentration of Triton X-100 = 1%). Mix by stirring for 30 minutes at 4°C.
52. Centrifuge the detergent extracted membrane lysate at 50,000 × *g* for 30 minutes at 4°C.
53. Save the supernatant for further analysis.

2.4.3.3 Sucrose gradient centrifugation

1. Use the sucrose gradient maker to make a linear (5%–20%) sucrose gradient.
2. Load the detergent extracted membrane lysate onto the linear sucrose gradient and centrifuge at 100,000 × *g* for 14–16 hours at 4°C with Beckman SW40 Ti swinging-bucket rotor.
3. Puncture the bottom of the centrifuge tube with an 18-gauge needle slightly to the side of the bottom. Collect the sucrose solution dropwise into Eppendorf tubes on ice, about 500 μL of sample per tube.
4. Measure the percentage of sucrose in each fraction with a refractometer.
5. Dialyze each sucrose gradient fraction against the extraction buffer with 1% Triton X-100.

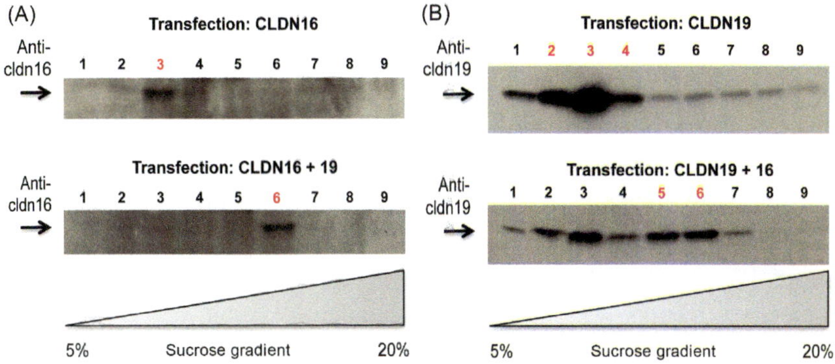

FIGURE 2.4.2 Sucrose gradient analysis of claudin assembly. Triton-soluble cell lysates from singly or doubly transfected claudin-expressing HEK293 cells were fractionated on 5%–20% of linear sucrose gradients and blotted with anti-Cldn16 (A) or anti-Cldn19 (B) antibody. Note that claudin monomers were recovered from sucrose fraction #3 (8% sucrose) whereas claudin oligomers were enriched in sucrose fraction #6 (14% sucrose). *Reproduced with permission from Gong, Y., Renigunta, V., Zhou, Y., Sunq, A., Wang, J., Yang, J., Renigunta, A., Baker, L. A., & Hou, J. (2015). Biochemical and biophysical analyses of tight junction permeability made of claudin-16 and claudin-19 dimerization.* Molecular Biology of the Cell, 26, *4333–46.*

6. Take 20 μL from each fraction and run Western blot with anti-Cldn16 and anti-Cldn19 antibodies to determine the fraction containing claudin oligomers (Fig. 2.4.2).

2.4.3.4 Chemical cross-linking

1. Add cross-linkers to the oligomeric claudin fraction.
   Glutaraldehyde: 1:100 dilution; final concentration 0.5%.
   BS(PEG)$_9$: 1:50 dilution; final concentration 2.5 mM.
   BS3: 1:100 dilution; final concentration 50 μg/mL.
   DMS: 1:50 dilution; final concentration 2.5 mM.
2. Incubate the samples on ice for 30 minutes with stirring.
3. Add glycine buffer to stop the cross-linking reaction.
   Glycine: 1:50 dilution; final concentration 50 mM.
4. Take 20 μL from each cross-linked sample and run Western blot with anti-Cldn16 or anti-Cldn19 antibody to determine the molecular size of claudin oligomer (Fig. 2.4.3).

2.4.4 Data analysis

Because the sedimentation rate in a linear sucrose gradient is largely based on the molecular mass, the claudin oligomer migrates to a higher sucrose density due to its larger sedimentation coefficient than that of the monomers

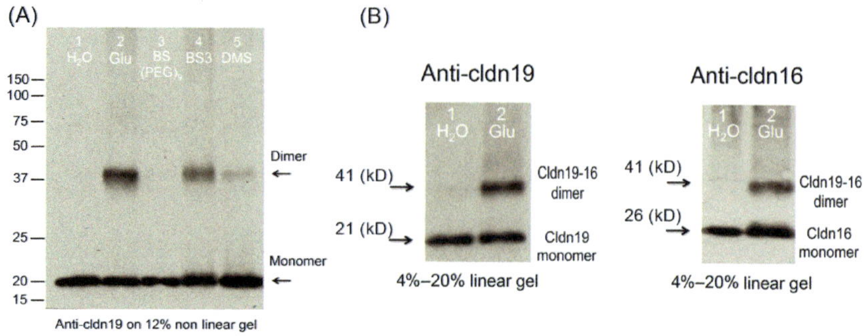

FIGURE 2.4.3 Cross-linking analysis of claudin oligomer from sucrose gradient fractionation. (A) Various cross-linkers were used to cross-link the claudin oligomers from sucrose fraction #6 (14% sucrose) in Fig. 2.4.2. (B) Linear SDS-PAGE was used to determine the molecular weight of claudin monomer and oligomer. *Reproduced with permission from Gong, Y., Renigunta, V., Zhou, Y., Sunq, A., Wang, J., Yang, J., Renigunta, A., Baker, L. A., & Hou, J. (2015). Biochemical and biophysical analyses of tight junction permeability made of claudin-16 and claudin-19 dimerization.* Molecular Biology of the Cell, 26, *4333–46.*

(Gong et al., 2015). The protein composition in claudin oligomer can be analyzed by Western blot using selective antibodies against each claudin species. The use of a linear SDS–polyacrylamide gel is important to accurately report the molecular mass of claudin oligomer. Silver staining can be used to detect the claudin oligomer after electrophoretic separation on a polyacrylamide gel. The claudin oligomer recovered from the silver-stained polyacrylamide gel can be analyzed by mass spectrometry to determine the stoichiometry of the oligomer.

2.4.5 Troubleshooting

2.4.5.1 Specificity of cross-linking

The specificity of cross-linking is the most critical variable in the use of this method to determine the proteins interacting with claudins. In one-step cross-linking procedures, where all the reactants are included in a single step, excessive cross-linking may occur under certain conditions, which lead to the conjugation of noninteracting proteins, that is, false positive results. For instance, cross-linking at pH values above 8 can promote deprotonation and activation of many classes of nucleophiles on the protein surface so that the protein effectively becomes a sponge for cross-linkers and noninteracting proteins that randomly collide with it. The same scenario occurs when the concentration of the cross-linker is in great excess of the concentration of the target protein. Reaction conditions that alter the protein structure may expose the buried regions in the claudin protein, leading to nonspecific cross-linking to other proteins. For example, carrier solvents such as DMSO, which are used to introduce water-insoluble hydrophobic cross-linkers into a reaction, can alter the protein structures to create false-positive results. The choice of buffer also affects the specificity of cross-linking. Buffers such as Tris, which contain free amines, may compete with the nucleophilic side chains in claudin for the cross-linker, especially when the buffer concentration is far higher than that of the protein. A two-step cross-linking process, which exploits the properties of the reactive groups in a bifunctional cross-linker, can be used to increase the cross-linking specificity. In these procedures, the claudin protein is first modified by the reactive group (e.g., maleimido or succinimidyl group) in a photoreactive bifunctional cross-linker. Then, unbound cross-linker molecules are washed off the protein. In the second step, the cross-linker-derivatized protein is mixed with its potential binding protein and exposed to an activating wavelength of light to induce cross-linking (Nadeau, Traxler, Fee, Baldwin, & Carlson, 1999).

2.4.5.2 Molecular ruler and nearest neighbor approaches

Because the side-chain nucleophiles cannot be cross-linked if they are separated by a distance greater than the length of the cross-linker, the efficacy of cross-linking depends upon the distance separating the reactive groups in the cross-linker (i.e., the length of spacer arm). A basic method to determine the maximal distance between the reactive amino acid residues in an oligomer, is to use a homologous series of cross-linkers with spacer arms of varying lengths, that is, the molecular ruler approach (Fig. 2.4.1). The distance between reacting residues determined using the molecular ruler approach is defined as maximal because the residues separated by a distance less than the extended length of the cross-linker may still be cross-linked, owing to its molecular flexibility (Green, Reisler, & Houk, 2001). The nearest neighbor approach is used to determine the topography of the interacting interface in a heterooligomer. The basic premise of this approach is that the proteins in an oligomer are cross-linked only if they are in sufficiently close proximity to accommodate the smallest span among the selected cross-linkers (Lambert, Boileau, Cover, & Traut, 1983).

2.4.5.3 In vitro versus in vivo cross-linking

The advantage of in vivo cross-linking is that the protein interactions occur under conditions that are close to the natural niche. Because of the complexity of the protein components in an in vivo system, short cross-linkers are used to ensure that only proximal proteins will be conjugated. For example, formaldehyde, a lipid-soluble cross-linker with a very short spacer arm (2–3 Å), has been used for in vivo cross-linking. Controlling the conditions to optimally cross-link the proteins within the TJ is a difficult process at best. Besides claudins, the cross-linker can react with numerous other proteins, including cell surface proteins, cell membrane proteins, and cytoskeletal proteins. Moreover, cellular metabolites are potential competitors for the cross-linker if they contain reactive groups such as amines, carboxylates, and thiols. In vitro cross-linking offers a far better controllability. Water or lipid-soluble cross-linkers can be used to respectively modify the hydrophilic or the hydrophobic groups in the interacting proteins. In vitro cross-linking also allows the level of purity to be optimized for claudins and their binding partners so that the cross-linking patterns can be easily interpreted. The disadvantage of in vitro cross-linking is that the interacting proteins are conjugated under nonphysiological conditions. When proteins are extracted from the cell membrane by detergents, the extraction process potentially disrupts protein–lipid and protein–protein interactions.

2.4.5.4 *cis* versus *trans* oligomer

The *cis* oligomers have been found for claudin-2, claudin-5, claudin-16, and claudin-19 (Gong et al., 2015; Rossa et al., 2014; Van Itallie, Mitic, & Anderson, 2011). The *trans* claudin oligomers, on the other hand, are difficult to capture, in part because the *trans* claudin interactions must be dynamic to allow ions and solutes of different sizes to permeate through the TJ. The cell culture condition is another factor. To promote *trans* interactions, cells expressing different claudin proteins have to be cocultured at high density to establish a sufficient number of cell–cell junctions. The membrane in the cell junction contains high levels of cholesterol and glycosphingolipids, which are hard to dissolve with mild detergents. Strong detergents such as SDS can solubilize the cell junctional membrane at the expense of breaking potential *trans* claudin interactions.

2.4.6 Concluding remarks

Cross-linking is a versatile technique for studying the oligomeric nature of protein complexes. Within the oligomeric complex, cross-linking can reveal the interacting subunits, determine the subunit stoichiometry, uncover the maximal distance separating the subunits, and identify the amino acid residues making the interaction interface. Cross-linking also provides corroboration for the results of yeast two-hybrid screens, in vitro pull-downs, and immunoprecipitations. Tight junction is a macromolecular architecture composed of over 400 proteins. Many TJ proteins, including claudins, can form oligomers. These oligomers then assemble into a functional TJ. From a biochemical point of view, molecular assembly of protein monomer to oligomer establishes the premise of TJ biology.

References

Furuse, M., Sasaki, H., & Tsukita, S. (1999). Manner of interaction of heterogeneous claudin species within and between tight junction strands. *The Journal of Cell Biology*, *147*, 891–903.

Gong, Y., Renigunta, V., Zhou, Y., Sunq, A., Wang, J., Yang, J., ... Hou, J. (2015). Biochemical and biophysical analyses of tight junction permeability made of claudin-16 and claudin-19 dimerization. *Molecular Biology of the Cell.*

Green, N. S., Reisler, E., & Houk, K. N. (2001). Quantitative evaluation of the lengths of homobifunctional protein cross-linking reagents used as molecular rulers. *Protein Science: a Publication of the Protein Society*, *10*, 1293–1304.

Hou, J., Renigunta, A., Konrad, M., Gomes, A. S., Schneeberger, E. E., Paul, D. L., ... Goodenough, D. A. (2008). Claudin-16 and claudin-19 interact and form a cation-selective tight junction complex. *The Journal of Clinical Investigation*, *118*, 619–628.

Lambert, J. M., Boileau, G., Cover, J. A., & Traut, R. R. (1983). Cross-links between ribosomal proteins of 30S subunits in 70S tight couples and in 30S subunits. *Biochemistry*, *22*, 3913–3920.

Nadeau, O. W., Traxler, K. W., Fee, L. R., Baldwin, B. A., & Carlson, G. M. (1999). Activators of phosphorylase kinase alter the cross-linking of its catalytic subunit to the C-terminal one-sixth of its regulatory alpha subunit. *Biochemistry*, *38*, 2551–2559.

Rossa, J., Ploeger, C., Vorreiter, F., Saleh, T., Protze, J., Gunzel, D., ... Piontek, J. (2014). Claudin-3 and claudin-5 protein folding and assembly into the tight junction are controlled by non-conserved residues in the transmembrane 3 (TM3) and extracellular loop 2 (ECL2) segments. *The Journal of Biological Chemistry*, *289*, 7641–7653.

Tsukita, S., Furuse, M., & Itoh, M. (2001). Multifunctional strands in tight junctions. *Nature Reviews Molecular Cell Biology*, *2*, 285–293.

Van Itallie, C. M., Mitic, L. L., & Anderson, J. M. (2011). Claudin-2 forms homodimers and is a component of a high molecular weight protein complex. *The Journal of Biological Chemistry*, *286*, 3442–3450.

Chapter 2.5

Yeast two-hybrid assay of claudin interaction

2.5.1 Background knowledge

2.5.1.1 Classic yeast two-hybrid assay

The classic yeast two-hybrid (Y2H) assay is based upon the concept that a transcriptional factor-Gal4 activates gene transcription in *Saccharomyces cerevisiae* in the presence of galactose (Fields & Song, 1989). The Gal4 protein contains two distinct domains: the N-terminal DNA-binding domain and the C-terminal transcription activation domain. When split into two halves, the N-terminal and the C-terminal fragments interact noncovalently to reconstitute a functional Gal4 protein. In the Gal4-based Y2H assay, the bait protein is expressed as a fusion to the Gal4 DNA-binding domain (DNA-BD), while the prey protein is expressed as a fusion to the Gal4 DNA-activation domain (DNA-AD). When bait and prey proteins interact, the DNA-BD and DNA-AD domains of Gal4 protein are brought into proximity to activate the transcription of the reporter genes (e.g., *lacZ*, *Ade2*, and *His3*) that are controlled by the Gal upstream activating sequences (Gal-UAS) (Fig. 2.5.1). Notably, the interaction between bait and prey proteins takes place in the nucleus.

2.5.1.2 Membrane yeast two-hybrid assay

The split-ubiquitin membrane yeast two-hybrid system (MbYTH) is based on the concept that ubiquitin can be experimentally separated into two moieties

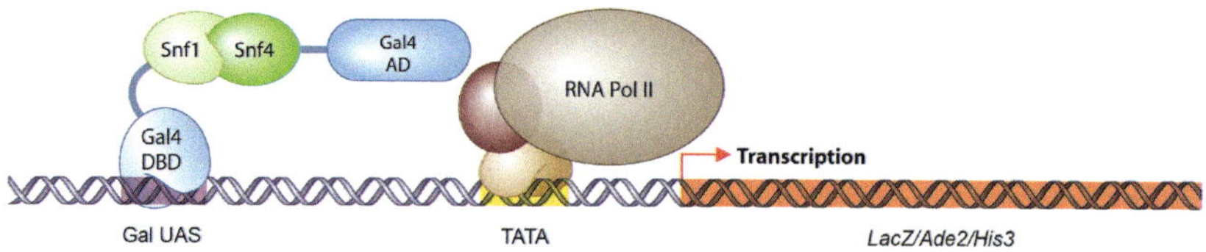

FIGURE 2.5.1 Classic yeast two-hybrid system. In the classic Y2H assay, the bait protein (X) is fused to the DNA-binding domain (DBD) of Gal4 and the prey protein is fused with the transcription activation domain (AD) of Gal4. The Gal4 DBD binds to the upstream activating sequences (UAS) of the Gal1 promoter. Interaction of bait with prey brings the Gal4 AD to the UAS, which recruits the basal transcriptional machinery to drive reporter gene transcription. *Reproduced with permission from Stynen, B., Tournu, H., Tavernier, J., & Van Dijck, P. (2012). Diversity in genetic in vivo methods for protein–protein interaction studies: From the yeast two-hybrid system to the mammalian split-luciferase system.* Microbiology and Molecular Biology Reviews: MMBR, 76, *331–382.*

that will reconstitute into a fully functional ubiquitin when brought into proximity (Johnsson & Varshavsky, 1994). In the MbYTH, the bait membrane protein is fused to the C-terminal half of ubiquitin (Cub), along with a transcriptional factor that consists of the bacterial LexA-DNA-binding domain and the *Herpes simplex* VP16 transactivator protein. The prey protein, either membrane-bound or cytosolic, is fused to the N-terminal half of ubiquitin (Nub). When bait and prey proteins interact, the two halves of ubiquitin reconstitute into a full ubiquitin that is recognized by the ubiquitin-specific proteases (UBPs) present in the cytoplasm of all eukaryotic cells. The released transcriptional factor then enters the nucleus and activates the transcription of the reporter genes (e.g., *lacZ*, *Ade2*, and *His3*) (Fig. 2.5.2).

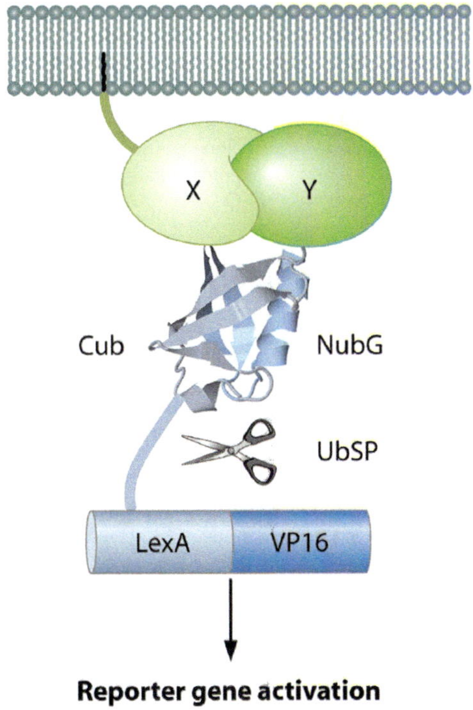

FIGURE 2.5.2 Split-ubiquitin membrane yeast two-hybrid system. Interaction between a membrane-bound bait X and prey Y leads to the reconstitution of a full ubiquitin and the cleavage of the transcription factor LexA-VP16. Released LexA-VP16 translocates to the nucleus to activate reporter genes. *Reproduced with permission from Stynen, B., Tournu, H., Tavernier, J., & Van Dijck, P. (2012). Diversity in genetic in vivo methods for protein-protein interaction studies: From the yeast two-hybrid system to the mammalian split-luciferase system.* Microbiology and Molecular Biology Reviews: MMBR *76, 331–382.*

2.5.2 Materials and reagents

2.5.2.1 *Saccharomyces cerevisiae* yeast strain

NMY51 yeast strain: MATa his3Δ200 trp1-901 leu2-3,112 ade2, LYS2::(lexAop)4-HIS3, ura3::(lexAop)8-lacZ, ade2::(lexAop)8-ADE2 GAL4; available from MoBiTec Inc.

2.5.2.2 Plasmids

pBT3 bait vector: expressing a fusion protein to Cub; available from MoBiTec Inc.

pPR3 prey vector: expressing a fusion protein to Nub; available from MoBiTec Inc.

2.5.2.3 Yeast growth media

YPAD (1 L)

10 g	Yeast extract
20 g	Peptone
40 mg	Adenine sulfate
900 mL	H_2O

Autoclave and store at 4°C

Add 100 mL of 10 × glucose solution before use

10 × Glucose

200 g	D-Glucose
1 L	H_2O

Filter, sterilize, and store at 4°C.

SD medium (1 L)

6.7 g	Yeast nitrogen base w/o amino acids
0.6 g	Dropout mix (-Ade, -His, -Leu, or -Trp)
900 mL	H_2O

Autoclave and store at 4°C.

Add 100 mL of 10 × glucose solution before use.

2.5.3 Experimental procedure

2.5.3.1 Transformation of bait and prey constructs into NMY51 yeast strain

1. Inoculate 50 mL of YPAD with a single colony of NMY51 taken from a fresh plate and grow overnight at 30°C with shaking.

2. Monitor the OD_{546} of the culture until it reaches 0.6–0.8. If the OD_{546} reading is above 1.0, dilute the culture to OD_{546} of 0.2 and regrow to $OD_{546} = 0.6$.
3. Pellet the 50 mL culture for 5 minutes at $2{,}500 \times g$ and resuspend in 2.5 mL of water.
4. Prepare the PEG/LiOAc master mix.

Component	Amount
50% PEG 3350 (Sigma P4338)	1.2 mL
1 M LiAc	180 μL
Single-stranded carrier DNA	125 μL

5. Set up the following reactions in reaction tubes.

Bait construct (pBT-Cldn16)	1.5 μg
Prey construct (pPR-Cldn19)	1.5 μg

6. Add 300 μL of PEG/LiAc master mix to each reaction tube, and vortex briefly.
7. Add 100 μL resuspended yeast cells from Step 3 to each reaction tube, and vortex 1 minute to thoroughly mix all components.
8. Incubate in a 42°C water bath for 45 minutes.
9. Pellet the reactions for 5 minutes at $2{,}500 \times g$.
10. Dissolve each pellet in 100 μL of 0.9% NaCl and plate each transformation onto the following plates.

Plasmid	SD/-Leu	SD/-Trp
Bait construct (pBT-Cldn16)	100 μL	
Prey construct (pPR-Cldn19)		100 μL

11. Incubate at 30°C for two days until colonies appear.

2.5.3.2 Verification of bait and prey protein expression by Western blot

1. Inoculate several colonies of bait and prey transformant strains into 10 mL of SD-Leu or SD-Trp and incubate overnight at 30°C. Grow to an OD_{546} of 0.6–1.
2. Centrifuge at $2{,}500 \times g$ for 5 minutes.
3. Wash the pellet once in 1 mM of EDTA.
4. Resuspend the pellet in 200 μL of 2 M NaOH, and incubate on ice for 10 minutes.
5. Add 200 μL of 50% trichloroacetic acid, vortex and incubate on ice for 2 hours.
6. Centrifuge at $14{,}000 \times g$ for 20 minutes.
7. Discard the supernatant.
8. Resuspend the pellet in 200 μL of ice-cold acetone.
9. Centrifuge at $14{,}000 \times g$ for 20 minutes.

10. Discard the supernatant and resuspend the pellet in 200 μL of 2 × Laemmli buffer.
11. Proceed to SDS-PAGE and Western blot.

2.5.3.3 Dual transformation and reporter gene expression assay

1. Inoculate a single NMY51 colony expressing the bait protein (Cldn16) into 50 mL of SD-leu and grow overnight at 30°C with shaking.
2. Monitor the OD_{546} of the culture until it reaches 0.6–0.8.
3. Pellet the 50 mL culture for 5 minutes at 2,500 × *g* and resuspend in 2.5 mL of water.
4. Prepare the PEG/LiOAc master mix.

Component	Amount
50% PEG 3350 (Sigma P4338)	1.2 mL
1 M LiAc	180 μL
Single-stranded carrier DNA	125 μL

5. Add 1.5 μg of prey construct (pPR-Cldn19) into a reaction tube.
6. Add 300 μL of PEG/LiAc master mix to each reaction tube, and vortex briefly.
7. Add 100 μL of resuspended yeast cells from Step 3 to the reaction tube, and vortex 1 minute to thoroughly mix all components.
8. Incubate in a 42°C water bath for 45 minutes.
9. Pellet the reactions for 5 minutes at 2,500 × *g*.
10. Dissolve the pellet in 100 μL of 0.9% NaCl and plate the transformation onto SD/-Leu/-Trp plates.
11. Incubate at 30°C for 2 days until colonies appear.
12. Streak colonies onto SD/-Leu/-Trp/-Ade and SD/-Leu/-Trp/-His plates to examine the growth reporter activities (Fig. 2.5.3).
13. Perform β-galactosidase assays to examine the *lacZ* reporter activities.

2.5.4 Data analysis

2.5.4.1 Protein topology

The working mechanism of the MbYTH system requires that the Cub-LexA-VP16 module of the bait and the Nub module of the prey must be located in the cytosol (Fig. 2.5.2). The crystal structure of claudin reveals that both the N-terminus and the C-terminus in the claudin molecule are located in the cytosol (Suzuki et al., 2014). The choice of the cytoplasmic domain in claudin for fusion to Cub-LexA-VP16 or Nub module is important for the interpretation of interaction data. From the study of claudin-16 and claudin-19 interaction, it is perceived that the strongest interaction was observed when Cub-LexA-VP16 was fused to the C-terminus of claudin-19 and Nub was fused to the N-terminus of claudin-16

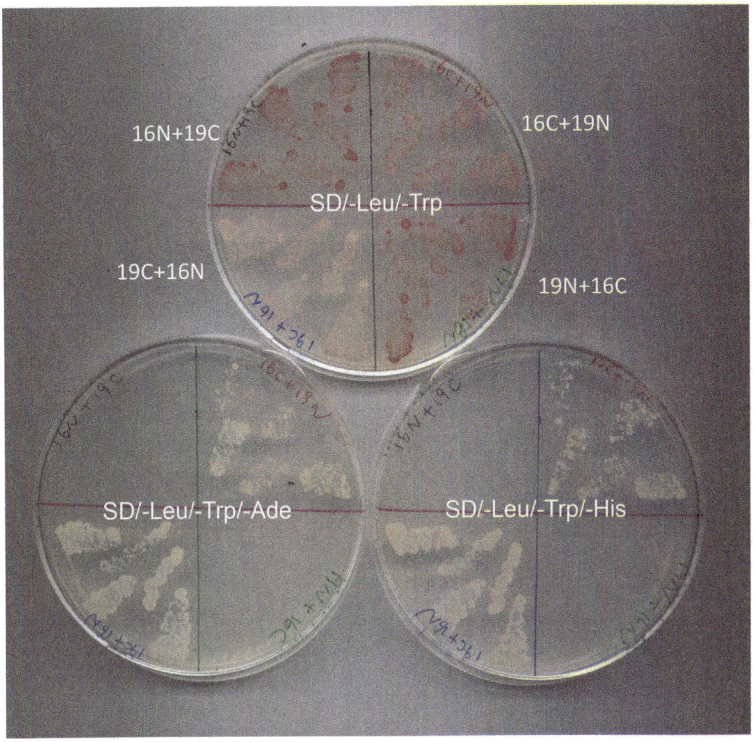

FIGURE 2.5.3 Split-ubiquitin membrane yeast two-hybrid assay of claudin-16 and claudin-19 interaction. Shown are plates with selective medium lacking leucine and tryptophan (SD/-Leu/-Trp), indicating the transforming of both bait and prey vectors (expressed as bait + prey); with SD/-Leu/-Trp/-Ade and SD/-Leu/-Trp/-His, indicating the expression of reporter genes Ade2 and His3, respectively. N stands for N-terminus; C stands for C-terminus.

(Fig. 2.5.3). Although the cause for this phenomenon has not been elucidated on the molecular level, one potential explanation is that the claudin-16 and claudin-19 interaction orientates the molecules in a certain way to bring the C-terminus of claudin-19 into proximity with the N-terminus of claudin-16.

2.5.4.2 Quantitative β-galactosidase assay

The *lacZ* reporter is used to assess the strength of interaction in a quantitative way. The *lacZ* gene encodes β-galactosidase, also known as β-gal, a glycoside hydrolase enzyme that catalyzes the hydrolysis of β-galactosides into monosaccharides. Many commercial kits offer sensitive colorimetric or fluorescent assays for β-galactosidase.

2.5.4.3 *Ade2* phenotype

When there is no interaction between bait and prey, the *Ade2* gene is not transcribed and the adenine synthesis pathway is blocked in the yeast cell, which causes the accumulation of a red-colored intermediate and gives the NMY51 colony pink to red colors. The presence of an interacting protein pair activates the *Ade2* gene, switches on the adenine synthesis pathway, and removes the red-colored intermediate. Therefore, the yeast cells harboring protein–protein interactions display faintly pink (weak interaction) to white colors (strong interaction).

2.5.4.4 *His3* reporter stringency test

Some bait proteins may show self-activation in the absence of prey proteins (vide infra). When the *His3* reporter gene is used as a selection marker, a chemical compound known as 3-aminotriazole (3-AT) can be included into the selection medium to reduce the background growth due to self-activation. The optimal 3-AT concentration is determined on the SD/-Leu/-Trp/-His selection plate supplemented with 1–10 mM 3-AT.

2.5.5 Troubleshooting

2.5.5.1 False-positive interaction

False-positive results can have multiple root causes. Some proteins might indeed be able to interact, but the interactions have no biological relevance (i.e., biological artifacts), because the interacting proteins are not present in the correct subcellular location in the mammalian cells. Other proteins may overcome the nutritional selection designed for *Ade2* or *His3* reporter. A particular case of a false-positive result is self-activation by the bait protein, that is, the bait can induce the expression of the reporter genes independently of the prey. Overexpression of the bait often causes self-activation. Some baits may display strong self-activation because they contain sequences that are targets of endogenous proteases.

2.5.5.2 Protein stability in *Saccharomyces cerevisiae*

In *S. cerevisiae*, protein stability depends on the nature of the N-terminal amino acid. Amino acids such as glycine, methionine, threonine, alanine, and cysteine stabilize the protein when they are present at its N-terminal end. In contrast, basic (e.g., lysine or arginine) or bulky hydrophobic amino acids at the N-terminus tend to promote protein degradation in a ubiquitin-dependent manner (Varshavsky, 1996). Protein stability is particularly important for the accuracy of the MbYTH assay. Degradation of the bait protein may release the Cub-LexA-VP16 module into the nucleus and induce the expression of the reporter genes. Degradation of the prey protein may reduce its "critical mass" level needed for interaction and result in a false-negative result.

2.5.6 Closing remarks

The TJ architecture is established upon the protein–protein interactions both within the TJ and between the TJ and the cytoskeleton. The TJ interactome can be systematically interrogated by the yeast two-hybrid assay. Because TJ is a membrane organelle, interactions need to be probed by the membrane-based yeast two-hybrid systems such as MbYTH. Notably, yeast cells do not make TJs, and yeast cell membrane differs from mammalian cell membrane in biochemical composition. The interaction data collected from yeast cell systems will have to be verified by independent biochemical approaches in mammalian cell systems.

References

Fields, S., & Song, O. (1989). A novel genetic system to detect protein-protein interactions. *Nature, 340*, 245–246.

Johnsson, N., & Varshavsky, A. (1994). Split ubiquitin as a sensor of protein interactions in vivo. *Proceedings of the National Academy of Sciences of the United States of America, 91*, 10340–10344.

Suzuki, H., Nishizawa, T., Tani, K., Yamazaki, Y., Tamura, A., Ishitani, R., … Fujiyoshi, Y. (2014). Crystal structure of a claudin provides insight into the architecture of tight junctions. *Science (New York, NY), 344*, 304–307.

Varshavsky, A. (1996). The N-end rule: Functions, mysteries, uses. *Proceedings of the National Academy of Sciences of the United States of America, 93*, 12142–12149.

Chapter 2.6

Recombinant claudin protein production in *Pichia pastoris*

2.6.1 Background knowledge

2.6.1.1 Recombinant protein expression system

2.6.1.1.1 Prokaryotes

In prokaryotes, there is no nuclear envelope or endoplasmic reticulum. The processes of mRNA transcription and protein translation occur simultaneously. The advantage of prokaryotic protein expression system is high production rate. The limitation is protein misfolding because prokaryotes lack the endoplasmic reticulum to perform posttranslational modification. Membrane protein folding is particularly difficult in prokaryotes. Misfolded proteins form inclusion bodies (IBs) in bacteria (Sabate, de Groot, & Ventura, 2010).

2.6.1.1.2 Eukaryotes

In eukaryotes, the processes of mRNA transcription and protein translation are temporospatially separated. The mRNA molecule is transcribed in the nucleus, and then translocated to the cytoplasm, where it is translated into protein. After translation, proteins are modified in various ways to complete their structure, designate their location, or regulate their activity in the cell. The advantage of eukaryotic protein expression system includes proper protein folding, bona fide protein modification, and correct protein localization. The limitation is low yield. The commonly used eukaryotic cell models include *P. pastoris* and *Saccharomyces cerevisiae* yeast cells, Sf9 and Sf21 insect cells, and HEK293 and CHO mammalian cells.

2.6.1.2 Chromatography

Chromatography is a separation technique based upon the differential partitioning of protein molecules between the mobile and stationary phases. There are four major categories of chromatographic techniques for protein purification: ion-exchange chromatography, hydrophobic interaction chromatography, affinity chromatography, and size-exclusion chromatography (Frey & Kang, 2005). Ion-exchange chromatography is based upon the reversible interaction between charged protein molecules in the solution and oppositely charged groups on the matrix. Hydrophobic interaction chromatography exploits the weak hydrophobic interaction between proteins and nonpolar stationary phase. Affinity chromatography exploits the natural specific recognition between protein molecules, for example, antigen and antibody. Size-exclusion chromatography separates proteins on the basis of their size. Unlike other chromatographic techniques, proteins are not retained by a size-exclusion column. On one hand, this feature is a strength, because fragile proteins can be separated using buffers compatible with physiological conditions. On the other hand, the lack of protein retention may limit the resolution of the chromatograph.

2.6.2 Materials and reagents

2.6.2.1 *Pichia pastoris* yeast strain

SMD1168 yeast strain: *his4*, *pep4*; available from Invitrogen.

2.6.2.2 Plasmids

pPICZ A, B, and C: expressing a PreScission protease cleavage site and a C-terminal GFP-His$_{10}$ tag; available from Invitrogen.

2.6.2.3 Yeast growth media

YPD (1 L)

10 g	Yeast extract
20 g	N–Z case plus
900 mL	H_2O

Autoclave and store at 4°C.

Add 100 mL of 10× glucose solution before use.

10× Glucose

200 g	D-glucose
1 L	H_2O

Filter, sterilize, and store at 4°C.

YPDS plates + Zeocin

5 g	Yeast extract
10 g	N–Z case plus
10 g	Agar
90 g	Sorbitol
450 mL	dH_2O

Autoclave and let cool to 50°C–60°C; add 50 mL of 10× glucose; add Zeocin to the final concentration of 100 μg/mL; pour plates.

Store plates at 4°C (*wrap plate with foil to protect against light; Zeocin is not stable; plate is stable for only 2 weeks at 4°C*).

50% (v/v) Glycerol

58.8 g	Glycerol
40 mL	dH_2O

Autoclave and store at room temperature.
10× YNB (yeast nitrogen base)

34 g	Yeast nitrogen base
100 g	Ammonium sulfate

Bring to 1 L with dH_2O and stir to dissolve.
Filter, sterilize, and store @4°C.
500× Biotin

20 mg	Biotin
100 mL	dH_2O

Filter, sterilize, and store @4°C.
1 M Potassium phosphate buffer (pH 6.0)

22.99 g	K_2HPO_4 (potassium phosphate dibasic)
118.13 g	KH_2PO_4 (potassium phosphate monobasic)

Bring to 1 L with dH_2O and adjust pH to 6.0.
Autoclave and store at room temperature.
BMG (buffered minimal glycerol medium)
Media used to grow yeast before induction—make fresh the day before use.

12 g	Glycerol
100 mL	1 M Potassium phosphate buffer (pH 6.0)
800 mL	dH_2O

Autoclave and cool to 4°C.
Add

100 mL	10× YNB
2 mL	500× Biotin

Store at 4°C.
BMM (buffered minimal methanol medium)
Media used to grow yeast for induction—make fresh the day before use.

200 mL	1 M Potassium phosphate buffer (pH 6.0)
700 mL	dH_2O

Autoclave and cool to 4°C.
Add

100 mL	10× YNB
2 mL	500× Biotin
7 mL	Pure methanol (to make 0.7% final concentration)

Store at 4°C.

2.6.2.4 *Pichia pastoris* EasyComp transformation kit

Solution I (store at 4°C).
Solution II (store at room temperature).
Solution III (store at 4°C); available from Invitrogen.

2.6.2.5 Chromatography

Bio-Scale Mini Nuvia IMAC Ni-charged cartridge; available from Bio-Rad.
Superdex 200 10/300 GL columns; available from GE Healthcare.

2.6.3 Experimental procedure

2.6.3.1 *Pichia pastoris* transformation

2.6.3.1.1 Prepare competent Pichia pastoris *cells*

1. Inoculate 10 mL of YPD with a streak of frozen *Pichia* cells (use sterile pipette tip). Grow overnight at 30°C in a shaking incubator (250–300 rpm).
2. Dilute cells from the overnight culture to an OD_{600} of 0.1–0.2 in a new 10 mL of YPD. Grow the cells at 30°C in a shaking incubator until the OD_{600} reaches 0.6–1.0. This will take approximately 4–6 hours.
3. Pellet the cells by centrifugation at 500 × *g* for 5 minutes at room temperature. Discard the supernatant.
4. Resuspend the cell pellet in 10 mL of *EasyComp Solution I.* No incubation time is required.
5. Pellet the cells by centrifugation at 500 × *g* for 5 minutes at room temperature. Discard the supernatant.
6. Resuspend the cell pellet in 1 mL of *EasyComp Solution I.* The cells are now competent.
7. Aliquot 100 μL of competent cells into labeled 1.5 mL of sterile Eppendorf microcentrifuge tubes.

   (*Note*: 100 μL of cells are used for each transformation. Cells can be thawed and refrozen several times without significant loss in transformation efficiency.)
8. At this point, the cells may be kept at room temperature and used directly for transformation or frozen for future use. To freeze cells, place tubes in a foam box and place in an −80°C freezer. It is important that the cells freeze down slowly. *Do not snap-freeze the cells in liquid nitrogen.*
9. Proceed to the transformation procedure.

2.6.3.1.2 Transformation

1. For each transformation, thaw one tube of competent cells at room temperature or use freshly prepared ones.

2. Add 2 μg linearized DNA to the competent cells.
3. Add 1 mL of *EasyComp Solution II* to the DNA/cell mixture and mix by vortexing or flicking the tube.
4. Incubate the transformation reactions for 1 hour at 30°C in an incubator. Mix the transformation reaction every 15 minutes by vortexing or flicking the tube.

   (*Failure to mix the transformation reaction every 15 minutes will result in decreased transformation efficiency.*)
5. Heat shock the cells in a 42°C heat block or water bath for 10 minutes.
6. Add 2 mL of YPD medium to each tube. Incubate the cells at 30°C with shaking for 1 hour to allow expression of Zeocin resistance gene.
7. Pellet the cells by centrifugation at 500 × *g* for 5 minutes at room temperature. Discard the supernatant.
8. Resuspend the cells in 1 mL of *EasyComp Solution III.*
9. Pellet the cells by centrifugation at 500 × *g* for 5 minutes at room temperature. Discard the supernatant.
10. Resuspend the cell pellet in 100 μL of *EasyComp Solution III.*
11. Plate the entire transformation on YPDS + Zeocin plates. Incubate the plates for 2–4 days at 30°C. Each transformation should yield approximately 50 colonies.

2.6.3.2 *Pichia pastoris* induction

1. Pick colonies (with sterile pipet tip) and inoculate 200 μL YPD + 100 μg/mL Zeocin in sterile 96-well plate.
2. Grow in 30°C incubator with shaking (250–300 rpm) for 24 hours.
3. Add to 25 mL of BMG medium (+100 μg/mL Zeocin) in a 250 mL flask. Grow at 24°C–27°C (heater off) in a shaking incubator (250–300 rpm) until culture reaches an OD_{600} = 2–6 (approximately 16–18 hours).
4. Use this 25 mL culture to inoculate 1 L of BMG (without Zeocin) in a 2 L *baffled* flask and grow at 24°C–27°C (heater off) with shaking (250–300 rpm) until the culture reaches log phase growth (OD_{600} = 2–6) (approximately 16–18 hours).
5. Harvest the cells using sterile centrifuge bottles by centrifuging at 1,000 × *g* for 5 minutes at room temperature.
6. To induce expression, decant the supernatant and resuspend cell pellet in equal volume of BMM (without Zeocin) (1 L of BMM corresponding to 1 L of BMG).
7. Cover the flasks with two layers of sterile cheesecloth and return to incubator. Continue to grow at 24°C–27°C (heater off) with shaking for 24 hours. (*If needing to grow for 48* hours, *add pure methanol to each flask at final concentration of 0.5% every 24 hours because of methanol evaporation*).

8. Harvest cells by centrifuge at 1,000 × *g* for 5 minutes at room temperature.
9. Decant the supernatant and freeze the pellet at −80°C for lysis.

2.6.3.3 HisGFP-claudin purification from *Pichia pastoris*

2.6.3.3.1 Lysis and extraction

1. Put dry and clean grinding jars w/grinding balls (part of Retsch Mixer Mill MM 400 grinding device) on dry ice.
2. Weigh 10 g of yeast cells for each grinding jar and chill on dry ice.
3. Cap the grinding jar gently and evenly.
4. Freeze the grinding jars containing recombinant yeast cells in liquid nitrogen for at least 5 minutes.
5. Load grinding jar symmetrically and start Retsch Mixer mill MM 400 (Program 10 Hz and 3 minutes intervals).
6. Unload grinding jar gently, freeze them in liquid nitrogen for 5 minutes, and start the second cycle of milling.
7. Repeat Step 6 for five or more times.
8. At the end of milling, use a jar wrench to open the grinding jar.
9. Use a dry and cold spoon to transfer yeast cells into a beaker containing 60 mL of 1 × lysis buffer (Section 2.6.2.3), and stir for 30 minutes.

   10 × Lysis buffer

Tris (pH 8.0)	500 mM
NaCl	1.5 M
EDTA (pH 8.0)	10 mM

10. Add *n*-dodecyl β-D-maltoside (DDM) to a final concentration of 2% (1.2 g for 60 mL).
11. Centrifuge at 30,000 × *g* for 1 hour at 4°C.
12. Keep supernatant at 4°C for further analysis.

2.6.3.3.2 Affinity chromatography

1. Connect Bio-Scale Mini Nuvia IMAC Ni-charged affinity column cartridge (5 mL) to FPLC system (Cutler, 2004).
2. Wash the affinity column with 10-bed volumes of dH_2O at a flow rate of 1 mL/min.
3. Equilibrate the affinity column with two-bed volumes of Talon binding buffer at a flow rate of 1 mL/min.

   Talon binding buffer

NaCl	150 mM
Tris pH 8.0	20 mM
β-Mercaptoethanol	5 mM
Imidazole	50 mM
DDM	2 mM

4. Load the supernatant from Step 12 (Section 2.6.3.3.1) at a flow rate of 1 mL/min.
5. Wash the affinity column with 10-bed volumes of Talon binding buffer until the protein concentration level returns to the baseline.
6. Elute HisGFP-claudin protein with elution buffer at a flow rate of 1 mL/min.

   Elution buffer

NaCl	150 mM
Tris pH 8.0	20 mM
β-Mercaptoethanol	5 mM
Imidazole	300 mM
DDM	2 mM

7. Collect eluted fractions and run 20 μL from each fraction on SDS-PAGE.
8. Stain SDS-PAGE gel with Coomassie blue (Fig. 2.6.1).

2.6.3.3.3 Size-exclusion chromatography

1. Connect GE Superdex 200 to FPLC system (Cutler, 2004).
2. Wash GE Superdex 200 with 10-bed volumes of dH_2O at a flow rate of 0.5 mL/min.
3. Equilibrate the GE Superdex 200 with 2-bed volumes of SEC buffer at a flow rate of 0.5 mL/min.

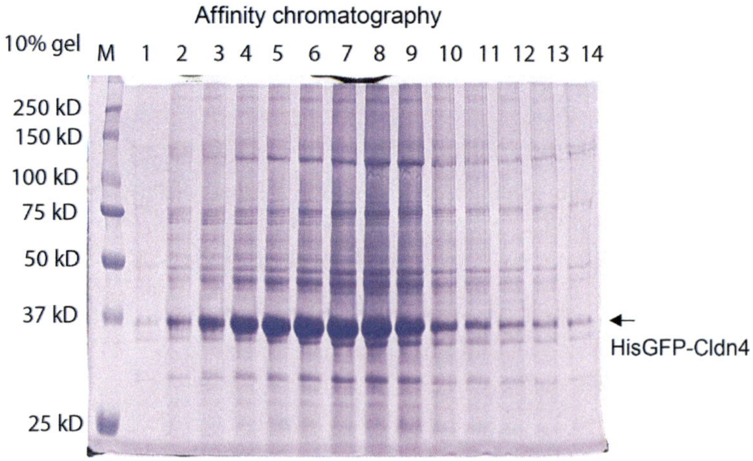

FIGURE 2.6.1 SDS-PAGE and Coomassie blue staining of eluted fractions from IMAC Ni-charged affinity chromatography. SDS-PAGE shows the enrichment of HisGFP-Cldn4 proteins in fractions eluted by imidazole at 300 mM.

SEC buffer

NaCl	150 mM
Tris pH 8.0	20 mM
β-Mercaptoethanol	5 mM
DDM	1 mM

4. Combine the fractions containing the target protein from Step 7 (Section 2.6.3.3.2) and concentrate it to 1.1 mL with Amicon Ultracentrifuge filter. Measure OD_{280} to estimate the protein yield.
5. Transfer the 1.1 mL of concentrated protein sample into a cold and clean 1.5 mL Eppendorf tube and centrifuge at 15,000 × *g* at 4°C for 10 minutes.
6. Use a 1 mL sample loop to load the protein sample into GE Superdex 200.
7. Elute the proteins at a flow rate of 0.35 mL/min.
8. Collect eluted fractions. Note that HisGFP-claudin appears in the second peak (Fig. 2.6.2).
9. Run 20 μL from each fraction on SDS-PAGE and stain SDS-PAGE gel with Coomassie blue (Fig. 2.6.3).

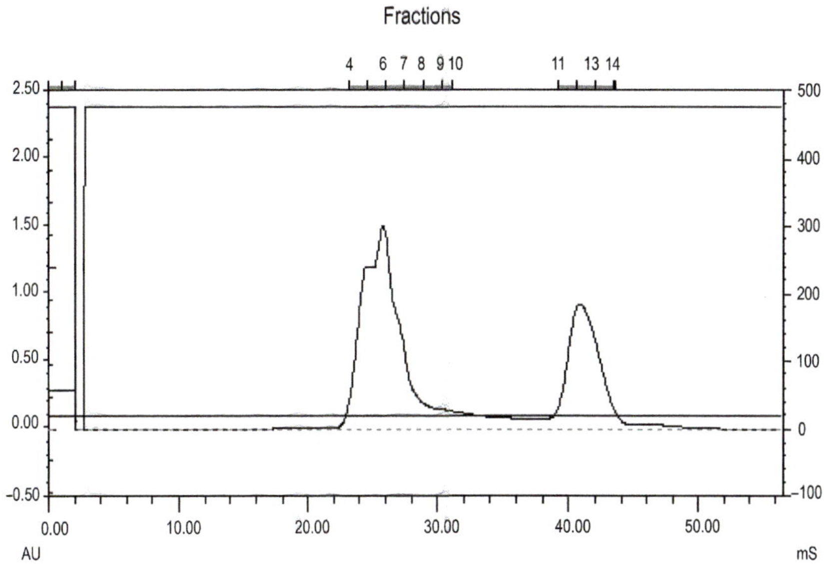

FIGURE 2.6.2 Size-exclusion chromatography. Size-exclusion chromatograph shows the separation of HisGFP-Cldn4 proteins (fractions 11−14) from endogenous *Pichia* cell membrane proteins (fractions 4−10). *Reproduced with permission from Belardi, B., Son, S., Vahey, M.D., Wang, J., Hou, J., & Fletcher, D. A. (2018). Claudin-4 reconstituted in unilamellar vesicles is sufficient to form tight interfaces that partition membrane proteins.* Journal of Cell Science, 132, *jcs221556.*

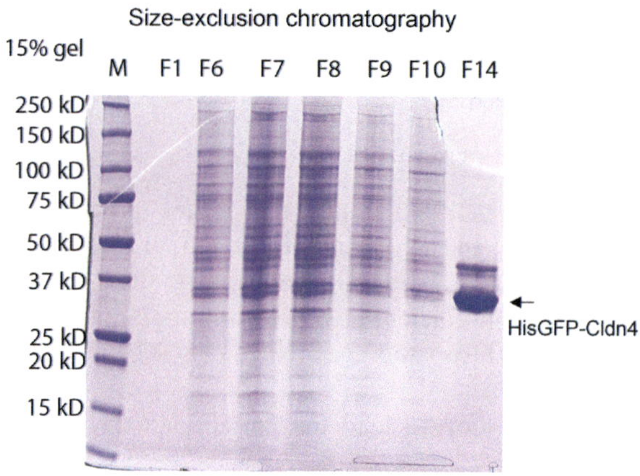

FIGURE 2.6.3 SDS-PAGE and Coomassie blue staining of eluted fractions from size-exclusion chromatography. SDS-PAGE shows the enrichment of HisGFP-Cldn4 proteins in fraction 14 (F14). *Reproduced with permission from Belardi, B., Son, S., Vahey, M.D., Wang, J., Hou, J., & Fletcher, D. A. (2018). Claudin-4 reconstituted in unilamellar vesicles is sufficient to form tight interfaces that partition membrane proteins.* Journal of Cell Science, 132, *jcs221556.*

2.6.4 Data analysis

The concentration and purity of recombinant claudin proteins can be easily determined by mass spectrometry. Functional analyses of purified claudin molecules require more advanced techniques. For example, the adhesion strength of homotypic claudin interactions can be measured in vitro by atomic force microscopy for recombinant claudin-1 proteins purified from *Escherichia coli* (Lim, Vedula, Kausalya, Hunziker, & Lim, 2008). The molecular assembly of claudin proteins can be monitored in reconstituted unilamellar vesicles for recombinant claudin-4 proteins purified from *P. pastoris* (Belardi et al., 2018).

2.6.5 Troubleshooting

2.6.5.1 Protein solubilization

Claudin proteins expressed in prokaryotic cell models such as *E. coli* may aggregate to form inclusion bodies, which must be solubilized by denaturants such as urea or guanidine-HCl. Restoration of the protein's native structure requires removing the denaturant by dialysis or size-exclusion chromatography. Claudin proteins expressed in eukaryotic cell models such as yeast, insect, and mammalian cells, are localized in the plasma membrane. Relatively mild detergents such as DDM and Triton X-100 are sufficient to extract claudin proteins from the membrane. These nonionic detergents can stabilize membrane proteins for biophysical and structural studies. Notably, the yeast cell contains a cell wall. A cryomilling step is needed to break the cell wall before the membrane can be solubilized.

2.6.5.2 Optimization of imidazole concentration for IMAC Ni-charged affinity chromatography

Most cellular lysates from which histidine-tagged (His_{10}) proteins are to be purified contain endogenous protein contaminants that bind to the IMAC absorbent. Usually, these proteins bind less strongly than the histidine-tagged proteins (His_{10}) because endogenous proteins only contain a short run of histidine (less than 10) in their sequences. Optimization of the imidazole concentration in the binding buffer and the elution buffer can effectively reduce or eliminate the coabsorption or coelution of protein contaminants with the target proteins. Optimization is carried out using a small amount of sample, a small IMAC column, and a gradient maker. The optimized conditions can then be used to design the purification protocol for a larger amount of sample on a larger column.

2.6.5.3 Reducing nonspecific protein interaction in IMAC Ni-charged affinity chromatography

Contaminating proteins may be connected to target proteins via hydrophobic interactions, electrostatic interactions or disulfide links. Including nonionic detergents (DDM or Triton X-100) or alcohols (20% ethanol or 50% glycerol) into the sample or the binding buffer reduces the hydrophobic interaction. Increasing NaCl concentration to 1 M in the binding buffer reduces the electrostatic interaction. Adding up to 20 mM β-mercaptoethanol to the binding buffer reduces the disulfide bond. Do not use a higher concentration of β-mercaptoethanol than required or use a more powerful reducing agent such as dithiothreitol or dithioerythritol, as the immobilized metal ion may be reduced to cause the loss of protein binding.

2.6.5.4 pH and EDTA levels in IMAC Ni-charged affinity chromatography

Low pH levels may reduce the binding of histidine-tagged (His_{10}) proteins to the IMAC Ni-charged affinity column. It is therefore important to adjust the pH level in the sample to 7.0–8.0 before loading the sample to the column. EDTA is a metal chelating reagent commonly used in the sample buffer to enhance protein solubility and prevent protein cleavage by endogenous metalloproteases. However, EDTA at 10 mM or higher concentrations strips away the chelated Ni^{++} ion and elute the bound proteins from the column.

2.6.5.5 Molecular weight and shape in size-exclusion chromatography

When using size-exclusion chromatography for protein purification, it is best to choose a gel with a selectivity curve where the molecular weight of the target protein is located in the middle of the linear range of the curve. A gel has different fractionation ranges for native proteins with tertiary and quaternary structures and denatured proteins in random coils. Therefore, protein molecules with the same weight but different shapes elute at different fractions (Heftmann, 2011). The size of the bead in the gel is an important parameter for the resolution of the protein peak. Medium bead diameters range from 4 to 300 μm. If high-resolution chromatography is needed, a smaller bead size and a narrower size distribution (10–34 μm) must be used.

2.6.5.6 Resolution in size-exclusion chromatography

There are several simple approaches to improve the resolution in the chromatogram, for example, increasing the column length by connecting two or three columns in series, reducing the flow rate, and reducing the sample volume. Protein aggregation is a key factor affecting the resolution in the chromatogram. Adding 10%–20% ethanol to the buffer prevents hydrophobic protein aggregates to form. Increasing the detergent concentration in the buffer higher than the CMC (critical micelle concentration) prevents size exclusion of micelle-associated protein aggregates.

2.6.6 Closing remarks

The TJ is a large-scale membrane organization consisting in over 400 proteins. Due to the complexity of the TJ architecture, the functional role of each TJ protein is best delineated by the in vitro approaches established upon fundamental chemical and structural principles, such as molecular interaction assay, liposome reconstitution assay, single-molecule imaging assay, and so on. These bottom-up approaches call for recombinant protein production. The purified TJ proteins can also be used as standards for various in vitro or in vivo applications or as immunogens to facilitate antibody production.

References

Belardi, B., Son, S., Vahey, M. D., Wang, J., Hou, J., & Fletcher, D. A. (2018). Claudin-4 reconstituted in unilamellar vesicles is sufficient to form tight interfaces that partition membrane proteins. *Journal of Cell Science*.

Cutler, P. (2004). *Protein purification protocols* (Vol. 244). Springer Science & Business Media.

Frey, D. D., & Kang, X. (2005). New concepts in the chromatography of peptides and proteins. *Current Opinion in Biotechnology*, *16*, 552–560.

Heftmann, E. (2011). *Chromatography: Fundamentals and applications of chromatographic and electrophoretic methods. Part A: Fundamentals and techniques*. Elsevier.

Lim, T. S., Vedula, S. R., Kausalya, P. J., Hunziker, W., & Lim, C. T. (2008). Single-molecular-level study of claudin-1-mediated adhesion. *Langmuir*, *24*, 490–495.

Sabate, R., de Groot, N. S., & Ventura, S. (2010). Protein folding and aggregation in bacteria. *Cellular and Molecular Life Sciences: CMLS*, *67*, 2695–2715.

Chapter 3

Biophysical approaches for tight junction

Chapter 3.1

Electrophysiology of epithelial transport

3.1.1 Electric potential, resistance, and capacitance of cell membrane

3.1.1.1 Membrane potential

The concentration differences of ions between intracellular compartment and extracellular compartment result in concentration gradients for each ion across the cell plasma membrane. Concentration gradients induce the diffusion of ions from higher to lower concentration. Because ions carry electric charges, the diffusion of ions will create electric potential differences across the cell plasma membrane, which counterbalance the concentration gradients. The membrane potential for each ion is the equilibrium potential at which there is no net flow of the ion across the cell plasma membrane, and can be calculated from the Nernst equation (Hille, 2001):

$$V_{\mathrm{eq}} = \frac{RT}{zF} \ln \frac{[ion]_{\mathrm{out}}}{[ion]_{\mathrm{in}}} \tag{3.1.1}$$

where V_{eq} is the equilibrium potential (relative to the extracellular space being regarded as zero), $[ion]_{\mathrm{out}}$ and $[ion]_{\mathrm{in}}$ are the outside and inside concentrations of the ion, respectively, R is the gas constant, T is the Kelvin temperature, F is the Faraday's constant, z is the valence of the ion, and ln is the logarithm to the base of e.

Because there are different ions each at different concentrations in intracellular and extracellular compartments, the overall membrane potential, also known as the resting membrane potential, is an average of the membrane

A Laboratory Guide to the Tight Junction. DOI: https://doi.org/10.1016/B978-0-12-818647-3.00003-9

potential for each individual ion weighted by its permeability, and can be calculated from the Goldman–Hodgkin–Katz equation (Hille, 2001):

$$V_{\mathrm{m}} = \frac{RT}{F}\ln\left(\frac{\Sigma_i^n P_{A_i}[A_i]_{\mathrm{out}} + \Sigma_j^m P_{B_j}[B_j]_{\mathrm{in}}}{\Sigma_i^n P_{A_i}[A_i]_{\mathrm{in}} + \Sigma_j^m P_{B_j}[B_j]_{\mathrm{out}}}\right) \tag{3.1.2}$$

where V_{m} is the resting membrane potential, P_{ion} is the permeability of the ion, A and B denote cation and anion, respectively, and $[ion]_{\mathrm{out}}$ and $[ion]_{\mathrm{in}}$ are the outside and inside concentrations of the ion, respectively.

Note that this equation only applies to monovalent ionic species. In most animal cells, the membrane permeability to potassium is much higher than the membrane permeability to other ions in the resting state. As a consequence, the resting membrane potential is usually close to the potassium equilibrium potential of -70 mV.

3.1.1.2 Membrane resistance

The membrane resistance (R_{m}) or its inverse quantity, the membrane conductance (G_{m}), reflects the ion's ability to penetrate the membrane. When there is no potential difference across the membrane, or the membrane is said to be short-circuited, the ionic flux occurs exclusively by diffusion, so the membrane resistance or conductance can be related to the membrane permeability (P_{m}) of the ion according to the following equation (Sten-Knudsen, 2002):

$$\frac{1}{R_{\mathrm{m}}} = G_{\mathrm{m}} = \frac{(zF)^2}{RT}\frac{[\mathrm{ion}]_{\mathrm{in}} - [\mathrm{ion}]_{\mathrm{out}}}{\ln([\mathrm{ion}]_{\mathrm{in}}/[\mathrm{ion}]_{\mathrm{out}})}P_m \tag{3.1.3}$$

In a mosaic membrane containing three different types of ion channels permeable to Na^+, K^+, and Cl^-, respectively, the resting membrane potential (V_{m}) can be related to the equilibrium potential of each ion and its membrane conductance by the Millman equation (Sten-Knudsen, 2002):

$$V_{\mathrm{m}} = \frac{G_{\mathrm{Na}}V_{\mathrm{Na}} + G_{\mathrm{K}}V_{\mathrm{K}} + G_{\mathrm{Cl}}V_{\mathrm{Cl}}}{G_{\mathrm{Na}} + G_{\mathrm{K}} + G_{\mathrm{Cl}}} \tag{3.1.4}$$

where V_{Na}, V_{K}, and V_{Cl} are the equilibrium potentials of Na^+, K^+, and Cl^-, respectively; G_{Na}, G_{K}, and G_{Cl} are the membrane conductances of Na^+, K^+, and Cl^-, respectively.

3.1.1.3 Membrane capacitance

The cell plasma membrane also acts as a capacitor. A capacitor stores charge. The amount of stored charge can be calculated by the following equation:

$$Q_{\mathrm{m}} = V_{\mathrm{m}}C_{\mathrm{m}} \tag{3.1.5}$$

where Q_m is the charge stored in the membrane, V_m is the resting membrane potential, and C_m is the membrane capacitance. The membrane capacitance imparts a delay to membrane potential alteration because the membrane potential must first overcome a change in stored charge.

3.1.2 Basic principles of cell membrane electrophysiology

3.1.2.1 Equivalent electrical circuit of cell membrane

Because the cell plasma membrane has potential, resistance, and capacitance, it can be described by an equivalent electrical circuit (Fig. 3.1.1). The electronic diagram symbols commonly used in a circuit are shown in Fig. 3.1.2.

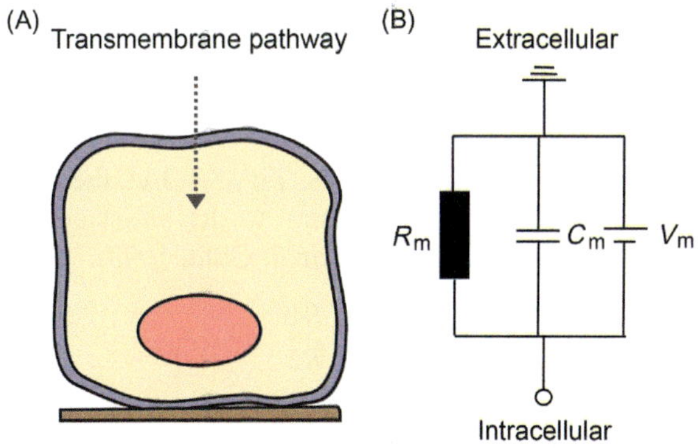

FIGURE 3.1.1 Equivalent electric circuit of cell plasma membrane. (A) Ions permeate through the transmembrane pathway. (B) Each element in the membrane is represented by an electric equivalent, that is, an equilibrium potential, V_m; a resistor, R_m; a capacitor, C_m.

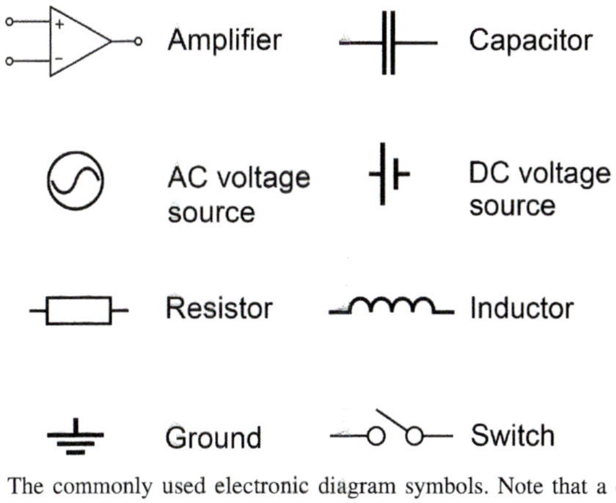

FIGURE 3.1.2 The commonly used electronic diagram symbols. Note that a different resistor symbol is sometimes used by other books.

3.1.2.2 Electric current

Electric current (I) is measured in amperes (A) and can range from picoamperes (pA) to microamperes (μA) in various cell or organ systems. There are two important rules for current measurements: (1) current is conserved at any branch point and (2) current only flows in a complete circuit.

3.1.2.3 Ohm's law

Ohm's law relates electric current, electric potential, and electric resistance. The potential difference (ΔV) between two points linked by a resistor with a resistance (R) is directly proportional to the current (I) flowing through the resistor.

$$\Delta V = IR \tag{3.1.6}$$

3.1.2.4 Voltage divider

When two resistors are connected in series (Fig. 3.1.3A), the potential difference across each resistor is proportional to its resistance and can be described by the following equation (Sherman-Gold, 1993):

$$\frac{\Delta V1}{\Delta V2} = \frac{R1}{R2} \tag{3.1.7}$$

where ΔV_1 is the potential difference across the first resistor with a resistance R_1 and ΔV_2 is the potential difference across the second resistor with a resistance R_2.

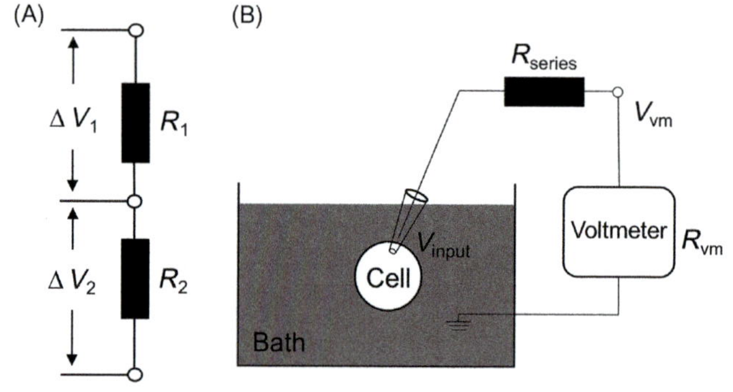

FIGURE 3.1.3 Voltage divider in theory and practice. (A) Equivalent electric circuit represents a voltage divider in which two resistors are connected in series. ΔV_1: the potential difference across the resistor R_1; ΔV_2: the potential difference across the resistor R_2. (B) Practical design of a voltmeter with extremely high resistance. V_{vm} is the potential difference across the voltmeter; R_{vm} is the resistance of the voltmeter; V_{input} is the input potential; R_{series} is the series resistance including the pipette resistance.

In a voltage measurement (Fig. 3.1.3B), the use of a voltmeter with an extremely large resistance (R_{vm}, $>$MΩ) allows the maximal recording accuracy because the potential difference across the voltmeter approximates the input potential, when R_{vm} predominates in the voltage divider according to the following equation:

$$V_{vm} = V_{input}\frac{R_{vm}}{R_{vm} + R_{series}} \tag{3.1.8}$$

where V_{vm} is the potential difference across the voltmeter, V_{input} is the input potential, and R_{series} is the series resistance including the pipette resistance.

3.1.2.5 Current through capacitor

The current (I) flowing through a capacitor with capacitance (C) is proportional to the potential change (ΔV) with time (Δt), which can be described by the following equation (Sherman-Gold, 1993):

$$I = C\frac{\Delta V}{\Delta t} \tag{3.1.9}$$

When a pulse of current is applied to a circuit consisting in a resistor arranged in parallel with a capacitor as in the cell plasma membrane (Fig. 3.1.4A), the current first charges the capacitor, and then charges the resistor. The potential difference ($V(t)$) across the circuit approaches the steady state along an exponential time course (Fig. 3.1.4B).

$$V(t) = V_{inf}(1 - e^{-(t/\tau)}) \tag{3.1.10}$$

The steady-state potential (V_{inf}, also known as the infinite-time potential) is determined by the current (I) and the resistance (R) according to Ohm's law.

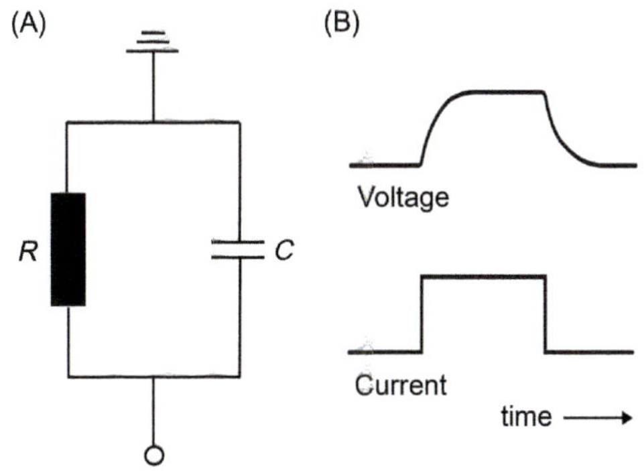

FIGURE 3.1.4 Capacitive current through cell plasma membrane. (A) Equivalent electric circuit represents the membrane as a resistor, R in parallel with a capacitor, C. (B) Voltage response of a parallel RC circuit to a current pulse.

$$V_{inf} = IR \tag{3.1.11}$$

The time constant (τ) is determined by the resistance (R) and the capacitance (C) according to the following equation:

$$\tau = RC \tag{3.1.12}$$

3.1.2.6 Electrode, liquid junction potential, and salt bridge

An electrode transforms the flow of electrons in a metal wire to the flow of ions in a solution. The most common electrode used in electrophysiological measurements is the Ag/AgCl electrode, which converts Ag atoms and Cl^- to AgCl and electrons. The Ag/AgCl electrode only works in solutions containing Cl^-. If two electrodes face different Cl^- concentrations (for instance, 3 M KCl inside a micropipette and 140 mM NaCl in a bathing solution), there will be a difference in the electrochemical potentials at the two electrodes, known as the liquid junction potential. The most common method of eliminating the liquid junction potential is to place a salt bridge consisting of 3 M KCl between the electrode and the solution (e.g., containing 140 mM NaCl). When a salt bridge is used, the ions in the salt bridge are present in large excess and carry almost all of the current across the liquid junction.

3.1.3 Electrophysiology of an epithelium

3.1.3.1 Equivalent electrical circuit of an epithelium

Transepithelial electric current includes contributions from two pathways: the transcellular pathway and the paracellular pathway (Fig. 3.1.5A). In the equivalent circuit of the epithelium (Fig. 3.1.5B), R denotes the electric resistance and V_{eq} denotes the equilibrium potential (also known as E, the zero-current potential). V_{eq} is caused by the electrogenic diffusion of permeant ions (also see Section 3.1.1.1). The subscripts denote: a = apical membrane; b = basolateral membrane; and p = paracellular pathway.

3.1.3.2 Transepithelial resistance

Because the transcellular pathway and the paracellular pathway are arranged in parallel, the transepithelial resistance (TER or R_{te}) can be related to the transcellular ($R_a + R_b$) and the paracellular (R_p) resistance by the following equation:

$$\frac{1}{R_{te}} = \frac{1}{R_a + R_b} + \frac{1}{R_p} \tag{3.1.13}$$

The degree of epithelial leakiness is best assessed by the ratio of transcellular resistance versus paracellular resistance, that is, $(R_a + R_b)/R_p$. The larger

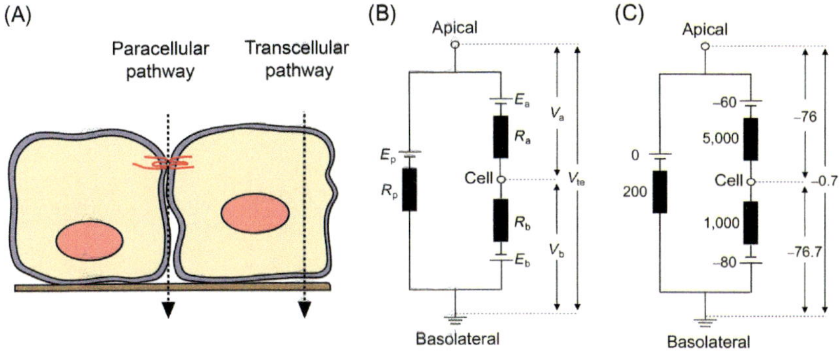

FIGURE 3.1.5 Equivalent electric circuit of an epithelium. (A) Ions can permeate through the tight junction (the paracellular pathway) or through the cell (the transcellular pathway) by crossing the apical and the basolateral membrane. (B) Each element in the circuit (a: apical membrane, b: basolateral membrane, p: paracellular pathway) is represented by an electric equivalent, that is, an equilibrium potential, E, in series with a resistor, R. (C) Membrane potentials (in mV) for a realistic epithelium with the assumed values of E (in mV) and R (in Ω/cm^{-2}). Note that the membrane potentials differ from the respective E values. *Reproduced with permission from Hou, J. (2018). The paracellular channel - biology, physiology and disease (Academic Press).*

this ratio is, the leakier the epithelium becomes. It is clear that the conductance (G) is proportional to area, and consequently that the resistance ($R = 1/G$) decreases with increasing area. This is the reason why R_{te} is expressed as Ω/cm^{-2} whereas G_{te} as S/cm^2 in epithelial transport physiology. R_{te} can be measured by applying a current pulse, I, across the epithelium and recording the resultant potential difference, ΔV. According to Ohm's law (Eq. 3.1.6), the resistance is calculated as the potential difference across the epithelium divided by the amplitude of the current pulse through the epithelium ($\Delta V/I$).

3.1.3.3 Transepithelial potential

The transepithelial potential depends upon E and R of all circuit elements. The reason is that there is an intraepithelial loop current due to the presence of the paracellular pathway. The loop current (I_e) is given by Hoffman and Jamieson (1997):

$$I_e = \frac{E_b + E_p - E_a}{R_a + R_b + R_p} \tag{3.1.14}$$

and the potential of each membrane element at $I \neq 0$ is given by:

$$V = E - I_e R \tag{3.1.15}$$

Inserting the expression for I_e into the membrane potential equation yields the following three equations for the apical membrane potential (V_a), the basolateral membrane potential (V_b), and the transepithelial potential (V_{te}) (Hoffman & Jamieson, 1997).

$$V_a = \frac{E_a(R_b + R_p) + (E_b + E_p)R_a}{R_a + R_b + R_p} \tag{3.1.16}$$

$$V_b = \frac{E_b(R_a + R_p) + (E_a - E_p)R_b}{R_a + R_b + R_p} \tag{3.1.17}$$

$$V_{te} = \frac{E_p(R_a + R_b) + (E_a - E_b)R_p}{R_a + R_b + R_p} \tag{3.1.18}$$

Fig. 3.1.5C shows the solution of Eqs. (3.1.16)–(3.1.18) for assumed values of E and R in realistic cell membranes. The potentials of the two membrane domains depend upon each other because they are connected by the paracellular pathway, which acts as an electric shunt. The changes in the properties of one of the membranes alter not only its own potential but also the potential across the opposite membrane and the potential across the entire epithelium. The electric shunting process requires a low paracellular resistance (R_p) due to the presence of paracellular channels. If the value of R_p approaches infinity, then the cell membrane potential approaches the value of the respective E, and I_e approaches zero.

3.1.4 Noise prevention and signal conditioning

3.1.4.1 External noise

The three most common types of external noise are hum noise, switch noise, and digital noise (Bretschneider & De Weille, 2018). Hum noise is caused by the alternating current (AC) in the power grid with a frequency of 60 Hz in North America. Switch noise is caused by the inductive current in the power grid when major devices are turned on or off. Digital noise is caused by the high-frequency clock pulse in digital devices, for example, monitors and computers. These external noises can be removed by grounding objects and shielding sensitive circuits. Ideal electrical ground is a large metal object lodged deep into the earth, which eliminates the noise by short-circuiting it via a low resistance route. Shielding in the form of a grounded metal cage (known as Faraday cage) is necessary to exclude the electromagnetic interference.

3.1.4.2 Intrinsic noise

The major type of intrinsic noise is thermal noise. It results from the random motion of thermally excited charge carriers in a conductor. The amplitude of intrinsic noise is measured by its root-mean-square (σ^2, rms) value. Noise can be classified by its power spectrum, the logarithm of σ^2 (on the y-axis)

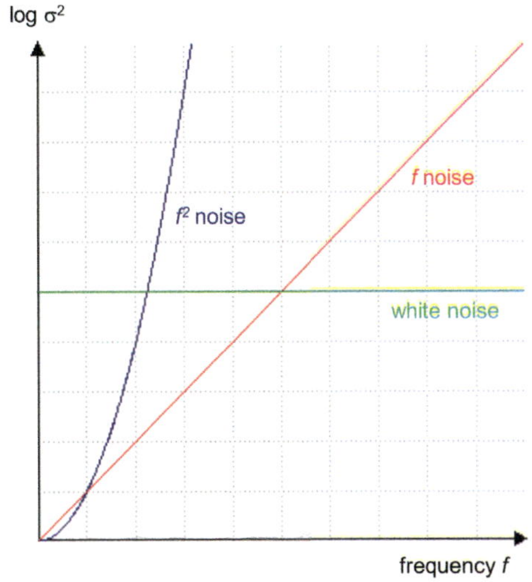

FIGURE 3.1.6 Power spectra of common intrinsic noises. The logarithm of σ^2 is on the y-axis and the frequency is on the x-axis.

at different frequencies (on the x-axis). The power spectrum of thermal noise may show an f signature or an f^2 signature or appear as a flat line (so-called white noise) (Fig. 3.1.6).

3.1.4.3 Filtering

A filter is a circuit that removes selected frequencies from the signal. Filtering is often performed to remove unwanted noises from the data. If the range of frequencies that contain the signal of interest is known, then all other frequencies can be filtered out. There are four major types of filters—low-pass, high-pass, band-pass, and notch filters (Fig. 3.1.7) (Bretschneider & De Weille, 2018). A low-pass filter shows little or no resistance at low frequencies while blocks the signals at high frequencies. A high-pass filter lets through the high-frequency signals but blocks the low-frequency signals. Band-pass and notch filters are combinations of a low-pass and a high-pass filter to selectively leave or block a range of frequencies. Low-pass filters are most commonly used in biological measurements because biological signals usually have an upper frequency limit.

3.1.5 Data acquisition and digitization

In electrophysiological experiments, data are most often presented in current and voltage waveforms whose magnitudes vary with time, so-called analog signals. In order for computers to document analog signals, a process of "analog-to-digital" conversion is undertaken to convert the analog data into a

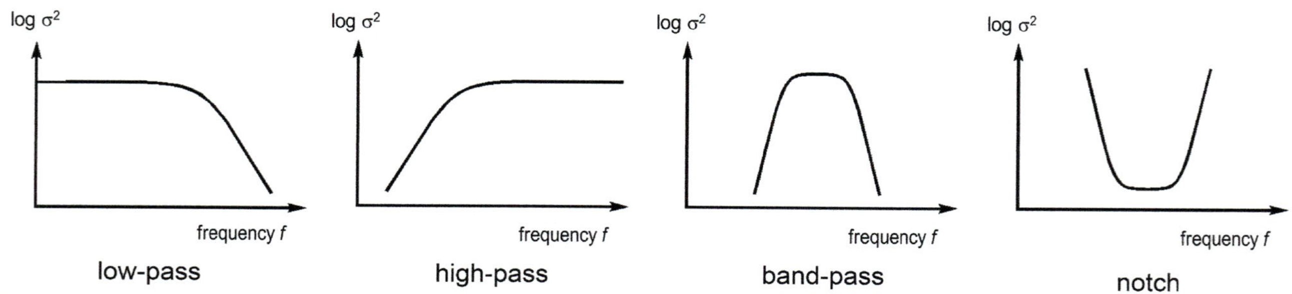

FIGURE 3.1.7 Characteristics of four common types of filters. Power spectra of low-pass, high-pass, band-pass, and notch filters indicate the output of the filter when the input is a flat signal.

digital format. Digitization involves dividing continuous analog data into discrete numbers by a process known as sampling. The quality of conversion depends upon the temporal and amplitude resolution of sampling. The temporal resolution is determined by the Nyquist theorem, which states that data must be sampled at a rate equal to or faster than twice the bandwidth of the signal in order to prevent a phenomenon known as aliasing (Marvasti, 2012). In practice, it is common to sample at a rate significantly faster than the minimum rate determined by the Nyquist theorem. When low-pass filters are used, sampling is typically performed at 2.5 times the filter bandwidth. For example, if the data are filtered at 10 kHz then they need to be sampled at 25 kHz or faster. The amplitude resolution is described by the sampling number at discrete time points. When the total measurement range is divided into a fixed number of possible values, the number of values is a power of two, referred to as the bits. For example, two bits can combine into four different values. There are 2^n values for an n-bit input in a digital device.

References

Bretschneider, F., & De Weille, J. R. (2018). *Introduction to electrophysiological methods and instrumentation*. Academic Press.

Hille, B. (2001). *Ion channels of excitable membranes* (Vol 507). Sunderland, MA: Sinauer.

Hoffman, J., & Jamieson, J. (1997). *Handbook of physiology: Section 14: Cell physiology*. Oxford University Press.

Marvasti, F. (2012). *Nonuniform sampling: theory and practice*. Springer Science & Business Media.

Sherman-Gold, R. (1993). *The axon guide for electrophysiology & biophysics laboratory techniques*. Axon Instruments.

Sten-Knudsen, O. (2002). *Biological membranes: Theory of transport, potentials and electric impulses*. Cambridge University Press.

Chapter 3.2

Epithelial cell cultures in Ussing chamber

3.2.1 Background knowledge

3.2.1.1 Theoretic considerations

3.2.1.1.1 Recording of transcellular transport

The Ussing chamber was first developed by Dr. Hans Ussing in the 1950s to study active transepithelial transport in the frog skin (Ussing & Zerahn, 1951). The need to distinguish the active transcellular pathway from the passive paracellular pathway led to the development of a dual chamber system that sandwiched and superfused the tissue sample with proper electrolyte solutions to maintain cellular viability. To eliminate the driving force for paracellular ion and water transport, identical electrolyte solutions of the same volume were added to each chamber. The electrogenic ionic current, known as the short-circuit current (I_{sc}), can be measured by clamping the transepithelial potential (V_{te}) to zero with an external current passing through the epithelium (Ussing, 1953). In the theory, I_{sc} is determined by V_{te} and the resistance of the epithelium (R_{te}), which includes the transcellular resistance ($R_a + R_b$) and the paracellular resistance (R_p).

$$I_{sc} = \frac{V_{te}}{R_{te}} \tag{3.2.1}$$

Because the transcellular pathway and the paracellular pathway are arranged in parallel (Fig. 3.1.5B), R_{te} can be related to $R_a + R_b$ and R_p by Eq. (3.1.13). Inserting the expression of R_{te} from Eq. (3.1.13) into Eq. (3.2.1) yields:

$$I_{sc} = \frac{V_{te}}{\frac{(R_a + R_b)R_p}{R_a + R_b + R_p}} \tag{3.2.2}$$

According to Eq. (3.1.18), V_{te} is caused by the equilibrium potential difference between the apical membrane (E_a) and the basolateral membrane (E_b), mindful that the equilibrium potential of the paracellular pathway (E_p) is now zero due to the saline condition applied to each chamber. Inserting the expression of V_{te} from Eq. (3.1.18) into Eq. (3.2.2) yields:

$$I_{sc} = \frac{E_a - E_b}{R_a + R_b} \tag{3.2.3}$$

Note that I_{sc} is not equivalent to the loop current (I_e), which is generated when the transcellular pathway and the paracellular pathway are arranged in series and can be described by Eq. (3.1.14). However, when $E_p = 0$, and $R_a + R_b >> R_p$, $I_{sc} \approx I_e$, albeit in opposite sign. Because E_a and E_b are theoretic values and difficult to measure, Eq. (3.2.3) is practically less meaningful. Instead, a simplified version of I_{sc}, known as the equivalent short-circuit current (I'_{sc}), is used more often in epithelial transport physiology.

$$I'_{sc} = \frac{V_{te}}{R_p} \tag{3.2.4}$$

I'_{sc} represents the paracellular current in the parallel circuit (Fig. 3.1.5B). When $R_a + R_b > R_p$, the paracellular pathway becomes the predominant pathway to carry the transepithelial current, and therefore, I'_{sc} approximates I_{sc}.

3.2.1.1.2 Recording of paracellular transport

The paracellular conductance (G_p) or its inverse value, R_p, is the most important measurement of paracellular ionic permeability. According Eq. (3.1.13), R_p equals R_{te} when $R_a + R_b$ approaches infinity, that is, when the transcellular pathway is pharmacologically inhibited. R_{te} is often measured by applying a current pulse, I, across the epithelium and recording the resultant potential difference, ΔV. Ohm's law dictates that $R_{te} = \Delta V/I$ (Eq. 3.1.6).

The ionic selectivity of an epithelium can be determined from the diffusion potential that builds up across the epithelium if an ionic gradient is applied. Because the transcellular transport is directional, the ionic gradient can be experimentally set up in the opposite direction to the transcellular pathway in order to allow a specific measurement of paracellular ionic selectivity. The Goldman–Hodgkin–Katz equation relates the ionic selectivity to the diffusion potential across the epithelium (Eq. 3.1.2). When Na^+ and Cl^- are the only ions present in the solution, then the Goldman equation is reduced to:

$$\eta = \frac{-\left(\varepsilon - e^{(V_{dif}/(RT/F))}\right)}{1 - \varepsilon e^{(V_{dif}/(RT/F))}} \tag{3.2.5}$$

where $\eta = P_{Na}/P_{Cl}$, the ratio of Na^+ to Cl^- permeability, $\varepsilon = C^{(b)}/C^{(a)}$, the ratio of basolateral to apical NaCl concentration, V_{dif} is the diffusion potential (relative to the apical space being regarded as zero), R is the gas constant, T is the Kelvin temperature, and F is the Faraday's constant (Hou, 2018).

In a simple situation where the epithelium is surrounded on both sides by the same salt and in equal concentrations, the conductance of an ion—j is related to its permeability by the following equation:

$$G_j = \frac{z_j^2 F^2}{RT} C_j P_j \tag{3.2.6}$$

where z is the valence of the ion and C is the concentration of the ion in either side of the epithelium (Sten-Knudsen, 2002). If the epithelium is placed in a NaCl buffered solution, then the permeabilities of Na^+ and Cl^- can be determined from R_{te} and η according to the Kimizuka–Koketsu equation.

$$P_{Na} = \frac{RT}{F^2}\frac{1}{CR_{te}}\frac{\eta}{(1+\eta)} \tag{3.2.7}$$

$$P_{Cl} = \frac{RT}{F^2}\frac{1}{CR_{te}}\frac{1}{(1+\eta)} \tag{3.2.8}$$

3.2.1.2 Practical applications

3.2.1.2.1 Classic Ussing chamber

The chamber apparatus originally designed by Dr. Ussing consists in two paired hemichambers between which an epithelial tissue can be mounted and superfused (Fig. 3.2.1). A CO_2/HCO_3^--buffered Ringer solution is used as the superfusate in the reservoir above each hemichamber. Oxygen and carbon dioxide tension are maintained in the Ringer solution by a circulating gas lift made of a mixture of 95% O_2 and 5% CO_2. Thermostated water is infused into the water jacket of each hemichamber to maintain a desired temperature. Salt bridges are used both to eliminate the liquid junction potential and to prevent exposure of the epithelium to toxic Ag atoms from the Ag–AgCl electrodes. The salt bridges typically consist of 3% agar melted in 3 M KCl solution that is congealed in the tubing. The classic Ussing chamber apparatus including various tissue adaptors, water-jacketed reservoirs, and Ag–AgCl electrodes is available from Physiologic Instruments (San Diego, CA), Warner Instruments (Hamden, CT), and World Precision Instruments (Sarasota, FL).

3.2.1.2.2 Self-contained Ussing chamber

Physiologic Instruments (San Diego, CA) and Warner Instruments (Hamden, CT) make self-contained Ussing chamber apparatus in which the superfusate reservoir and the gas lift are integrated into each hemichamber (Fig. 3.2.2). The temperature in the chamber apparatus can be controlled by an external thermostat via direct connection to a metal heating block. When HEPES-buffered superfusate is used, the tubing system that provides gas lift circulation can be removed from the chamber apparatus.

3.2.1.2.3 Transwell permeable supports

Transwell permeable supports are essential for epithelial cells to form tight junctions, gain apicobasal polarity, and develop TER. They were

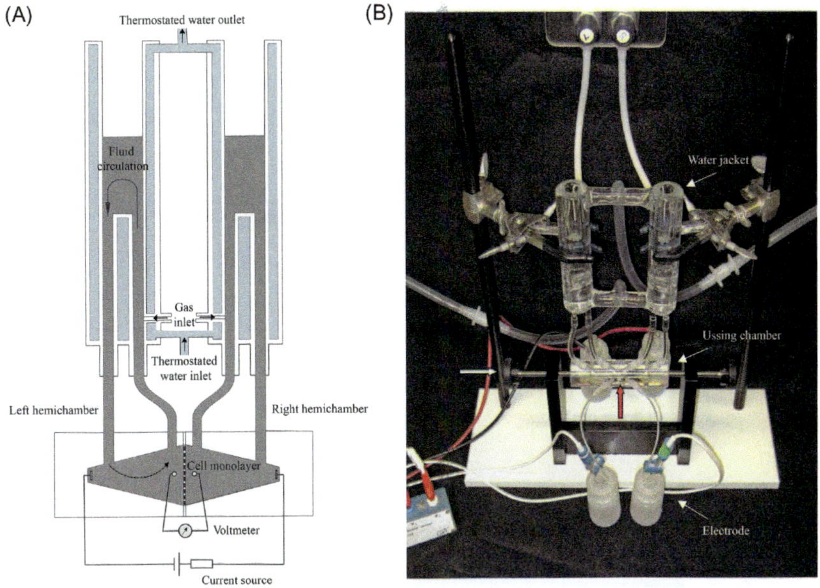

FIGURE 3.2.1 Classic Ussing chamber. (A) Schematic diagram of the Ussing chamber. The Ussing chamber is named after the Danish physiologist Dr. Hans Ussing and consists in two fluid-filled hemichambers separated by an epithelial tissue in the center. Two fluid reservoirs with gas lifts provide circulation and equilibration with O_2 and CO_2. Electric current is applied via two current electrodes connected to a current generator. Transepithelial potential is recorded by two voltage electrodes connected to a voltmeter. (B) Fully assembled Ussing chamber apparatus. The key components of Ussing chamber apparatus include the Ussing chamber itself, the water jacket, and the electrodes that record the electrophysiological properties of the epithelial tissue. *Reproduced with permission from Hou, J. (2018).* The paracellular channel—Biology, physiology and disease. *Academic Press and Clarke, L. L. (2009). A guide to Ussing chamber studies of mouse intestine.* American Journal of Physiology Gastrointestinal and Liver Physiology, 296, *G1151–G1166.*

developed in the 1970s to study in vitro epithelial cell transport (Cereijido, Robbins, Dolan, Rotunno, & Sabatini, 1978; Misfeldt, Hamamoto, & Pitelka, 1976). Corning Inc. (Lowell, MA) commercialized the culture system on permeable supports as Transwell (Fig. 3.2.3). The Transwell system consists of Transwell inserts of different diameters in standard tissue culture wells or dishes (Table 3.2.1). The bottom of the Transwell insert is made of a microporous membrane that allows fluid exchange. The microporous membrane is available in three materials: polycarbonate, polyester, and polytetrafluoroethylene. The pore size in the membrane varies from 0.4 to 3.0 μm. Cells are seeded on top of the membrane in the Transwell insert. The Transwell insert is then placed into a tissue culture well or dish. Culture media or superfusates can be separately added to the interior and the exterior compartments of the Transwell insert. After the cells form tight

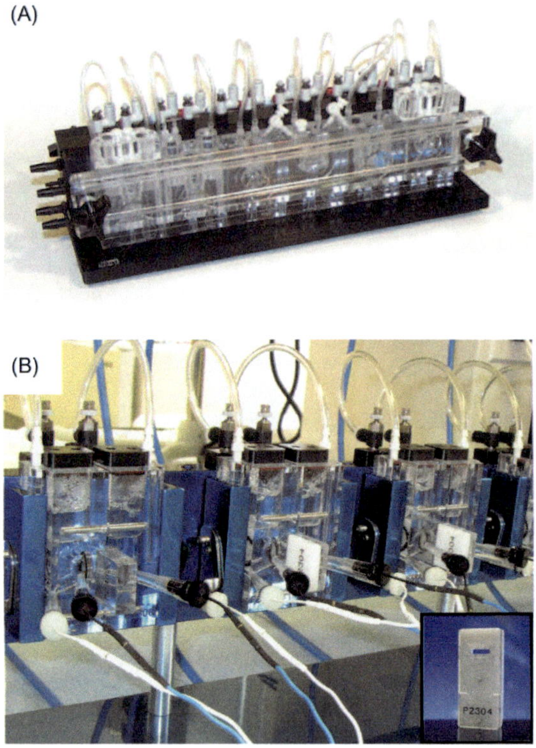

FIGURE 3.2.2 Self-contained Ussing chamber. Two commercial suppliers offer the self-contained Ussing chamber apparatus. (A) Navicyte vertical Ussing system from Warner Instruments. (B) P2250 6-chamber system from Physiologic Instruments. *Inset*: "slider" with pins and aperture for mounting cell cultures and tissue preparations. *Reproduced with permission from Clarke, L. L. (2009). A guide to Ussing chamber studies of mouse intestine.* American Journal of Physiology Gastrointestinal and Liver Physiology, 296, *G1151–G1166.*

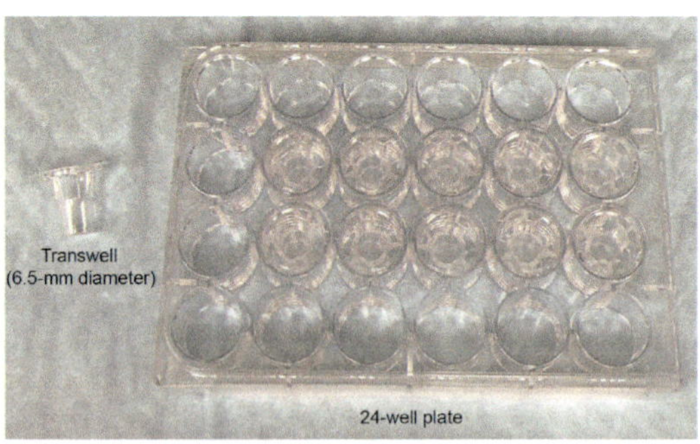

FIGURE 3.2.3 Transwell permeable supports from Corning Inc. Transwell inserts of 6.5 mm diameter in a 24-well plate are shown.

TABLE 3.2.1 Transwell permeable support diameter and growth area.

Transwell insert diameter	Multiple well plate style	Insert membrane growth area
6.5 mm	24 well	0.33 cm^2
12 mm	12 well	1.12 cm^2
24 mm	6 well	4.67 cm^2
75 mm	100 mm dish	44 cm^2

junctions on the Transwell membrane, the membrane can be excised and mounted onto the Ussing chamber apparatus for electrophysiological recording.

3.2.2 Materials and instrumentation

3.2.2.1 Electrophysiological rig

The electrophysiological rig is assembled from the following components (Fig. 3.2.4):

- National Instruments PCI-6014 Multifunction I/0 Device (16-bit, 200 kS/s) receives analog signals.
- National Instruments Data Acquisition (NI-DAQmx) software converts analog signals to digital signals.
- Strathclyde Electrophysiology Software (WinWCP) analyzes voltage and current signals from voltage clamp amplifiers.
- Warner Instruments Epithelial Voltage Clamp amplifier (EC-800) records voltage and current levels of the sample.
- A-M Systems Isolated Pulse Stimulator (Model 2100) provides constant voltage or current pulses to the sample.

3.2.2.2 Ussing chamber assembly

The self-contained Ussing chamber system is assembled from the following components (Fig. 3.2.5):

- Warner Instruments U9926 single-channel Ussing chamber base assembly includes a base plate, two water-jacketed hemichambers, a plug-in cartridge for mounting the sample, two fluid reservoir covers, two drain valves, two clamp screws and four air/gas Luer connectors.
- Warner Instruments U9975A electrode set including two Ag/AgCl pellets and two Ag wires detects the electric potential differences across and passes electric currents to the sample respectively.

FIGURE 3.2.4 Epithelial voltage clamp rig. A fully assembled rig to perform epithelial voltage and current measurements consists in a voltage clamp amplifier, a voltage or current pulse generator and a data acquisition system.

- Warner Instruments U9565SC electrode bridge fitting kit connects the electrodes to the Ussing chamber.
- Warner Instruments U9924T-06 Transwell insert adapter of 6.5 mm diameter allows a Transwell insert to fit into the plug-in cartridge of the Ussing chamber.
- Warner Instruments TC120 thermocirculator provides thermostated water to the water jacket of the Ussing chamber.

3.2.2.3 Superfusate

The superfusates that bathe the sample derive from the Ringer solution (Table 3.2.2). HEPES is used as a buffering agent. Glucose is included as an energy source. Mannitol is used to adjust the osmolality of the superfusates.

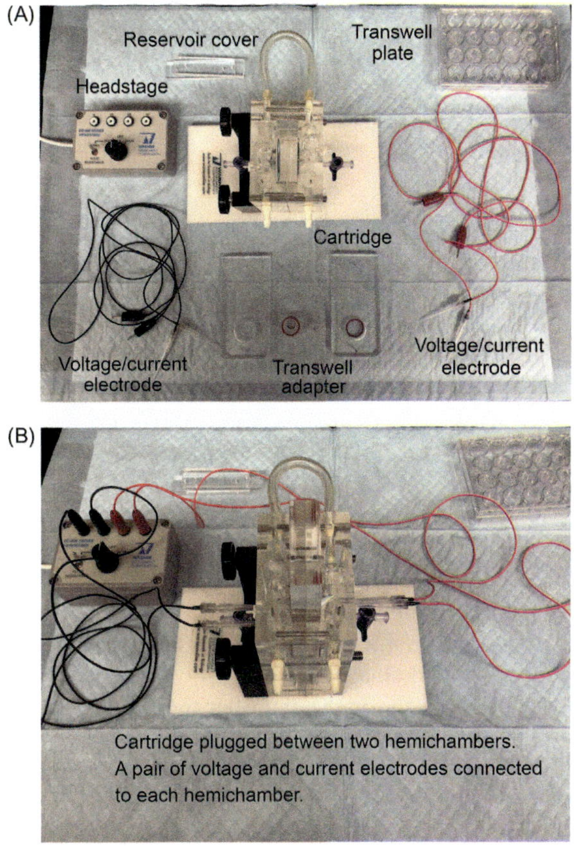

FIGURE 3.2.5 Self-contained Ussing chamber system. Exploded view (A) and assembled view (B) of the Ussing chamber apparatus connected to the headstage of epithelial voltage clamp amplifier via voltage and current electrodes.

3.2.2.4 Cell culture

The Madin-Darby canine kidney (MDCK) strain II cells are cultured as described in Chapter 7.2. After passage, the MDCK-II cells are seeded onto the Transwell inserts (6.5 mm diameter, Table 3.2.1) to allow forming tight junctions.

3.2.3 Experimental procedure

3.2.3.1 Setting up Ussing chamber

1. Assemble the self-contained Ussing chamber system without the Transwell insert in the plug-in cartridge (Fig. 3.2.5). Activate the circulating water flow to the water jacket in the Ussing chamber.

TABLE 3.2.2 Superfusates in Ussing chamber system.

Superfusate	Physio-buffer	Diluting-buffer
Calculated osmolality (mosm/kg)	300	300
NaCl	145	80
$CaCl_2$	2	2
$MgCl_2$	1	1
Glucose	10	10
Mannitol	0	130
HEPES	10	10

Concentration is in mmol/L; pH is adjusted to 7.4 by NaOH or HCl.

2. Insert the voltage and current electrodes into the electrode ports on each hemichamber. Connect a salt bridge made of 3% agar melted in 3 M KCl solution to the tip of each electrode.
3. Fill the chamber reservoirs with 10 mL Physio-buffer prewarmed to 37°C (Table 3.2.2). Make certain the valve at the bottom of each reservoir is closed to prevent fluid drainage.
4. Toggle the MODE selector on the amplifier to AMPLIFY.
5. Use the INPUT OFFSET switch to offset the electric potential bias including the liquid junction potential between the two voltage electrodes ($V_1 - V_2$). Watch on the computer screen that V (voltage) goes to zero (Fig. 3.2.6).

   Note the output V is the differential voltage ($V_1 - V_2$) and in mV. V_1 is placed into the apical solution and V_2 is placed into the basolateral solution.
6. Use the FLUID RESISTANCE panel to offset the series resistance of the bathing solution. A 25 μA electric current is injected to the I1 electrode. The adjustment is made by depressing the PUSH TO ADJUST button and turning the dial until V reaches zero on the computer screen (Fig. 3.2.7).

   Note the output I is the current flowing from I1 to I2 and in μA. I1 is placed into the apical solution and I2 is placed into the basolateral solution.

3.2.3.2 Measuring the short-circuit current

7. Place a Transwell insert seeded with the MDCK-II cells into the plug-in cartridge. Assemble the cartridge into the Ussing chamber.
8. Toggle the MODE selector on the amplifier to V. CLAMP (voltage clamp mode). On the COMMANDS panel, set the VOLTAGE switch to OFF position.

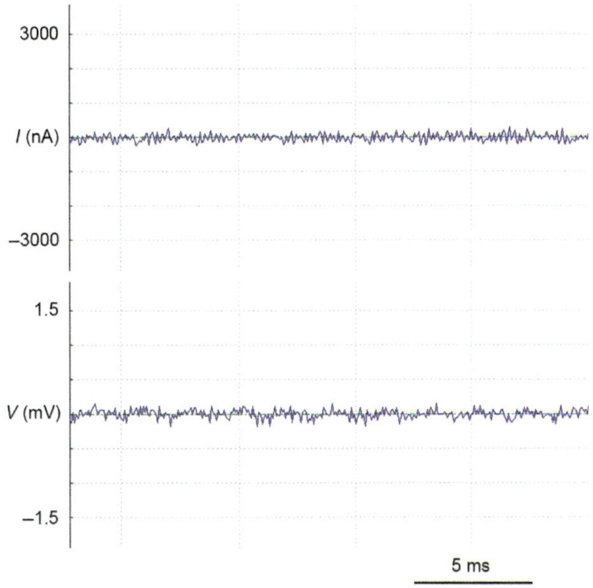

FIGURE 3.2.6 Current and voltage traces during input offset. The electric potential bias between V_1 and V_2 electrodes is automatically offset by the amplifier.

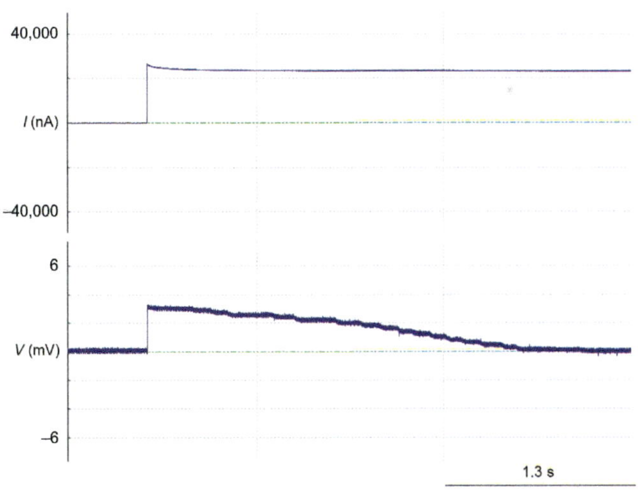

FIGURE 3.2.7 Current and voltage traces during fluid resistance offset. The series resistance of the bathing solution is automatically offset by the amplifier.

9. Record the trace of V on the computer screen.
   Note the recorded voltage is the spontaneous transepithelial potential (V_{te}).

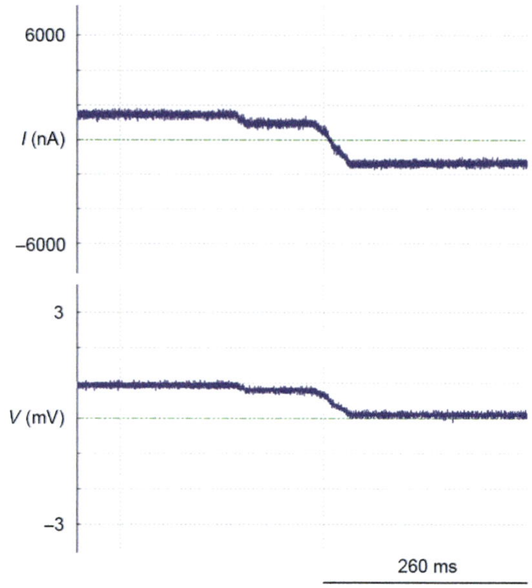

FIGURE 3.2.8 Transepithelial potential and short-circuit current in cultured MDCK-II cells. Note that the MDCK-II cells develop a spontaneous lumen-positive transepithelial potential of ~1 mV and a short-circuit current of ~2 μA due to electrogenic ion transport across the apical and basolateral membranes.

10. Flip the VOLTAGE switch to " − " or " + " position and turn the dial of potentiometer to clamp V to zero.
11. Watch on the computer screen that I (current) changes accordingly (Fig. 3.2.8).

    Note the recorded current is the short-circuit current (I_{sc}) but in opposite sign.
12. Add 1 mM ouabain to the basolateral side of the sample and watch I_{sc} return to zero (Fig. 3.2.9).

    Note ouabain inhibits the Na^+/K^+-ATPase in the basolateral membrane and the transcellular transport pathway.

3.2.3.3 Measuring the paracellualr conductance

13. Toggle the MODE selector on the amplifier to C. CLAMP (current clamp mode). On the COMMANDS panel, set the CURRENT switch to OFF position.
14. Watch the trace of I on the computer screen. If I≠0, then use the CURRENT switch on the COMMANDS panel to clamp I to zero.
15. Turn on the pulse stimulator. Inject continuous current pulses of 10 μA at 1 Hz for 20 ms to the I1 electrode.

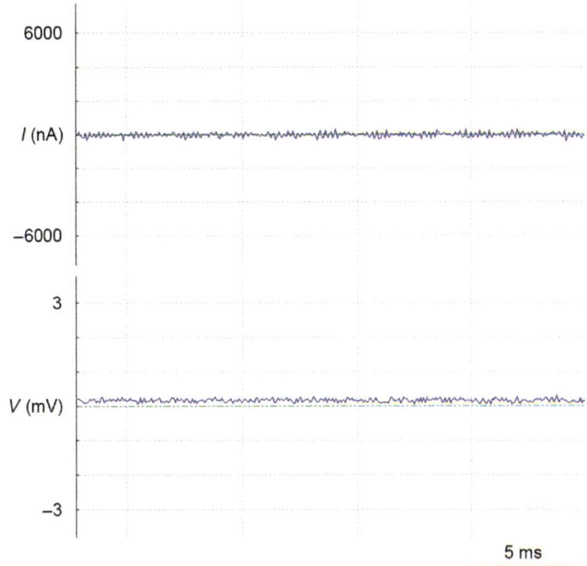

FIGURE 3.2.9 Effects of ouabain on short-circuit current in cultured MDCK-II cells. Note that ouabain inhibits the Na^+/K^+-ATPase in the basolateral membrane and the electrogenic ion transport of the transcellular pathway.

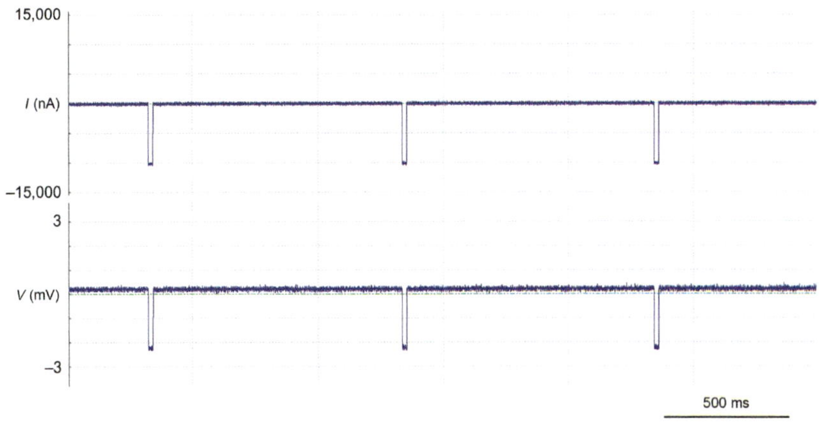

FIGURE 3.2.10 Paracellular resistance in cultured MDCK-II cells. Note that a current pulse of 10 μA elicits a voltage deflection of ~2 mV across the epithelial sample.

16. Record the trace of V on the computer screen (Fig. 3.2.10).

Note the voltage deflection (ΔV) divided by I reflects the paracellular resistance (R_p), an inverse measure of the paracellular conductance (G_p). Because conductance is proportional to area, resistance is expressed as Ω/cm^{-2} and R_p is the product of the value of $\Delta V/I$ and the area of Transwell insert, 0.33 cm^2 in this case (Table 3.2.1).

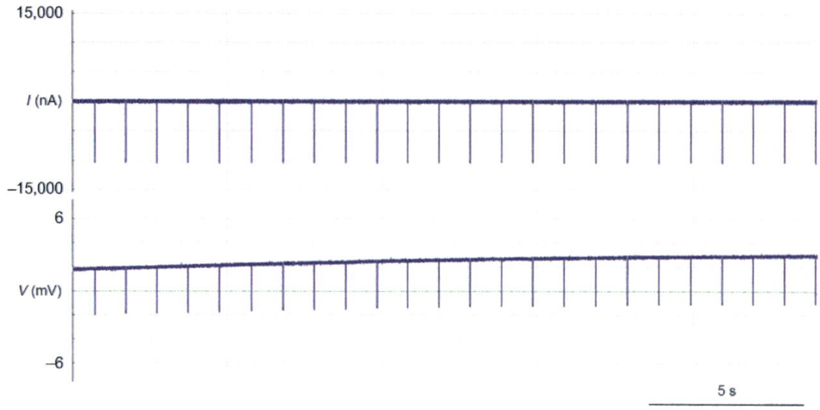

FIGURE 3.2.11 Diffusion potential in cultured MDCK-II cells. Note that a lumen-positive diffusion potential develops when the apical NaCl concentration is reduced indicating the presence cation selective paracellular channels.

Also note that the Transwell insert membrane has an intrinsic resistance, which needs to be measured with a blank insert and subtracted from R_p.

3.2.3.4 Measuring the paracellular ion selectivity

17. Continue to clamp the sample at $I = 0$.
18. Perfuse the apical side of the sample with the diluting buffer (Table 3.2.2).
19. Watch on the computer screen that V changes accordingly (Fig. 3.2.11).

    Note the recorded voltage is the diffusion potential (V_{dif}).

3.2.4 Data analysis

3.2.4.1 Baseline and peak amplitude

The baseline is set to a section of the trace where there is no activity. An event is defined to be a significant deviation from the baseline. There are three important measurements of an event—peak amplitude, peak time, and slope (Fig. 3.2.12). Because the baseline may drift over time during an experiment, each event must be relative to the most proximal baseline region it precedes or succeeds.

3.2.4.2 *I–V* curve

The *I–V* curve for paracellular channels can be obtained by plotting the voltage deflection (ΔV) against a series of current steps from $-50\,\mu A$ to

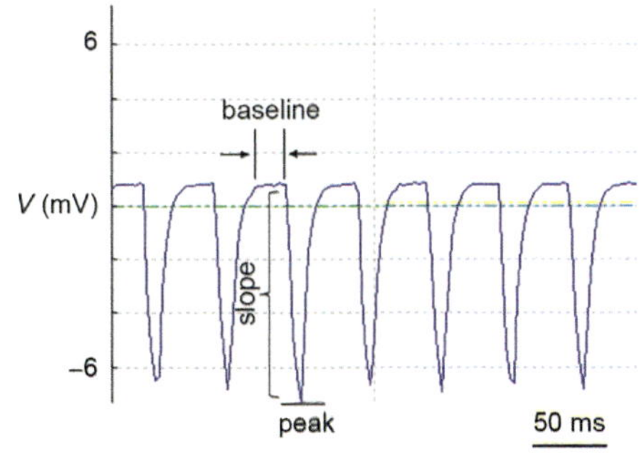

FIGURE 3.2.12 Intended statistical baseline and region positions. The peak amplitude, peak time, and slope can be measured from the representative voltage trace.

+ 50 μA in 10 μA increments at a holding current of 0 μA. The paracellular channel conductance is constant if its $I-V$ curve is a straight line. The slope of the $I-V$ curve equals the conductance of the channel.

3.2.4.3 MATLAB functions

Eqs. (3.2.5), (3.2.7), and (3.2.8) can be written into MATLAB (MathWorks) programming functions to calculate paracellular ion selectivity ($\eta = P_{Na}/P_{Cl}$), paracellular Na^+ permeability (P_{Na}), and paracellular Cl^- permeability (P_{Cl}), respectively.

```
function yeta=Na_Cl_permeability_ratio (dilution_potential)

% elph=basal_con/apical_con
% yeta=Na_permeability/Cl_permeability
% v=dilution_potential/(kT/e)
% dilution potential is regarded as 0 at the apical side

k=1.3805*10^(-23);  % k in J/K
T=273+37; % temperature in K
eValue=1.602*10^(-19); % in C
v=dilution_potential*10^(-3)*eValue/(k*T);

apical_con=145*10^(-3);
basal_con=80*10^(-3);
elph=basal_con/apical_con;

yeta=-(elph-exp(v))./(1-elph*exp(v));
```

3.2.5 Troubleshooting

3.2.5.1 Quality of cell monolayer

Damages to the cell monolayer may reduce the R_{te}, increase the I_{sc}, and dissipate the V_{te}, all of which induce system errors in the measurement of paracellular permeabilities. The quality of cell culture is also important. As elaborated in Chapter 7.2, the membrane material in Transwell inserts, the passage number of cell strains, the loading density of cells, and the culture condition including culture medium and culture duration, all affect the development of tight junction and the value of R_{te}. Therefore recordings must be performed on multiple independent Transwell inserts for each cell strain or cell clone. If genetic manipulation has been introduced to cells, then multiple independent clones must be analyzed.

3.2.5.2 Quality of electrodes

A bubble or break in the salt bridge that connects the Ussing chamber to voltage and current electrodes may cause an interruption in the electrical circuit that senses electric potential difference or inject electric current. Care must be taken to avoid the intense bubbling of gas mixture through the bathing medium, which may introduce noise to the electric potential measurement. The electric potential bias between two voltage electrodes is another major source of noise (Fig. 3.2.13). Even though the amplifier can automatically offset the bias, it is highly recommended to neutralize the bias by

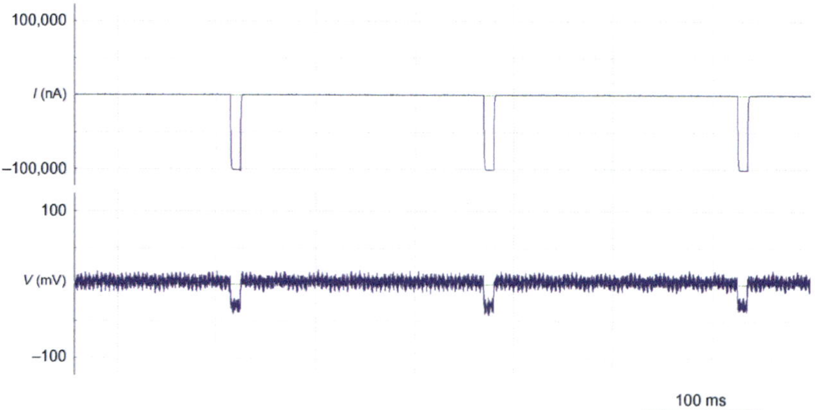

FIGURE 3.2.13 Illustration of electric noise in paracellular resistance measurement. The same MDCK-II cells as in Fig. 3.2.10 were recorded with a pair of flawed voltage electrodes. Note that a current pulse of 100 μA has to be applied to elicit a voltage deflection larger than the baseline noise level.

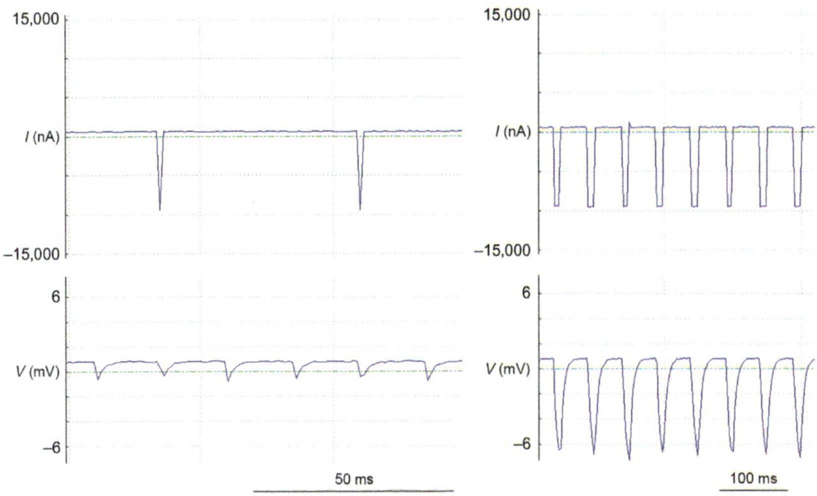

FIGURE 3.2.14 Effect of current pulse duration on paracellular resistance measurement. Human Colon Carcinoma Cell Line 2 (Caco2) cells were recorded with two different current pulses: 10 μA, 20 Hz, 1 ms duration (left) and 10 μA, 20 Hz, 10 ms duration (right). Note that the 1 ms current pulse is not sufficient to charge the cell membrane or elicit a voltage deflection.

connecting the two electrodes and dipping the Ag/AgCl pellet into 3 M KCl for at least 24 hours.

3.2.5.3 Electric pulse

The cell membrane capacitance in epithelial cells ranges from 1 to 100 $\mu F/cm^2$. In the epithelial cells with high membrane capacitance (C_m) and high paracellular resistance (R_{te}), the duration of current pulse must be long enough, that is, greater than the time constant (τ, the product of C_m and R_{te}) because the current first charges the capacitor and then charges the resistor. Transient current pulses may cause underestimation of V_{te} and R_{te} in some epithelial cells or tissues (Fig. 3.2.14).

3.2.6 Closing remarks

The Ussing chamber technique is a robust and reliable approach to record epithelial transport properties. Its in vitro use, largely based upon cell cultures, facilitates the quantitative analyses of pharmacological inhibitors of ion channels and transporters. More advanced techniques can be added to the Ussing chamber system, such as the patch clamp technique, the confocal microscopy, and the electron microscopy. Genetic manipulation can also be applied to cell cultures in the Ussing chamber. The Ussing chamber technique is not limited to the study of epithelial cell biology. Endothelial cells, neural cells, and even stem cells can all be cultured in the Ussing chamber and interrogated by the voltage and current clamp circuitry.

References

Cereijido, M., Robbins, E. S., Dolan, W. J., Rotunno, C. A., & Sabatini, D. D. (1978). Polarized monolayers formed by epithelial cells on a permeable and translucent support. *The Journal of Cell Biology*, *77*, 853–880.

Hou, J. (2018). *The paracellular channel—Biology, physiology and disease*. Academic Press.

Misfeldt, D. S., Hamamoto, S. T., & Pitelka, D. R. (1976). Transepithelial transport in cell culture. *Proceedings of the National Academy of Sciences of the United States of America*, *73*, 1212–1216.

Sten-Knudsen, O. (2002). *Biological membranes: Theory of transport, potentials and electric impulses*. Cambridge University Press.

Ussing, H. H. (1953). Transport through biological membranes. *Annual Review of Physiology*, *15*, 1–20.

Ussing, H. H., & Zerahn, K. (1951). Active transport of sodium as the source of electric current in the short-circuited isolated frog skin. *Acta Physiologica Scandinavica*, *23*, 110–127.

Chapter 3.3

Epithelial tissues in Ussing chamber

3.3.1 Background knowledge

3.3.1.1 Intestinal epithelium

The Ussing chamber technique has been widely used to study intestinal transport physiology. In the small intestine, the transcellular Na^+-dependent absorption of glucose via the Na^+/glucose cotransporter 1 (SGLT1) is coupled to the paracellular backflux of Na^+ via claudin-2 and claudin-15 (Hou, 2018). The large intestine absorbs Na^+ through a luminal Na^+ channel known as the epithelial Na^+ channel (ENaC) and secretes Cl^- via the cystic fibrosis transmembrane conductance regulator (CFTR) in the luminal membrane (Kunzelmann & Mall, 2002). The claudin-2–dependent paracellular backflux of Na^+ facilitates the development of diarrhea, which is important for colonic clearance of bacteria (Tsai et al., 2017). Applying the Ussing chamber technique to living intestinal tissues requires a specially designed cartridge for mounting the sample (Fig. 3.3.1). Oxygenating the perfusate to $Po_2 > 400$ mmHg is necessary to overcome tissue hypoxia due to the lack of hemoglobin delivery when the intestinal tissue is removed from the circulation. The electrogenic Na^+ absorption generates a lumen-negative spontaneous transepithelial potential (V_{te}) across the intestinal epithelium, which is less than 5 mV in amplitude for human jejunum and ileum, but ranges from

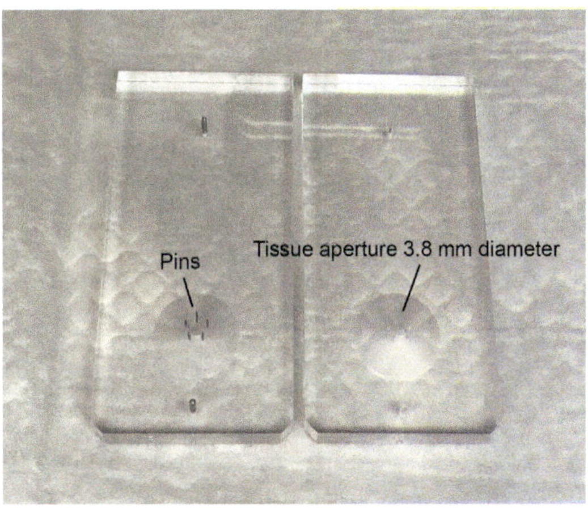

FIGURE 3.3.1 Tissue mounting cartridge for Ussing chamber. The tissue sample is impaled onto a circle of pins outside the aperture of the half-cartridge. The pins enter the opposing blunt end holes when the two half-cartridges are assembled and inserted into the Ussing chamber.

25 to 50 mV in amplitude for human colon (Giebisch, 1979). The R_{te} varies from 50 to 200 Ω/cm^{-2} in human intestines depending upon the segmental location (Giebisch, 1979).

3.3.1.2 Renal epithelium

The Ussing chamber recording principle can be applied to the study of renal transport physiology in perfused single renal tubules. After nearly 50 years since its invention, renal tubule perfusion remains to be a challenging task and can only be performed by a handful of laboratories. To perfuse renal tubules ex vivo, they are first manually dissected from the kidney, then one end is connected to a glass micropipette to perfuse the tubular lumen, and finally the free end is connected to another micropipette to collect the perfused fluid (Figs. 3.3.2 and 3.3.3). In this fashion, the tubular lumen is effectively separated from the bath. Electrophysiological probes can be inserted into the perfusing micropipette to record V_{te} and R_{te}. Cellular morphology can also be examined either during perfusion by light microscopy or after perfusion by electron microscopy. Due to its technical difficulty and limited utility, renal tubule perfusion will not be elaborated in this chapter. Detailed description of renal tubule perfusion can be found in these references (Burg, 1972; Burg, Issaacson, Grantham, & Orloff, 1968; Greger, 1990; Greger & Hampel, 1981).

3.3.2 Materials and instrumentation

3.3.2.1 Electrophysiological rig

The electrophysiological rig is assembled from the components described in Chapter 3.2.2.1.

3.3.2.2 Ussing chamber assembly

The self-contained Ussing chamber system is assembled according to Chapter 3.2.2.2. Warner Instruments U9924B-04 tissue mounting insert of 3.8 mm diameter with pins allows the tissue to fit into the plug-in cartridge of the Ussing chamber.

3.3.2.3 Superfusate

The superfusates that bathe the sample derive from the Ringer solution (Table 3.2.2).

3.2.2.4 Mouse colon preparation

Following euthanasia, the mouse distal colon is dissected by insertion of a glass stick in the lumen and the muscularis externa is removed mechanically.

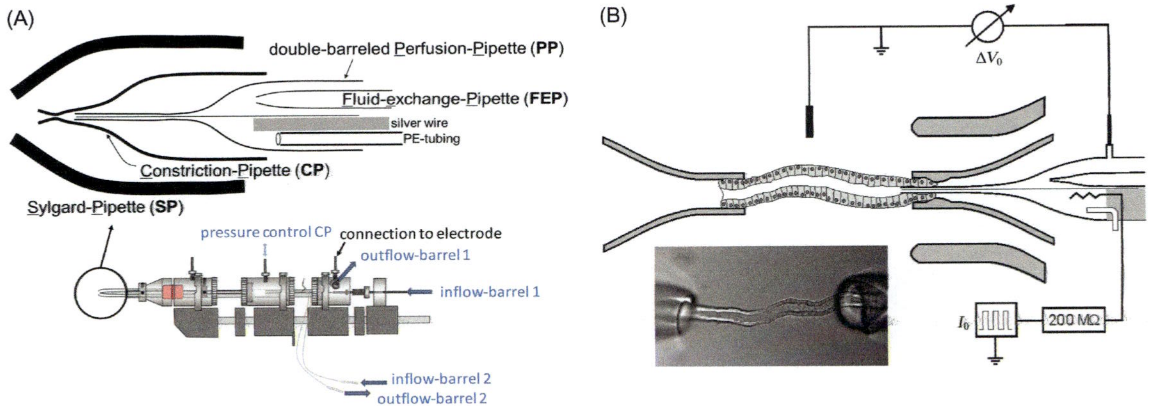

FIGURE 3.3.2 Concentric glass pipette system for renal tubule perfusion. (A) Schematic graph of perfusion pipette. The perfusion pipette (PP) is double barreled with an outer diameter of 10−12 μm. One barrel is used for perfusion, fluid exchange, and transepithelial potential measurement. Other barrel is used for constant current injection. The constriction pipette (CP) consists of a glass pipette with an inner diameter of 15−20 μm. The Sylgard pipette (SP) consists of a glass pipette with an inner diameter of 45 μm and provides electrical insulation. The rate of perfusion is 10-20 nL/minute. (B). Schematic graph of equivalent electrical circuit of perfusion system. A 13-nA current pulse (I_0) is injected to the circuit, which causes a transepithelial voltage deflection (ΔV_0). Cable equation is then used to calculate the transepithelial resistance (R_{te}) from I_0 and ΔV_0. *Inset*: a mouse thick acending limb tubule in perfusion.

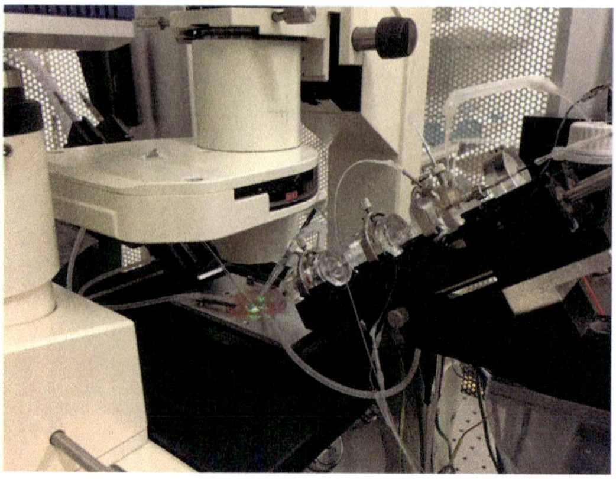

FIGURE 3.3.3 Renal tubule perfusion rig. A fully assembled rig to perform renal tubule perfusion consists in an inverted microscope, a concentric glass pipette system for fluid perfusion (right) and a concentric glass pipette system for fluid collection (left), and microelectrodes to sense voltage and current differences across the tubular epithelium.

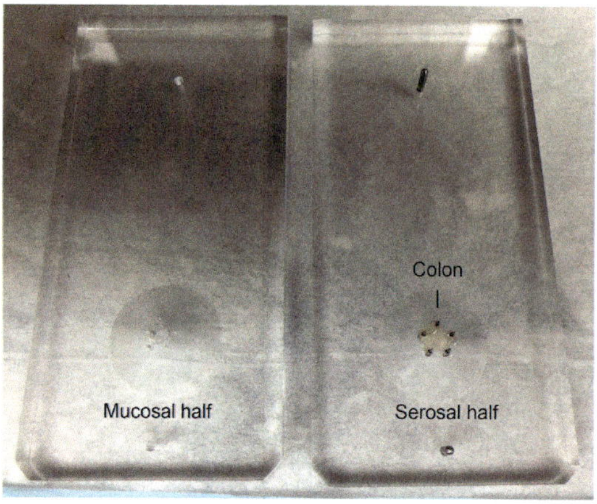

FIGURE 3.3.4 Murine colon preparation mounted on pins of the serosal half of an acrylic plug-in cartridge. Note corresponding holes for pins in the mucosal half and guideposts for assembly into the Ussing chamber.

The intestinal section is made longitudinally along the mesenteric attachment so that the antimesenteric mucosa is situated in the aperture of the Ussing chamber. The seromusculature layers (including serosa, longitudinal, and circular smooth muscle) are relatively thin in the mouse intestine, so the whole mouse distal colon sheet is mounted onto the plug-in cartridge with an exposed area of 0.11 cm^2 (Fig. 3.3.4).

3.3.3 Experimental procedure

3.3.3.1 Setting up Ussing chamber

1. Assemble the self-contained Ussing chamber system without the colon tissue in the plug-in cartridge as described in Chapter 3.2.3.1. Activate the circulating water flow to the water jacket to maintain the temperature of 37°C.
2. Insert the voltage and current electrodes into the electrode ports on each hemichamber. Connect a salt bridge made of 3% agar melted in 3 M KCl solution to the tip of each electrode.
3. Fill the chamber reservoirs with 10 mL Physio-buffer prewarmed to 37°C (Table 3.2.2). Make certain the valve at the bottom of each reservoir is closed to prevent fluid drainage.
4. Turn on the gas lift to provide a mixture of 95% O_2 and 5% CO_2 (also known as carbogen) to the superfusate in the chamber reservoir. The gas lift pressure is set so that the individual bubbles of gas are just discernible to the naked eye.
5. Turn on the EC-800 amplifier and adjust INPUT OFFSET and FLUID RESISTANCE according to Chapter 3.2.3.1.

3.3.3.2 Measuring the short-circuit current

6. Assemble the plug-in cartridge mounted with the colon tissue into the Ussing chamber.
7. Toggle the MODE selector on the amplifier to V. CLAMP (voltage clamp mode). On the COMMANDS panel, set the VOLTAGE switch to OFF position.
8. Record the trace of V on the computer screen (Fig. 3.3.5).

   Note the recorded voltage is the spontaneous transepithelial potential (V_{te}).
9. Flip the VOLTAGE switch to " − " or " + " position and turn the dial of potentiometer to clamp V to zero.
10. Watch on the computer screen that I (current) changes accordingly (Fig. 3.3.6).

    Note the recorded current is the short-circuit current (I_{sc}) but in opposite sign.
11. Add 50 μM amiloride to the luminal side of the sample and watch I_{sc} return to zero (Fig. 3.3.7).

    Note amiloride inhibits the ENaC channel in the luminal membrane and the transcellular transport pathway.

3.3.3.3 Measuring the paracellualr conductance

12. Toggle the MODE selector on the amplifier to C. CLAMP (current clamp mode) and use the CURRENT switch on the COMMANDS panel to clamp I to zero.

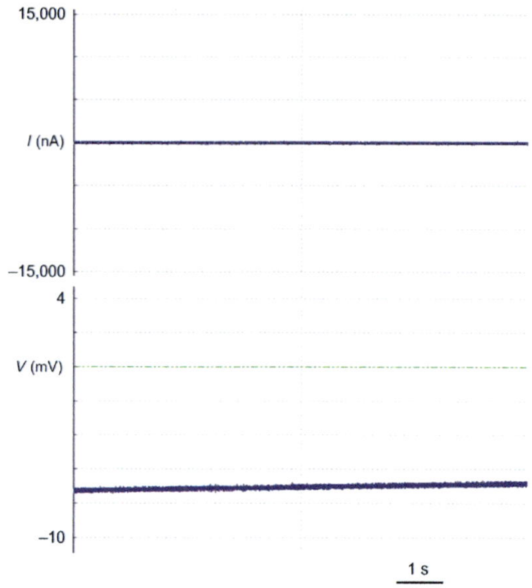

FIGURE 3.3.5 Spontaneous transepithelial potential in mouse colon. Note that the mouse colon develops a spontaneous lumen-negative transepithelial potential of ~ −7 mV due to electrogenic Na^+ absorption and Cl^- secretion.

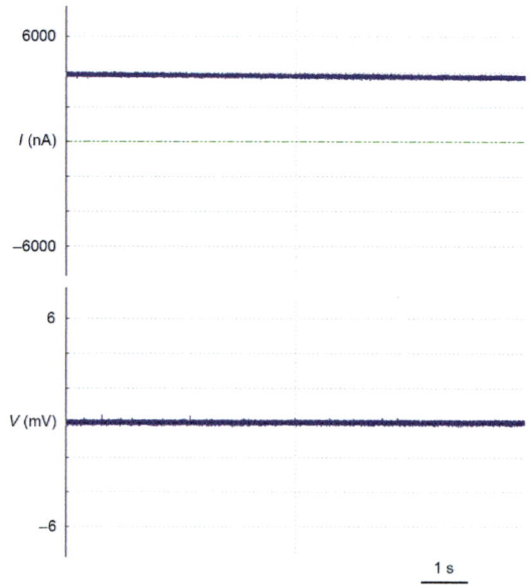

FIGURE 3.3.6 Short-circuit current in mouse colon. Note that the mouse colon develops a short-circuit current of ~4 μA due to the spontaneous transepithelial potential.

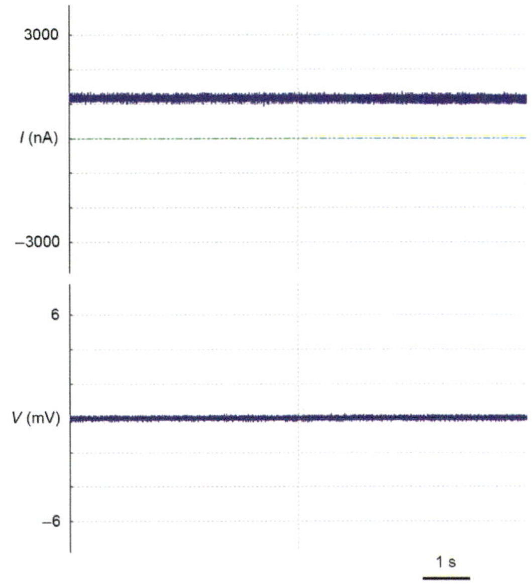

FIGURE 3.3.7 Effects of amiloride on short-circuit current in mouse colon. Note that amiloride inhibits ENaC in the luminal membrane and the electrogenic ion transport of the transcellular pathway. The short-circuit current is ~1 μA in the presence of 50 μM amiloride, suggesting that there exists an additional electrogenic pathway via CFTR-mediated Cl^- secretion.

13. Turn on the pulse stimulator. Inject continuous current pulses of 1 μA at 1 Hz for 50 ms to the I1 electrode.
14. Record the trace of V on the computer screen (Fig. 3.3.8).
    Note the voltage deflection (ΔV) divided by I reflects the paracellular resistance (R_p), an inverse measure of the paracellular conductance (G_p). Because conductance is proportional to area, resistance is expressed as Ω/cm^{-2} and R_p is the product of the value of $\Delta V/I$ and the area of aperture in the mounting cartridge.
15. Continue to clamp the sample at $I = 0$.
16. Perfuse the basolateral side of the sample with the diluting buffer (Table 3.2.2).
17. Watch on the computer screen that V changes accordingly (Fig. 3.3.9).
    Note the recorded voltage is the diffusion potential (V_{dif}).

3.3.4 Data analysis

3.3.4.1 MATLAB functions

The calculation of paracellular ion selectivity ($\eta = P_{Na}/P_{Cl}$), paracellular Na^+ permeability (P_{Na}), and paracellular Cl^- permeability (P_{Cl}) can be made by using the MATLAB programming functions written for Chapter 3.2.4.3.

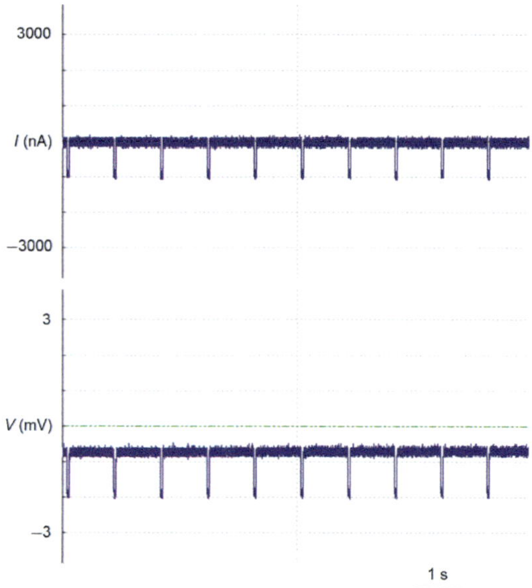

FIGURE 3.3.8 Paracellular resistance in mouse colon. Note that a current pulse of $I = 1\ \mu A$ elicits a voltage deflection (ΔV) of ~ 1 mV across the epithelial sample. Multiplying $\Delta V/I$ with the area of aperture (0.11 cm^2) gives the paracellular resistance of $\sim 110\ \Omega/cm^{-2}$.

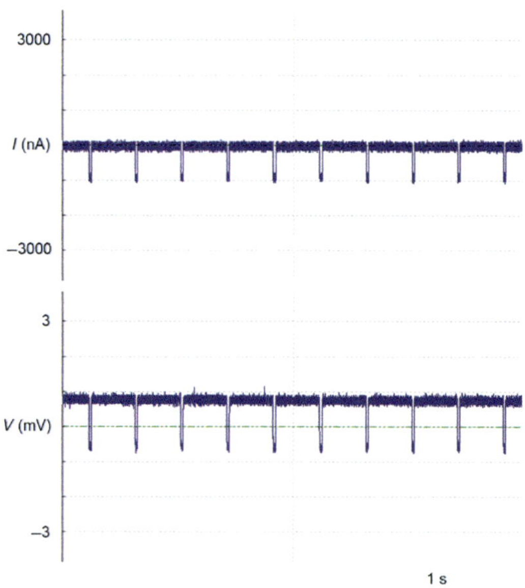

FIGURE 3.3.9 Diffusion potential in mouse colon. Note that a lumen-positive diffusion potential develops when the basolateral NaCl concentration is reduced indicating the presence anion selective paracellular channels.

3.3.4.2 Electrical contribution from multiple tissue layers

The seromusculature layers represent a significant diffusion barrier, which retains its own electric resistance and needs to be separately recorded and subtracted from the resistance of the epithelium (R_{te}). Seromusculature stripping removes the serosa (visceral peritoneum) and the longitudinal/circular muscle layers, leaving only the mucosal layer, that is, remnants of muscle, the extracellular matrix, and the epithelium. The damage to the mucosal layer from seromuscular stripping may induce phospholipase C or A2 production, resulting in liberation of arachidonic acid and subsequent eicosanoid generation, which in turn activates ENaC and CFTR channel activities (Smith, Marks, Lieberman, & Marks, 2005). The integrity of the mucosal layer is also vital to the R_{te} measurement. Gaps or holes in the mucosal layer will reduce the R_{te} level. Therefore whole intestine recording followed by seromusculature recording and electrical subtraction provides a safer approach for the functional study of paracellular channels in the mucosal layer.

3.3.5 Troubleshooting

3.3.5.1 Edge damage

Edge damage refers to the extrusion of a small portion of the crushed mucosa into the plug-in cartridge along the outer perimeter of the aperture when the two half-cartridges are clamped together (Fig. 3.3.10). Since the crushed mucosa is a shunt pathway for electric current to pass between the cartridge halves, the ΔV is reduced and the R_{te} is increased to include this artifact. Previous studies have shown that the erroneous effect of edge

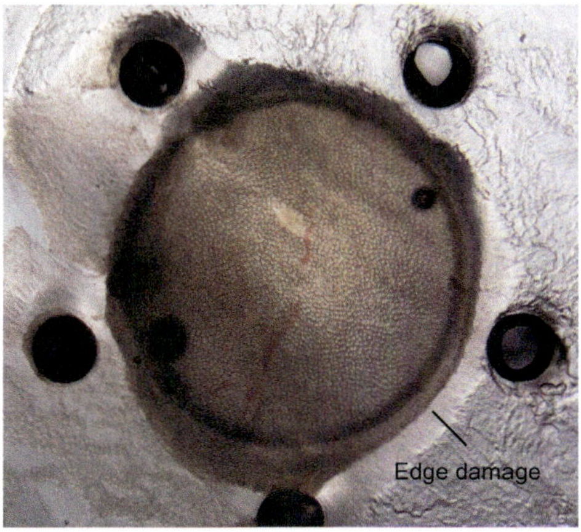

FIGURE 3.3.10 Edge damage to mouse colon. A post-recording plug-in cartridge is exposed to show the edge damage to the mucosa due to pressure clamping.

damage increased proportionally with a reduction in the aperture diameter (Dobson & Kidder, 1968). Thus using a relatively larger aperture to accommodate the excised mucosa will reduce the recording error caused by edge damage.

3.3.5.2 Tissue viability and variability

Upon removal from the animal, the ex vivo organ preparation has limited viability. Viability decreases with longer experimental duration. The R_{te} itself is a good indicator of viability as tissue integrity deteriorates faster than metabolism. Tissue viability may cause variability in recording. Thus control and treatment groups should be matched not only for animal sex, age, and genotype but also for tissue preparation step, incubation time, and culture condition.

3.3.5.3 Electric pulse

Because the aperture for mounting the tissue sample is small, that is, $0.11\ cm^2$, the electric resistance across the tissue sample in the aperture must be large, ranging from 500 to 2000 Ω, knowing that R_{te} varies from 50 to $200\ \Omega/cm^{-2}$ among different intestinal tissues. Therefore current pulses of large magnitude, for example, $> 10\ \mu A$ should be avoided, as they elicit voltage deflections exceeding 10 mV, which may depolarize the cell membrane.

3.3.6 Closing remarks

The ex vivo Ussing chamber recording technique provides a short-term organ culture approach to enable the real-time measurement of paracellular transport pathways in living epithelial tissues, including the skin, the stomach, the intestines, and the bladder. Compared to in vitro cultures, the orderly epithelial cell layer of the native tissue produces greater effect size in paracellular transport function, simply because the number of available epithelial cells within the same gross surface area is greater. When combined with animal transgenesis to knock-out or knock-in genes, the ex vivo Ussing chamber recording technique allows interrogating genetic effects on paracellular transport function.

References

Burg, M. B. (1972). Perfusion of isolated renal tubules. *The Yale Journal of Biology and Medicine*, *45*, 321–326.

Burg, M. B., Issaacson, L., Grantham, J., & Orloff, J. (1968). Electrical properties of isolated perfused rabbit renal tubules. *The American Journal of Physiology*, *215*, 788–794.

Dobson, J. G., Jr., & Kidder, G. W., III (1968). Edge damage effect in in vitro frog skin preparations. *The American Journal of Physiology*, *214*, 719–724.

Giebisch, G. H. (1979). *Membrane transport in biology: Transport organs. 2 v* (Vol 4). Springer.

Greger, R. (1990). An electrophysiological approach to the study of isolated perfused tubules. *Methods in Enzymology*, *191*, 289–302.

Greger, R., & Hampel, W. (1981). A modified system for in vitro perfusion of isolated renal tubules. *Pflugers Archiv: European Journal of Physiology*, *389*, 175–176.

Hou, J. (2018). *The paracellular channel—Biology, physiology and disease*. Academic Press.

Kunzelmann, K., & Mall, M. (2002). Electrolyte transport in the mammalian colon: Mechanisms and implications for disease. *Physiological Reviews*, *82*, 245–289.

Smith, C., Marks, A., Lieberman, M., & Marks, D. (2005). *Metabolism of the eicosanoids. Basic medical biochemistry* (pp. 654–667). Philadelphia, PA: Lippincott, Williams and Wilkins.

Tsai, P. Y., Zhang, B., He, W. Q., Zha, J. M., Odenwald, M. A., Singh, G., ... Yeruva, S. (2017). IL-22 upregulates epithelial claudin-2 to drive diarrhea and enteric pathogen clearance. *Cell Host & Microbe*, *21*, 671–681, .e4.

Chapter 3.4

Epithelial ohmmeter

3.4.1 Background knowledge

3.4.1.1 "Chopstick" electrode system

The R_{te} is the most important measurement of epithelial transport property. A simplified Ussing chamber system, consisting in an ohmmeter and a pair of "chopstick" electrodes, is widely used to measure R_{te} for various epithelial and endothelial cell models (Fig. 3.4.1). Each "chopstick" electrode contains a Ag/AgCl pellet to sense electric potential and a Ag wire to pass electric current. The advantage of the "chopstick" system is that R_{te} can be rapidly and repeatedly interrogated when cells are in different culturing stages. The disadvantage of this system is that the electric field generated by the two "chopstick" electrodes is not uniform, which can cause significant variations in the measurement of R_{te}.

3.4.1.2 Current clamp and Ohm's law

The "chopstick" electrode system is driven by a bipolar current (I) at 2 Hz in the form of square wave and with the amplitude of 10 μA. This current, when flowing through the epithelial sample, will cause a voltage deflection, ΔV (Fig. 3.4.2). R_{te} equals $\Delta V/I$, according to Ohm's law (Eq. 3.1.6).

3.4.2 Materials and instrumentation

- World Precision Instruments EVOM2 Epithelial Voltohmmeter
- World Precision Instruments STX2 chopstick electrodes

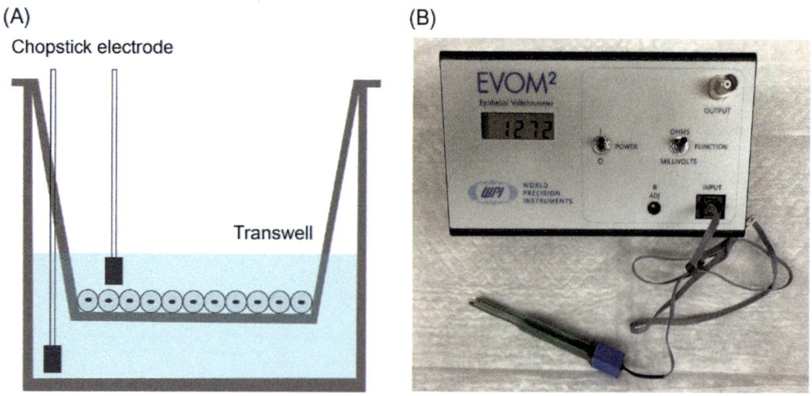

FIGURE 3.4.1 Epithelial ohmmeter. (A) Schematic diagram of the "chopstick" electrodes in a Transwell chamber. (B) A commercial epithelial ohmmeter made by World Precision Instruments.

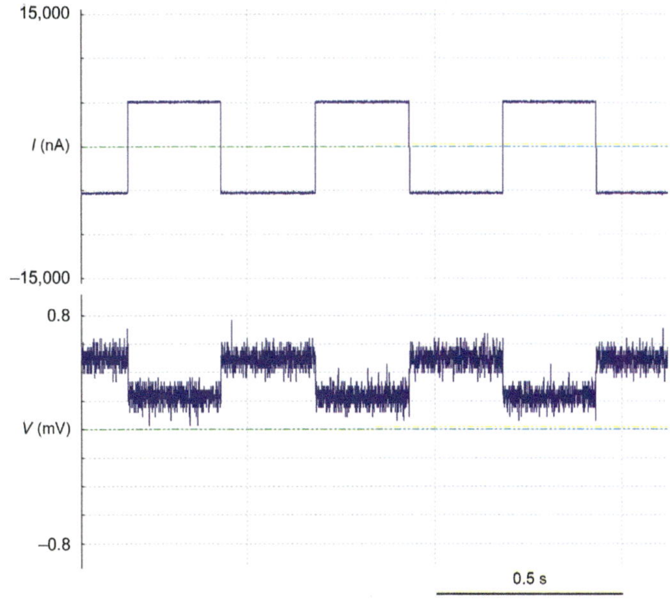

FIGURE 3.4.2 Current and voltage traces from epithelial ohmmeter. A bipolar square wave current (10 μA) elicits a voltage deflection (0.2 mV) across a blank Transwell insert with a resistance level of 20 Ω.

- Physio-buffer (Chapter 3.2, Table 3.2.2)
- Corning Transwell inserts (12 mm diameter) (Chapter 3.2, Table 3.2.1)
- MDCK strain II cells

3.4.3 Experimental procedure

1. Turn the EVOM2 **Power** on (*I*).
2. Sterilize the electrodes with 70% ethanol.
3. Connect the electrodes to the meter.
4. Precondition the electrodes in Physio-buffer.
5. Set the **Function** switch to **Ohms**.
6. Measure the blank Transwell insert resistance and record the value.
7. Perform the measurements with cells cultured on Transwell inserts. To obtain the actual epithelial tissue reading, subtract the blank resistance value.
8. Clean the electrodes with 70% ethanol, dry and store the electrode.

3.4.4 Data analysis

The measured electrical resistance is inversely proportional to the membrane area of the Transwell insert. The R_{te} value is often normalized to the Transwell insert membrane area and expressed as Ω/cm^{-2}. A digital display on the epithelial ohmmeter registers the readings of resistance values. The current and voltage traces are not displayed by the epithelial ohmmeter. Instead, it provides analog signals to the digitizer and computer (see Chapter 3.2 for details).

3.4.5 Troubleshooting

3.4.5.1 Nonuniform electric field

In the theory, a pair of "chopstick" electrodes should be placed above and beneath the center of the sample, respectively, in order to generate a uniform electric field. When the electrodes are placed near the edge of the sample, the current passes the Transwell membrane mainly near its edge and, to a lesser degree, through the center. If one "chopstick" electrode is moved relative to the other electrode from 0° to 180° around the circumference of the Transwell membrane, the measured R_{te} value progressively decreases (Fig. 3.4.3) and the noise increases (Fig. 3.4.4) (Jovov, Wills, & Lewis, 1991).

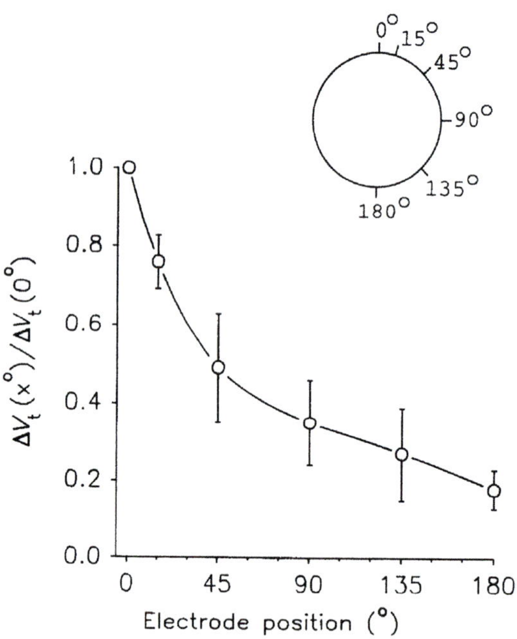

FIGURE 3.4.3 Effect of electrode position on epithelial resistance. The voltage deflection (ΔV_t) induced by a square current pulse was measured by the voltage-sensing electrodes positioned at different locations around the circumference of the Transwell membrane (*inset above*). The voltage deflection, reflecting the R_{te}, measured at a position x° (($\Delta V_t(x^\circ)$)) is expressed as fractional ratio relative to the voltage deflection measured at position 0° (($\Delta V_t(0^\circ)$)). *Reproduced with permission from Jovov, B., Wills, N. K., & Lewis, S. A. (1991). A spectroscopic method for assessing confluence of epithelial cell cultures.* The American Journal of Physiology, 261(6 Pt 1), *C1196–C1203.*

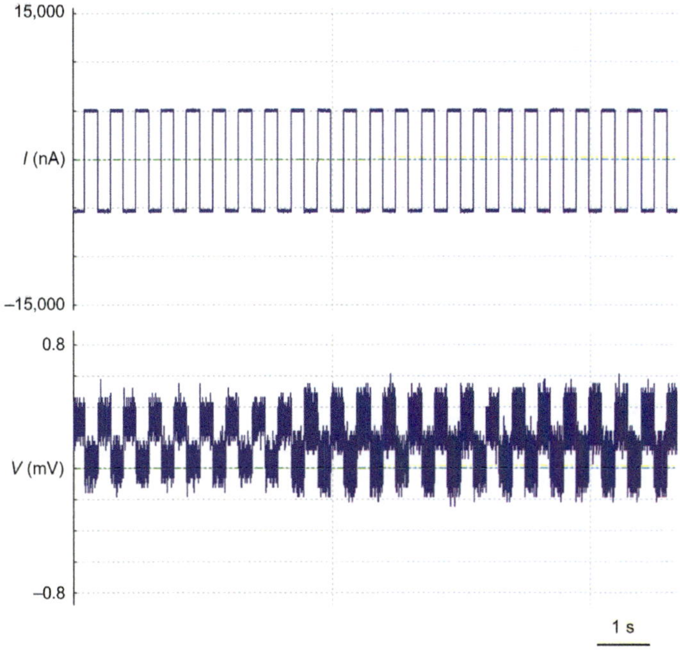

FIGURE 3.4.4 Electrode position and noise. Current trace shows a bipolar square wave current (10 μA) passes a blank Transwell insert with a resistance level of 20 Ω. Voltage trace shows the voltage deflection across the Transwell insert. Note that when the voltage-sensing electrodes were separated by 180°, the noise significantly increased (compared to Fig. 3.4.2).

3.4.5.2 Cell capacitance and underestimation of R_{te}

The epithelial ohmmeters apply AC at frequencies ranging from 2 to 20 Hz. Because the cell membrane is made of parallel resistor-capacitor circuit (Fig. 3.5.1A), a part of the applied current will charge the membrane capacitor, causing systematic underestimations of ΔV and R_{te}. The higher the AC frequency is, the larger the systematic error becomes.

3.4.6 Closing remarks

The "chopstick"-electrode epithelial ohmmeter offers a simple and user-friendly platform to assess the epithelial resistance in a wide variety of cultured epithelial or endothelial cells. Several new developments are currently under way to improve the accuracy of epithelial ohmmeter. For example, a new design has been devised to embed the voltage-sensing and the current-passing electrodes into a double-layer poly(dimethylsiloxane) microfluidic chip in which the upper and the lower channel are separated by a semiporous membrane (Douville et al., 2010). This system allows the real-time measurement of R_{te} across the cells grown on the membrane. Advances in three-dimensional cell culture techniques have called for the development of new electrical recording devices to facilitate the measurement of epithelial resistance in microfluidic organ-on-chip models. An electrolyte-gated transistor, organic electrochemical transistors (OECTs), can transduce biological signals into electrical outputs using very low operating voltages. The OECT is fabricated into a thin layer of organic semiconductor film whose potential is modulated by a gate electrode when the film is in contact with an electrolyte (biological medium) (Pitsalidis et al., 2018). Cells growing on the film impede the ionic flux between the electrolyte and the film, which causes a potential difference on the gate electrode that can then be converted to the epithelial resistance.

References

Douville, N. J., Tung, Y. C., Li, R., Wang, J. D., El-Sayed, M. E., & Takayama, S. (2010). Fabrication of two-layered channel system with embedded electrodes to measure resistance across epithelial and endothelial barriers. *Analytical Chemistry*, *82*(6), 2505–2511.

Jovov, B., Wills, N. K., & Lewis, S. A. (1991). A spectroscopic method for assessing confluence of epithelial cell cultures. *The American Journal of Physiology*, *261*(6 Pt 1), C1196-203.

Pitsalidis, C., Ferro, M. P., Iandolo, D., Tzounis, L., Inal, S., & Owens, R. M. (2018). Transistor in a tube: A route to three-dimensional bioelectronics. *Science Advances*, *4*(10), eaat4253.

Chapter 3.5

Impedance measurement in Ussing chamber

3.5.1 Background knowledge

3.5.1.1 Concept of impedance measurement

Chapter 3.2 described an approach to measure the R_{te} with a direct current (DC) circuit. The cell membrane capacitance (C_m) was not taken into account. Impedance measurements make use of the concept that cell membranes act like capacitors and that the capacitive currents strongly depend on the frequency in an AC circuit (Fig. 3.5.1A). In the theory, an AC (I) with an angular frequency (ω) generates an oscillating potential (V) across the epithelium with the same frequency but different phases. The impedance (Z_{te}), deriving from V/I, reflects the R_{te} when ω approaches zero (Fig. 3.5.1B).

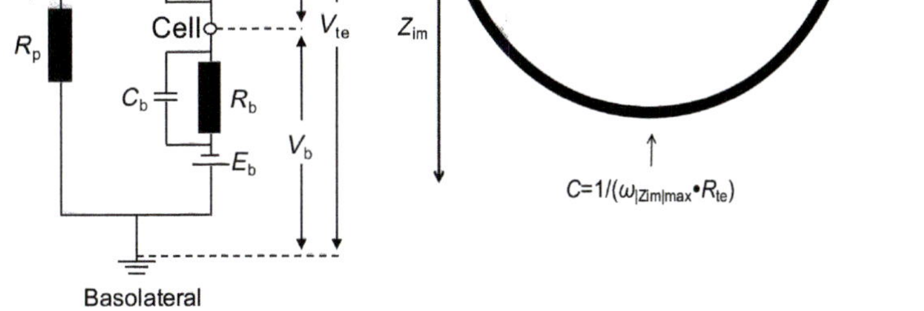

FIGURE 3.5.1 Equivalent electric circuit for impedance measurement. (A) An AC circuit takes cell membrane capacitance into account. Each element in the circuit (a: apical membrane, b: basolateral membrane, p: paracellular pathway) is represented by an electric equivalent, that is, an electromotive force, E, in series with a resistor, R, and with a capacitor, C. (B) Nyquist diagram (plot of the real and the imaginary axes of the impedance, Z_{re}, Z_{im}). At low frequencies ($\omega \to 0$), Z_{re} approaches R_{te}. The capacitance C can be calculated from the frequency at which $|Z_{im}|$ reaches a maximum value: $C = 1/(\omega|Z_{im}|_{max} \bullet R_{te})$. *Reproduced with permission from Hou, J. (2018).* The paracellular channel—Biology, physiology and disease. *Academic Press.*

3.5.1.2 Sinusoidal current waveform

When sinusoidal current passes through an electrical circuit made of resistors and capacitors, for example, an epithelial sample, the voltage across the circuit adopts a similar sinusoidal waveform:

$$i(t) = I_{\mathrm{m}}\cos(\omega t + \psi) \quad (3.5.1)$$

$$v(t) = V_{\mathrm{m}}\cos(\omega t + \theta) \quad (3.5.2)$$

where I_{m} and V_{m} are the amplitudes, ψ and θ are the phase angles, and ω is the angular frequency ($\omega = 2\pi f$, where f is the frequency in Hertz). Using Euler's relationship, Eqs. (3.5.1) and (3.5.2) can be written into:

$$i(t) = \mathrm{Re}[Ie^{j\omega t}] \quad (3.5.3)$$

$$v(t) = \mathrm{Re}[Ve^{j\omega t}] \quad (3.5.4)$$

where Re[] denotes the real part, $j = \sqrt{-1}$, and I and V are the imaginary part of the complex quantity. I and V, termed as the phasors, express the magnitude and phase angle of the sinusoids at angular frequency ω according to the following equations:

$$I = I_{\mathrm{m}}e^{j\omega\psi} \quad (3.5.5)$$

$$V = V_{\mathrm{m}}e^{j\omega\theta} \quad (3.5.6)$$

The impedance is a complex quantity and defined as the ratio of the voltage phasor to the current phasor at a given frequency.

$$Z = \frac{V}{I} = \frac{V_{\mathrm{m}}}{I_{\mathrm{m}}}e^{j\omega(\theta-\psi)} \quad (3.5.7)$$

3.5.1.3 Impedance of resistor and capacitor

The impedance of a resistor (Z_{r}) equals its resistance value (R).

$$Z_{\mathrm{r}} = R \quad (3.5.8)$$

The impedance of a capacitor (Z_{c}) is related to its capacitance by the following equation:

$$Z_{\mathrm{c}} = -j\frac{1}{\omega C} \quad (3.5.9)$$

The impedance of a series resistor-capacitor circuit is given by the following equation:

$$Z = Z_{\mathrm{r}} + Z_{\mathrm{c}} = R - j\frac{1}{\omega C} \quad (3.5.10)$$

The impedance of a parallel resistor-capacitor circuit (i.e., a biological membrane) is given by the following equation:

$$Z = \frac{1}{\frac{1}{Z_r} + \frac{1}{Z_c}} = \frac{R}{1 + (\omega RC)^2} - j\frac{\omega R^2 C}{1 + (\omega RC)^2} \tag{3.5.11}$$

From Eqs. (3.5.10) and (3.5.11), it is clear that the magnitude of the impedance equals R under the DC condition ($\omega = 0$), and progressively decreases as ω increases, since at higher frequency more current flows through the capacitor instead of the resistor. A simple "lumped" model of equivalent electric circuit (Fig. 3.5.1A) is often used to represent a unilayered epithelium with open lateral space (Lewis & de Moura, 1984; Schifferdecker & Fromter, 1978). In this model, the epithelium can be described as a parallel circuit made of an epithelial resistor (R_{te}) and a membrane capacitor (C_m), mindful that R_{te} includes the resistance of the apical membrane (R_a), the basolateral membrane (R_b), and the tight junction (R_p).

3.5.1.4 Nyquist plot

The Nyquist plot displays the amplitude of the real part (abscissa, Z_{re}) of impedance against the amplitude of the imaginary part (ordinate, Z_{im}) of impedance for each angular frequency (ω) (Fig. 3.5.1B). The frequencies (f, $\omega = 2\pi f$) employed in biological experiments typically range from 1 Hz to 100 kHz and appear in a discrete pattern (Fig. 3.5.2) (Gitter, Bendfeldt, Schulzke, & Fromm, 2000). The input waveform can either be current or voltage. The output waveform is then measured with a voltage or current amplifier.

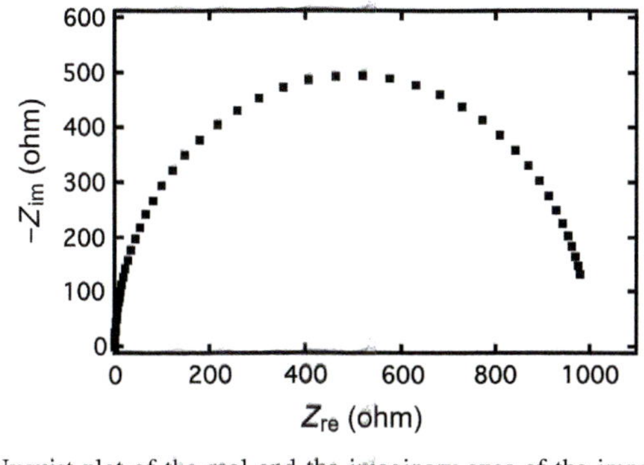

FIGURE 3.5.2 Nyquist plot of the real and the imaginary axes of the impedance of a parallel RC circuit. The circuit is made of a resistor of 1000 Ω and a capacitor of 1 μF. Measurements were performed with a sinusoidal signal of 500 mV (peak-to-peak) at 100 discrete frequencies from 1 Hz to 100 kHz.

3.5.2 Materials and instrumentation

3.5.2.1 Electrophysiological rig

The electrophysiological rig is assembled from the following components (Fig. 3.5.3).

- CH Instruments 600E Electrochemical Workstation
- Ag/AgCl working electrode, Ag/AgCl reference electrode, Pt counter electrode
- Faraday cage

3.5.2.2 Buffer

The saline buffer that bathes the sample derives from the Ringer solution—Physio-buffer (see Table 3.2.2)

3.5.2.3 Cell culture

The MDCK strain II cells are cultured as described in Chapter 7.2. After passage, the MDCK-II cells are seeded onto the Transwell inserts (12 mm diameter, Table 3.2.1) to allow forming tight junctions.

3.5.3 Experimental procedure

1. Assemble the working electrode, the reference electrode, and the counter electrode in a Faraday cage.

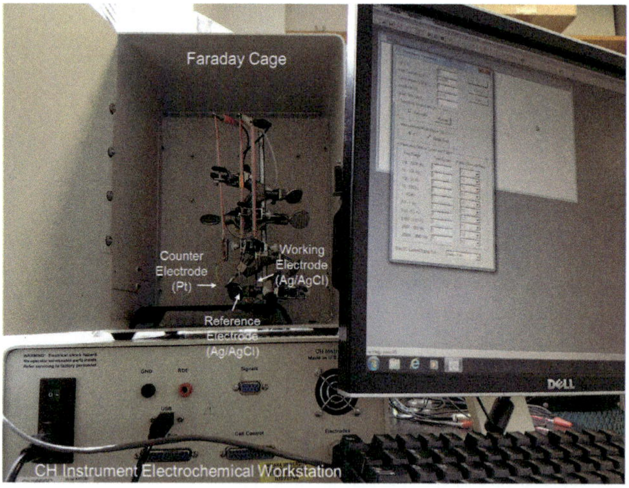

FIGURE 3.5.3 Electrochemical impedance spectroscopy rig. A fully assembled rig to perform impedance measurements consists in an electrochemical workstation, three electrodes and a Faraday cage.

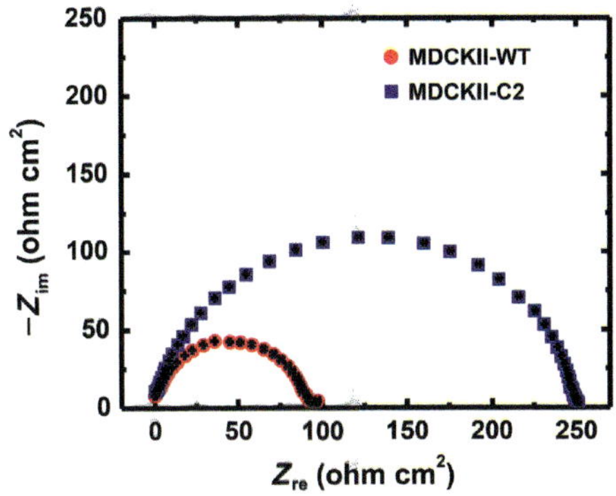

FIGURE 3.5.4 Nyquist plot of the real and the imaginary axes of the impedance of MDCK-II cells samples. Both wildtype (WT) and claudin-2 knockdown (C2) cells were recorded with impedance spectroscopy. Claudin-2 proteins function as a paracellular cation channel. Genetic deletion of claudin-2 proteins from the MDCK-II cells reduces the paracellular permeability and increases the R_{te}. *Reproduced with permission from Chen, C. C., Zhou, Y., Morris, C. A., Hou, J., & Baker, L. A. (2013). Scanning ion conductance microscopy measurement of paracellular channel conductance in tight junctions.* Analytical Chemistry, 85(7), *3621–3628.*

2. Place Transwell chamber into the Faraday cage. Replace the medium with the Physio-buffer (Table 3.2.2).
3. Place the working electrode into the apical chamber of the Transwell. Place the reference electrode and the counter electrode into the basolateral chamber of the Transwell.
4. Open CH Instruments graphic interface on computer. Click "Technique" and choose "AC Impedance."
5. Choose "Initial E (V)" value—0, "High Frequency (Hz)" value—100,000, "Low Frequency (Hz)" value—1, and "Amplitude (V)" value—0.02.
6. Click "Run" to start the measurement.

3.5.4 Data analysis

The Nyquist plots for MDCK-II cell samples are automatically generated by the CH Instruments graphic software (Fig. 3.5.4) (Chen, Zhou, Morris, Hou, & Baker, 2013). The R_{te} can be determined by extrapolation of the low and high frequency ends of the semicircular impedance curve to the abscissa. The resistance of a blank Transwell insert must be measured separately and subtracted from the R_{te}.

3.5.5 Troubleshooting

3.5.5.1 Sample-electrode distance

The distance between the voltage sensing electrodes and the epithelial sample must be as small as possible (within a few millimeters) in order to

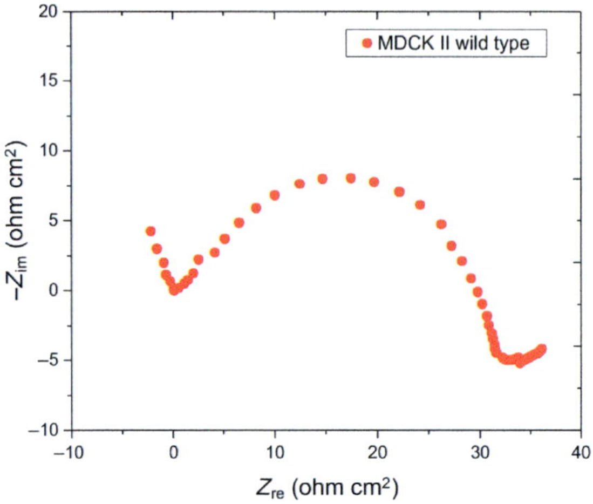

FIGURE 3.5.5 Nyquist plot of a phase shifted measurement. Wildtype MDCK-II cells were recorded with impedance spectroscopy. Note the significant phase shift and the lower than normal R_{te} value in this measurement.

minimize the contribution of the impedance of saline solution. The current passing electrode (i.e., the counter electrode) must be placed above or beneath the center of the sample so that there is uniform current density over the surface of the sample.

3.5.5.2 Phase shift

The impedance of the saline filled chamber and a blank Transwell insert (prior to mounting the epithelial sample) must be measured to ensure that there is no phase shift on voltage or current electrode (Fig. 3.5.5). Salt bridges are not connected to the electrodes to avoid abrupt changes in salt concentration from the saline buffers to the salt bridges, which may introduce "stray" capacitance to the circuit.

3.5.5.3 Paracellular versus transcellular pathway

Impedance measurement is not able to discriminate the paracellular resistance from the transcellular resistance on its own. One way to shut off the transcellular pathway is by using the pharmacological inhibitors for various membrane ion channels. The capacitance value from the Nyquist plot is important to estimate the cell plasma membrane area. Changes in membrane area can be related to changes in membrane transport rates due to exocytosis or endocytosis of ion channels. A good example of cell membrane expansion is when the frog stomach is stimulated by histamine to secrete protons (Clausen, Machen, & Diamond, 1983).

3.5.6 Closing remarks

Compared to the DC circuit, the AC circuit is particularly useful to separate the resistance of the epithelial tissue (made of resistors and capacitors, Fig. 3.5.1B) from the subepithelial tissue (made of resistors), for example, the extracellular matrix, the smooth muscle, and the blood vessels. In some inflammatory diseases, a reduction in R_{te} might be masked by a simultaneous increase in subepithelial tissue resistance when interrogated by the DC circuit (Burgel et al., 2002; Zeissig et al., 2007). The AC circuit also allows determining the total cell membrane capacitance, which is often neglected in studies using the DC circuit. The membrane capacitance may change in response to pharmacological treatments or after transgenic manipulation of paracellular channel proteins. Thus normalizing the R_{te} to the membrane capacitance in Ω/F^{-1} should be accepted as a new standard of documenting the epithelial tissue resistance.

References

Burgel, N., Bojarski, C., Mankertz, J., Zeitz, M., Fromm, M., & Schulzke, J. D. (2002). Mechanisms of diarrhea in collagenous colitis. *Gastroenterology*, *123*(2), 433–443.

Chen, C. C., Zhou, Y., Morris, C. A., Hou, J., & Baker, L. A. (2013). Scanning ion conductance microscopy measurement of paracellular channel conductance in tight junctions. *Analytical Chemistry*, *85*(7), 3621–3628.

Clausen, C., Machen, T. E., & Diamond, J. M. (1983). Use of AC impedance analysis to study membrane changes related to acid secretion in amphibian gastric mucosa. *Biophysical Journal*, *41*(2), 167–178.

Gitter, A. H., Bendfeldt, K., Schulzke, J. D., & Fromm, M. (2000). Trans/paracellular, surface/crypt, and epithelial/subepithelial resistances of mammalian colonic epithelia. *Pflugers Archiv: European Journal of Physiology*, *439*(4), 477–482.

Lewis, S. A., & de Moura, J. L. (1984). Apical membrane area of rabbit urinary bladder increases by fusion of intracellular vesicles: An electrophysiological study. *The Journal of Membrane Biology*, *82*(2), 123–136.

Schifferdecker, E., & Fromter, E. (1978). The AC impedance of Necturus gallbladder epithelium. *Pflugers Archiv: European Journal of Physiology*, *377*(2), 125–133.

Zeissig, S., Burgel, N., Gunzel, D., Richter, J., Mankertz, J., Wahnschaffe, U., et al. (2007). Changes in expression and distribution of claudin 2, 5 and 8 lead to discontinuous tight junctions and barrier dysfunction in active Crohn's disease. *Gut*, *56*(1), 61–72.

Chapter 3.6

Flux assay in Ussing chamber

3.6.1 Background knowledge

3.6.1.1 Fick's law

Fick's law relates the diffusive flux of a tracer molecule j to the concentration (C) of the tracer at a physical location x.

$$J(x) = -D\frac{dC(x)}{dx} \tag{3.6.1}$$

where J is the flux or the number of tracer molecules that pass a unit area per unit time; D is called the diffusion coefficient of the tracer (Sten-Knudsen, 2002).

Assuming the concentration profile of the tracer is linear in the epithelium, then the derivative of C with respect to x is given by:

$$\frac{dC(x)}{dx} = \frac{C^{(a)} - C^{(b)}}{h} \tag{3.6.2}$$

where $C^{(a)}$ and $C^{(b)}$ are the concentration of the tracer in the bathing buffer from the apical and basal sides, respectively, and $C^{(b)} > C^{(a)}$; h is the thickness of the epithelium.

The permeability P is related to the diffusion coefficient by the following equation:

$$D = Ph \tag{3.6.3}$$

Inserting the expression of dC/dx from Eq. (3.6.2) and the expression of D from Eq. (3.6.3) into Eq. (3.6.1) yields:

$$J = P(C^{(b)} - C^{(a)}) \tag{3.6.4}$$

If there is no tracer in the apical compartment at $t = 0$ or $C^{(a)}_{(0)} = 0$, then the flux is related to the accumulation of the tracer in the apical compartment by the following equation:

$$J = \frac{V^{(a)}C^{(a)}_{(t)}}{At} \tag{3.6.5}$$

where $V^{(a)}$ is the volume of the apical solution, $C^{(a)}_{(t)}$ is the concentration of the tracer in the apical compartment after the time t, and A is the surface area of the epithelium.

Inserting the expression of J from Eq. (3.6.5) into Eq. (3.6.4) yields:

$$P = \frac{V^{(a)}C_{(t)}^{(a)}}{AC_{(t)}^{(b)}t} \tag{3.6.6}$$

where $C_{(t)}^{(b)}$ is the concentration of the tracer in the basal compartment after the time t, assuming that diffusion is a slow process or $C_{(t)}^{(b)} \approx C_{(0)}^{(b)} >> C_{(t)}^{(a)}$.

If $C^{(a)} > C^{(b)}$ and $C_{(0)}^{(b)} = 0$, then the permeability is given by:

$$P = \frac{V^{(b)}C_{(t)}^{(b)}}{AC_{(t)}^{(a)}t} \tag{3.6.7}$$

where $V^{(b)}$ is the volume of the basal solution.

3.6.1.2 Radioisotope

The use of radioisotope allows accurate measurement of flux and permeability of ions and solutes across the epithelium (Ussing, 1980). The side of the epithelium to which radioisotope is added is referred to as the "hot" side; the opposite side of the epithelium is referred to as the "cold" side. The following equation relates the flux to the accumulation of radioactivity in the "cold" side:

$$J = \frac{dpm_{(t)}^{(cold)}}{(S^*)At} \tag{3.6.8}$$

where $dpm_{(t)}^{(cold)}$ (abbreviation for disintegrations per minute) is the radioactivity in the "cold" side after the time t, S^* is the specific activity that indicates the amount of radiolabeled mass in a sample, A is the surface area of the epithelium, and t is the reaction time.

Inserting the expression of J from Eq. (3.6.8) to Eq. (3.6.4) yields:

$$P = \frac{J}{C^{(hot)}} = \frac{J}{\frac{dpm_{(t)}^{(hot)}}{(S^*)V^{(hot)}}} = \frac{dpm_{(t)}^{(cold)}V^{(hot)}}{dpm_{(t)}^{(hot)}At} \tag{3.6.9}$$

where $C^{(hot)}$ is the concentration of radioisotope in the "hot" side, $V^{(hot)}$ is the volume of the solution in the "hot" side, $dpm_{(t)}^{(hot)}$ is the radioactivity in the "hot" side after the time t, assuming that diffusion is a slow process or $dpm_{(t)}^{(hot)} >> dpm_{(t)}^{(cold)}$.

3.6.2 Materials and instrumentation

3.6.2.1 Buffer

The saline buffer that bathes the sample derives from the Ringer solution—Physio-buffer (see Table 3.2.2).

3.6.2.2 Cell culture

The MDCK strain II cells are cultured as described in Chapter 7.2. After passage, the MDCK-II cells are seeded onto the Transwell inserts (12 mm diameter, Table 3.2.1) to allow forming tight junctions.

3.6.2.3 Liquid scintillation counter

Beckman/PerkinElmer LS 3801 liquid scintillation counter

3.6.2.4 Radioisotope

American Radiolabeled Chemicals [^{3}H]-mannitol

- "Cold" buffer: Physio-buffer with no radioisotope
- "Hot" buffer: Physio-buffer with 1 μCi/mL of [^{3}H]-mannitol

3.6.3 Experimental procedure

1. Aspirate medium from each Transwell.
2. Add 500 μL of "cold" buffer to the inner chamber of Transwell.
3. Add 1.5 mL of "hot" buffer to the outer chamber of Transwell.
4. Incubate at 37°C for 30 minutes
5. Remove 100 μL of "cold" buffer and 10 μL of "hot" buffer, and mix with 4 mL of ScintiVerse BD mixture (Fisher Scientific).
6. Count dpm with liquid scintillation counter.

3.6.4 Data analysis

Eqs. (3.6.8) and (3.6.9) can be written into MATLAB (MathWorks) programming functions to calculate the flux and the permeability of radioisotope labeled tracer, respectively.

```
function per=Na_permeability_NaCl (ratio, resistance)

% ratio=Na_permeability/Cl_permeability
% C is the concentration in which resistance is measured
% resistance is the total resistance in om.cm2
% SI_Unit_trans: from a combination of complex units to cm/s

R=8.3143;  % in J(K-1)(mol-1)
T=273+37; % temperature in K
F=96488; % in C(mol-1)
C=145; % in mM
SI_Unit_trans=1e6; %x SI_Unit_trans
per=SI_Unit_trans*R*T*ratio./(F*F*C*resistance.*(ratio+1));
```

3.6.5 Troubleshooting

3.6.5.1 Radiation safety

Radiation safety is the primary consideration when designing assays involving radioisotopes. Although Transwell is not a *bona fide* Ussing chamber, it offers an enclosed system in which radioisotopes can be contained, counted and disposed. Because the assay is carried out in the HEPES-buffered saline (Table 3.2.2), a radioisotope-dedicated CO_2 incubator will not be necessary. When performing multiple assays, it is important that pipettes, tips, or plasticwares are not reused to prevent radioisotope contamination. Contamination is not only a safety concern but also causes variation in the assays.

3.6.5.2 Differentiation of paracellular from transcellular pathway

Ussing derived a flux-ratio equation to differentiate paracellular from transcellular transport across an epithelium (Ussing, 1949).

$$\frac{J^{(ab)}}{J^{(ba)}} = \frac{C^{(a)}}{C^{(b)}} e^{\left(\frac{zV_{te}F}{RT}\right)} \tag{3.6.10}$$

where J denotes the flux (superscript ab refers to the flux from the apical to the basal compartment; superscript ba refers to the flux from the basal to the apical compartment), $C^{(a)}$ and $C^{(b)}$ are the concentration of the tracer in the apical and basal solutions, respectively, V_{te} is the transepithelial potential, z is the valence of the tracer, R is the gas constant, T is the Kelvin temperature, and F is Faraday's constant.

When the tracer carries no charge, that is, $z=0$ or $V_{te}=0$, the flux-ratio equation can be reduced to:

$$\frac{J^{(ab)}}{J^{(ba)}} = \frac{C^{(a)}}{C^{(b)}} \tag{3.6.11}$$

If the experimentally determined flux ratio fits the equation, then the transport is paracellular. If the flux ratio does not fit the equation, then active transcellular transport must be involved. The paracellular permeability can only be determined when the transcellular pathway is pharmacologically inhibited and the flux-ratio relationship is fully satisfied.

3.6.6 Closing remarks

The flux assay is a simple and reliable approach to measure transepithelial transport levels. The use of radioisotope provides a direct means to quantify the number of tracer molecules that pass the epithelium. Modern techniques, such as fluorophore labeling and enzyme-linked immunosorbent assay, albeit matching the sensitivity of radioisotope, utilize indirect algorithms to calculate the number of tracer molecules in the flux assay. The conversion from fluorescent or luminescent signals to molecular numbers is a major source of error. Compared to electrophysiological approaches, the flux assay reflects the accumulative nature of epithelial transport pathways. Cellular viability, metabolic status, and intracellular trafficking processes may influence the measurement during the assay. Therefore, meaningful conclusion can only be drawn when the flux assay is used in combination with other recording approaches able to capture transient permeability alterations such as electric resistance measurement and impedance measurement.

References

Sten-Knudsen, O. (2002). *Biological membranes: Theory of transport, potentials and electric impulses*. Cambridge University Press.

Ussing, H. H. (1949). The active ion transport through the isolated frog skin in the light of tracer studies. *Acta Physiologica Scandinavica*, *17*, 1–37.

Ussing, H. H. (1980). Life with tracers. *Annual Review of Physiology*, *42*, 1–16.

Chapter 3.7

Measurement of water permeability in Ussing chamber

3.7.1 Background knowledge

3.7.1.1 Transepithelial water permeability

The volume flow driven by hydrostatic pressure and osmotic pressure obey the following equation:

$$\frac{dV}{dt} = AP_{\text{os}}\Delta\pi + AP_{\text{hydro}}\Delta p \tag{3.7.1}$$

where A is the surface area of the epithelium, V is the volume in either side of the epithelium in an enclosed system, P_{os} is the osmotic permeability coefficient of water, P_{hydro} is the hydrostatic permeability coefficient of water (note that $P_{\text{hydro}} \neq P_{\text{os}}$), $\Delta\pi$ is the osmotic pressure difference between the apical and basal solutions, and Δp is the hydrostatic pressure difference between the apical and basal compartments (Kedem & Katchalsky, 1958).

When no hydrostatic pressure difference exists between the apical and basal compartments, Eq. (3.7.1) can be reduced to:

$$\frac{dV}{dt} = AP_{\text{os}}\Delta\pi \tag{3.7.2}$$

Inserting $\pi = RTC$ to Eq. (3.7.2) gives:

$$\frac{dV}{dt} = ARTP_{\text{os}}\Delta C \tag{3.7.3}$$

where C is the osmotic concentration and denotes the sum of the concentrations of all the solutes whether the epithelium is permeable to them or not, R is the gas constant, and T is the Kelvin temperature.

If the apical solution only contains nonpermeable solutes, N_{m}, and the basal solution is continuously perfused to maintain its osmotic concentration, then $C^{(\text{b})}$ is a constant, and $C^{(\text{a})} = N_{\text{m}}^{(\text{a})}/V^{(\text{a})}$, where $N_{\text{m}}^{(\text{a})}$ is also a constant. The above equation can be written as:

$$\frac{dV^{(\text{a})}}{dt} = ARTP_{\text{os}}\left(\frac{N_{\text{m}}^{(\text{a})}}{V^{(\text{a})}} - C^{(\text{b})}\right) \tag{3.7.4}$$

Since solving this differential equation requires advanced mathematics, calculation of P_{os} from direct measurement of $V^{(\text{a})}$ is not an ideal solution.

3.7.1.2 A simplified model

If the apical solution only contains a nonpermeable fluorophore, N_f, then $V^{(a)} = N_f^{(a)}/C^{(a)}$. At low concentrations, the fluorescence intensity (F_λ) approaches a linear relationship with the concentration:

$$F_\lambda = kC \tag{3.7.5}$$

where k is a constant unique to each fluorophore and λ is the emission wavelength (Guilbault, 1967).

$V^{(a)}$ is therefore inversely related to $F_\lambda^{(a)}$:

$$V^{(a)} = \frac{kN_f^{(a)}}{F_\lambda^{(a)}} \tag{3.7.6}$$

Inserting the expression of $V^{(a)}$ from Eq. (3.7.6) into Eq. (3.7.2) yields:

$$P_{os} = \frac{kN_f^{(a)}}{\Delta\pi\Delta F_\lambda^{(a)} At} \tag{3.7.7}$$

where $\Delta F_\lambda^{(a)}$ is the fluorescence intensity difference in the apical solution before and after water permeation, $\Delta\pi$ is considered to be constant assuming that water permeation is a slow process, and t is the reaction time.

The expression of $kN_f^{(a)}$ can be determined from $V_{(0)}^{(a)}$ and $F_{\lambda(0)}^{(a)}$ at $t = 0$ according to Eq. (3.7.6):

$$kN_f^{(a)} = V_{(0)}^{(a)} F_{\lambda(0)}^{(a)} \tag{3.7.8}$$

Inserting $kN_f^{(a)}$ from Eq. (3.7.8) into Eq. (3.7.7) yields:

$$P_{os} = \frac{V_{(0)}^{(a)} F_{\lambda(0)}^{(a)}}{\Delta\pi\Delta F_\lambda^{(a)} At} \tag{3.7.9}$$

3.7.2 Materials and instrumentation

3.7.2.1 Ussing chamber perfusion rig

The Ussing chamber perfusion rig is assembled from the following components (Fig. 3.7.1):

- Warner Instruments NaviCyte horizontal Ussing system with Snapwell adapter
- Corning Snapwell insert
- Harvard Apparatus PLI-100 compressed air driven Pico-injector
- Gravity perfusion rig made of a 50-mL syringe cylinder connected to the tubing and a flow regulator.

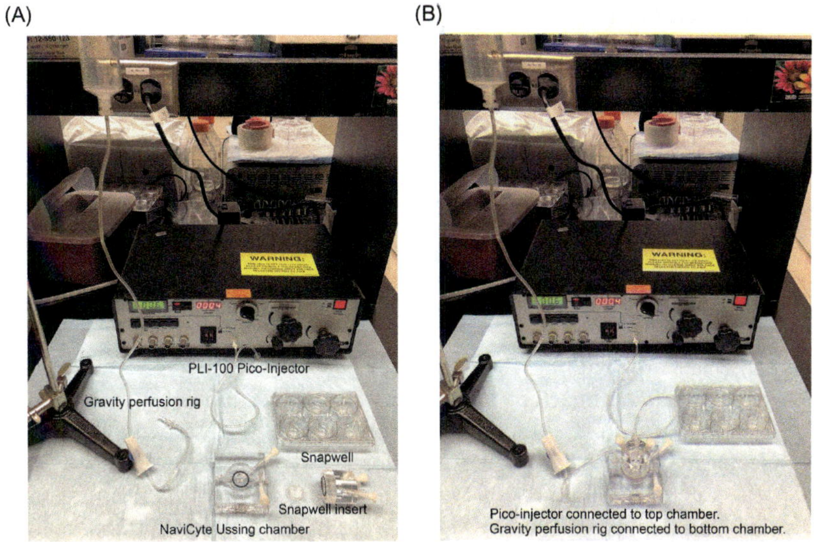

FIGURE 3.7.1 Ussing chamber perfusion rig. Exploded view (A) and assembled view (B) of the Ussing chamber apparatus connected to the compressed air driven Pico-injector and the gravity driven perfusion cylinder.

TABLE 3.7.1 Superfusates in Ussing chamber system.

Superfusate	FITC-buffer	Hypertonic-buffer
Calculated osmolality (mosm/kg)	300	400
NaCl	145	145
$CaCl_2$	2	2
$MgCl_2$	1	1
Glucose	10	10
Mannitol	0	100
HEPES	10	10
150 kDa FITC-dextran	0.001	0

Concentration is in mmol/L; pH is adjusted to 7.4 by NaOH or HCl.

3.7.2.2 Superfusate

The saline superfusate that bathes the sample derives from the Ringer solution and can be supplemented with mannitol to alter the osmolality level (see Table 3.7.1)

3.7.2.3 Cell culture

The MDCK strain II cells are cultured as described in Chapter 7.2. After passage, the MDCK-II cells are seeded onto the Snapwell inserts (12.7 mm diameter, 1.12 cm^2 area) to allow forming tight junctions.

3.7.3 Experimental procedure

3.7.3.1 Cancelation of hydrostatic pressure

1. Adjust the height of the perfusion cylinder to achieve water flow rate of 2 mL/minute.
2. Stop water flow with the flow regulator.
3. Connect the Pico-injector to the gravity perfusion cylinder.
4. Resume water flow. Adjust the air pressure ($\sim$0.2 psi) with the Pico-injector to counterbalance the water flow.
5. Stop water flow with the flow regulator. Disconnect the Pico-injector from the gravity perfusion cylinder.

3.7.3.2 Measuring transepithelial water permeability

6. Place a Snapwell insert seeded with the MDCK-II cells into the bottom half of the horizontal Ussing chamber. Assemble the top half of the horizontal Ussing chamber.
7. Connect the Pico-injector to the top chamber. Connect the gravity perfusion cylinder to the bottom chamber.
8. Fill the top chamber reservoir with 2 mL FITC-buffer prewarmed to 37°C (Table 3.7.1).
9. Place the assembled Ussing chamber on a thermostated hotplate at 37°C.
10. Fill the gravity perfusion cylinder with Hypertonic-buffer prewarmed to 37°C (Table 3.7.1). Start perfusion at a flow rate of 2 mL/minute. Make certain the valve at the bottom chamber is closed to prevent fluid drainage.
11. Turn on the Pico-injector to counterbalance the hydrostatic pressure caused by perfusion.
12. After $t = 10$ min, stop the perfusion and turn off the Pico-injector.
13. Aspirate the fluid from the top chamber to measurement its fluorescence intensity at excitation 488 nm and emission 520 nm. Also measure the fluorescence intensity in stock FITC-buffer as $t = 0$.

3.7.4 Data analysis

Eq. (3.7.9) can be written into MATLAB (MathWorks) programming function to calculate the transepithelial water permeability.

```
function per=Cl_permeability_NaCl (ratio, resistance)

% ratio=Na_permeability/Cl_permeability
% C is the concentration in which resistance is measured
% resistance is the total resistance in om.cm2
% SI_Unit_trans: in cm/s

R=8.3143;   % in J(K-1)(mol-1)
T=273+37; % temperature in K
F=96488; % in C(mol-1)
C=145; % in mM
SI_Unit_trans=1e6; %x SI_Unit_trans
per=SI_Unit_trans*R*T./(F*F*C*resistance.*(ratio+1));
```

3.7.5 Troubleshooting

3.7.5.1 Differentiating paracellular from transcellular water pathway

Similar to membrane ion channels, the use of chemical inhibitors to block membrane water channels, that is, aquaporins, is a good way to differentiate paracellular from transcellular water channels. In the proximal tubule of the rabbit kidney, an aquaporin inhibitor is able to reduce the apical membrane permeability to water by 77% and the basolateral membrane permeability to water by 92% (Carpi-Medina & Whittembury, 1988). Genetic deletion of aquaporins in cell and animal models provides a new method to more efficaciously inhibit the transcellular water pathway. There is no pharmacologic approach to block the paracellular water pathway. The molecular identity of paracellular water channel remains largely unknown, despite some sketchy evidence that claudin-2 may permeate water as a byproduct of cation permeation (Rosenthal et al., 2010).

3.7.5.2 Effect of transepithelial voltage

Many types of epithelial cells develop a spontaneous transepithelial voltage (V_{te}) that drives Na^+ and Cl^- permeation through the paracellular pathway. There are two consequences of the spontaneous ionic transport. First, the transport of salt will alter the osmolality on both sides of the epithelium, which affects the value of $\Delta\pi$ and renders it variable. Second, water flux through the claudin-2 channel appears to be augmented when there is concomitant Na^+ flux. While the extent of V_{te} influence on transepithelial water permeability is not clear, it is highly recommended to eliminate V_{te} with ouabain included in the basal solution (see Chapter 3.2.3.2).

3.7.5.3 Proton permeability

The use of tritiated water ([3]H_2O) or deuterium oxide ([2]H_2O) does not allow differentiating water permeability from proton permeability. While protons are strictly excluded from aquaporins by a large electrostatic barrier present in their protein structures (de Groot & Grubmuller, 2005), claudins have been shown to permeate protons via a favorable electrostatic field established by the extracellular loop (Angelow, Kim, & Yu, 2006; Yu et al., 2009). From a methodological point of view, tritiated water or deuterium oxide is only suitable to transcellular water flux measurement.

3.7.5.4 Limitation in direct measurement of volume

According to Eq. (3.7.2), water permeability can be measured from volume change over reaction time given that $\Delta\pi$ is constant. A modified Ussing chamber system has been described to facilitate such measurement. In the system, each hemichamber is connected to glass tube instead of gas lift. The height of meniscus in the glass tube is monitored with a video optic system to record the volume change over time (Rosenthal et al., 2010). Nevertheless, it is worth pointing out that, because epithelial water flux is generally low, in the range of μL/hour/cm^2, the recording process must last several hours in order to discern the meniscus shift in the glass tube. Longer experimental duration often causes reduction in cellular viability.

3.7.6 Closing remarks

The molecular nature of paracellular water pathway is an important but unsolved topic in physiology. To date, the only way to measure paracellular water permeability is with Ussing chamber and in a polarized epithelium. Because the molecular components of tight junctions, including bicellular and tricellular tight junctions, have all been identified, genetic deletion or addition tools, in combination with Ussing chamber recording techniques, will reveal how water is permeated through the paracellular pathway.

References

Angelow, S., Kim, K. J., & Yu, A. S. (2006). Claudin-8 modulates paracellular permeability to acidic and basic ions in MDCK II cells. *The Journal of Physiology*, *571*, 15–26.

Carpi-Medina, P., & Whittembury, G. (1988). Comparison of transcellular and transepithelial water osmotic permeabilities (Pos) in the isolated proximal straight tubule (PST) of the rabbit kidney. *Pflugers Archiv: European Journal of Physiology*, *412*, 66–74.

de Groot, B. L., & Grubmuller, H. (2005). The dynamics and energetics of water permeation and proton exclusion in aquaporins. *Current Opinion in Structural Biology*, *15*, 176–183.

Hou, J. (2018). *The paracellular channel - biology, physiology and disease*. Academic Press.

Guilbault, G. G. (1967). *Fluorescence: Theory, instrumentation, and practice*. E. Arnold.

Kedem, O., & Katchalsky, A. (1958). Thermodynamic analysis of the permeability of biological membranes to non-electrolytes. *Biochimica et Biophysica Acta*, *27*, 229–246.

Rosenthal, R., Milatz, S., Krug, S. M., Oelrich, B., Schulzke, J. D., Amasheh, S., … Fromm, M. (2010). Claudin-2, a component of the tight junction, forms a paracellular water channel. *Journal of Cell Science*, *123*, 1913–1921.

Yu, A. S., Cheng, M. H., Angelow, S., Gunzel, D., Kanzawa, S. A., Schneeberger, E. E., … Coalson, R. D. (2009). Molecular basis for cation selectivity in claudin-2-based paracellular pores: Identification of an electrostatic interaction site. *The Journal of General Physiology*, *133*, 111–127.

Chapter 4

Histological approaches for tight junction

Chapter 4.1

Fixation and fixatives

4.1.1 Classification of fixatives

Chemical fixatives are classified into three major categories:

1. Cross-linking agents
   Formaldehyde, glutaraldehyde, acrolein, glyoxal, Mercuric chloride, and carbodiimide
2. Oxidizing agents
   Osmium tetroxide, potassium permanganate, and potassium dichromate
3. Precipitating agents

   Acetic acid, acetone, methyl alcohol, and ethyl alcohol

4.1.2 Mechanism of fixation

4.1.2.1 Protein cross-linking and denaturation

4.1.2.1.1 Cross-link formation

Aldehyde crosslinks are formed between protein molecules, in particular with the basic amino acid—lysine (Fig. 4.1.1). Only those lysine residues on the exterior of the protein molecule react with aldehydes. The reaction between aldehyde and protein is pH-dependent, progressing more rapidly at higher pH values. The reaction with formaldehyde is reversible over the first 24 hours of reaction time. With glutaraldehyde, on the other hand, the

A Laboratory Guide to the Tight Junction. DOI: https://doi.org/10.1016/B978-0-12-818647-3.00004-0

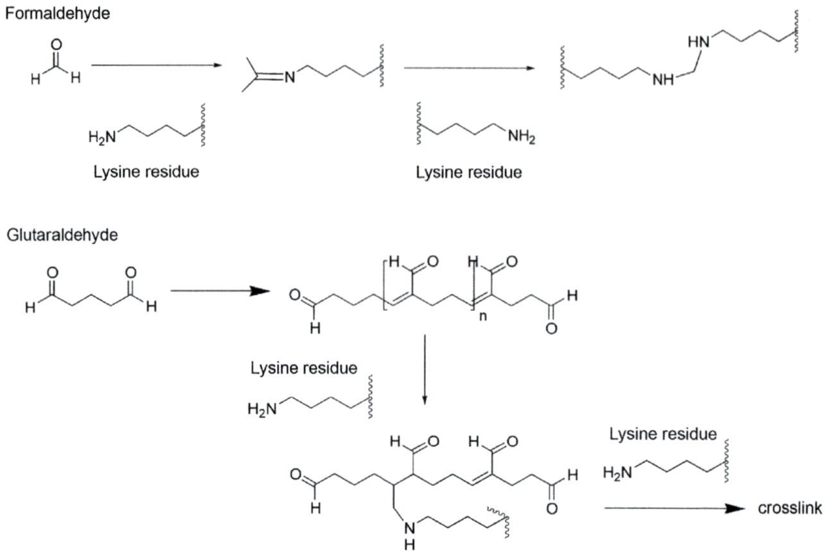

FIGURE 4.1.1 Mechanisms for cross-linking proteins by formaldehyde and glutaraldehyde via Maillard reaction.

reaction becomes irreversible. Cross-linking denatures proteins to some extent. Glutaraldehyde causes a loss of ~30% of the α-helix structure in the protein molecule, while formaldehyde markedly less. Because protein is a universal constituent of the cell, present not only in the cytosol but also in the cytoskeleton and the plasma membrane, soluble proteins can be cross-linked by aldehydes to each other and to the proteins throughout the cell. Therefore the proteins in the cell form a single interlocking meshwork. If some proteins were not cross-linked, they would nevertheless be trapped in the meshwork (Bancroft & Gamble, 2008).

4.1.2.1.2 Denaturation

Dehydrants such as alcohols and acetone remove and replace free water in cells and tissues and cause a change in the tertiary structure of proteins by destabilizing the hydrophobic interaction and the hydrogen bonds. Hydrophobic areas, frequently found on the inside of protein molecules, are exposed due to the repulsion of water. Hydrophilic areas on the outside of protein molecules, which are loosely bound by hydrogen bonds, are destabilized by the removal of water. The conformational changes in the protein molecules may cause water soluble proteins to become insoluble, a change that is largely irreversible if the protein is returned to an aqueous environment (Eltoum, Fredenburgh, Myers, & Grizzle, 2001).

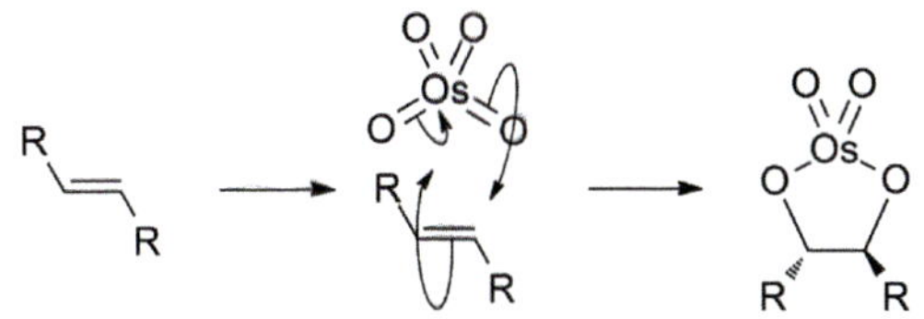

FIGURE 4.1.2 Mechanism for oxidizing alkenes in lipids by osmium tetroxide.

4.1.2.2 Lipid oxidization and cross-linking

Osmium tetroxide is a symmetrical compound with a molecular weight of 254.2 and contains four double-bonded oxygen molecules. Osmium tetroxide reacts primarily with lipid moieties by oxidizing the unsaturated bonds in the fatty acids (Fig. 4.1.2). The reduced osmium metal adds density and contrast to the biological sample.

Aldehydes can not only cross-link but also oxidize lipids. Phospholipids that contain amino groups, such as phosphatidyl ethanolamine, are fixed by aldehydes via cross-linking reactions. The double bonds in unsaturated fatty acids are oxidized by aldehydes into 1:3 glycols and 1:3 dioxans (Bancroft & Gamble, 2008).

4.1.2.3 Reaction of fixatives with nucleic acids

Ethanol and methanol are commonly used in the fixation of nucleic acids. Physical and chemical measurements have shown that DNA is largely collapsed in ethanol and methanol. When the denatured DNA is rehydrated, there is substantial reversion to the original form. For RNA, the precipitant fixatives ethanol and acetone give the best results, when frozen tissues are used.

4.1.3 Concentration of fixatives

Experiments have been performed to reveal the effects of different concentrations of aldehyde fixatives on the resulting tissue morphology. Over a 10-fold range, varying the concentrations of formaldehyde from 2% to 20% in a fixative solution had little effect on the cytoplasmic volume in one-centimeter blocks of rat kidney (Fox, Johnson, Whiting, & Roller, 1985). Only at very high (37%–40%) concentrations was there a significant decrease in the cytoplasmic volume. Formaldehyde is normally used as a 3.7% solution; glutaraldehyde is normally used as a 3% solution. The use of low concentrations of aldehydes, for example, formaldehyde (1%) or glutaraldehyde (0.25%), has been found ideal for immunolabeling purposes, including immunofluorescence microscopy and immuno-electron microscopy.

4.1.4 Osmolality of fixative solution

The osmolality of fixatives has a major effect on tissue morphology. Hypertonic solutions give rise to cell shrinkage. Isotonic fixatives produce swollen cells and poor fixation, as do hypotonic fixatives. The best results were obtained using slightly hypertonic solutions (440–450 mOsm). The vehicle osmolality is more important than the total osmolality of the fixative. Ideally, it should be isotonic with tissues in their normal environment. For aldehyde fixatives, vehicle osmolality should be about 300 mOsm with sodium chloride as added substance (Collins, Arborgh, & Brunk, 1977).

4.1.5 Penetration of fixatives

The depth (d) of fixative penetrated into a tissue is proportional to the square root of time (t), which can be written as follows:

$$d = K\sqrt{t}$$

where the constant (K) is the diffusion coefficient of the fixative (Medawar, 1941). Its unit is millimeter per hour. The K values of commonly used fixatives were measured in uniform tissues such as the liver or in gelatin pessaries (Table 4.1.1) (Baker, 1970). Because the values obtained in tissues are usually low, the blocks taken should be thin and small (e.g., 1 mm^3), to achieve satisfactory fixation.

4.1.6 Temperature of fixation

By tradition, fixation is carried out at room temperature. For electron microscopy and immunohistological applications, the temperature range is 0°C – 4°C. The argument in favor of the lower temperature range is that tissue autolysis is slowed down, as is diffusion of various cellular components. Against this is the fact that chemical reactions involved in fixation are more rapid at higher temperatures. In practice, the duration of fixation needs to be extended when lower temperature is used.

TABLE 4.1.1 Diffusion coefficient (*K*) for commonly used fixatives.

Fixative	Concentration	*K* value (mm/h)
Acetic acid	5	1.2
Formaldehyde	4	0.78
Ethanol	100	1
Glutaraldehyde	4	0.34
Mercuric chloride	0.78–0.84	2.2
Methanol	100	1.45[a]
Osmium tetroxide	0.5–2.0	0.29–0.58

All values were measured in the liver at room temperature.
[a]*Value measured in gelatine pessaries.*

4.1.7 Duration of fixation

Long fixation in aldehyde solutions is known to severely inhibit enzyme activity and immunological reactions. This can be attributed to limitations in substrate diffusion and cross-link formation between protein molecules. Prolonged fixation in formaldehyde ($>$ 24 hours) has been shown to cause tissue shrinkage and hardening. In the case of fixation with glutaraldehyde for electron microscopy, oxidizing fixatives such as osmium tetroxide and potassium permanganate are likely to degrade the tissues over time by oxidative cleavage of proteins (Bancroft & Gamble, 2008).

4.1.8 Fixation artifacts

Tissues often change in volume during fixation. Several mechanisms are involved, including inhibition of respiration, changes in membrane permeability, and swelling of extracellular matrix. Tissues fixed in formaldehyde and embedded in paraffin wax shrink by approximately 30% (Baker, 1970). The nuclei in frozen sections are usually larger than those in paraffin sections. Injured cells swell or shrink to a different extent from normal cells during fixation (Penttila, McDowell, & Trump, 1975). Materials may diffuse within or out of the tissue to give false localization. For example, unfixed hemoglobin molecules diffused to the periphery of spleen blocks because of the slow penetration of aldehyde fixatives. Finally, aldehydes will induce autofluorescence in tissues, which needs to be corrected by subsequent labeling techniques.

References

Baker, J. R. (1970). *Principles of biological microtechnique*.

Bancroft, J. D., & Gamble, M. (2008). *Theory and practice of histological techniques*. Elsevier Health Sciences.

Collins, V. P., Arborgh, B., & Brunk, U. (1977). A comparison of the effects of three widely used glutaraldehyde fixatives on cellular volume and structure. A TEM, SEM, volumetric and cytochemical study. *Acta Pathologica et Microbiologica Scandinavica Section A, Pathology*, *85a*, 157–168.

Eltoum, I., Fredenburgh, J., Myers, R. B., & Grizzle, W. E. (2001). Introduction to the theory and practice of fixation of tissues. *Journal of Histotechnology*, *24*, 173–190.

Fox, C. H., Johnson, F. B., Whiting, J., & Roller, P. P. (1985). Formaldehyde fixation. *The Journal of Histochemistry and Cytochemistry: Official Journal of the Histochemistry Society*, *33*, 845–853.

Medawar, P. B. (1941). III.—The rate of penetration of fixatives. *Journal of the Royal Microscopical Society*, *61*, 46–57.

Penttila, A., McDowell, E. M., & Trump, B. F. (1975). Effects of fixation and postfixation treatments on volume of injured cells. *The Journal of Histochemistry and Cytochemistry: Official Journal of the Histochemistry Society*, *23*, 251–270.

Chapter 4.2

Fixation

4.2.1 Introduction

For many purposes, fixation can be achieved by simple immersion of small tissue pieces into the fixative solution. A more rapid and uniform fixation is usually obtained by perfusion via the vascular system, either through the heart or through the abdominal aorta. Perfusion fixation is preferred over immersion fixation. For most tissues, glutaraldehyde at 1% is sufficient for general ultrastructural preservation. For immunohistochemical studies, formaldehyde at 4% plus glutaraldehyde at 0.1% is preferable, but glutaraldehyde must be omitted if it compromises the antigenicity (Eltoum, Fredenburgh, Myers, & Grizzle, 2001).

4.2.2 Materials and reagents

Anesthetic.

Scissors, forceps, scalpels, and clamps for surgical procedures.

Gauze swabs.

Thin razor blades.

Operating table.

Gloves.

Short-beveled syringe needle for perfusion of abdominal aorta (length, 50 mm; outer diameter, 1.5 mm).

Blunt syringe needle for perfusion of heart (length, 100 mm; outer diameter, 2.0 mm).

Perfusion set with drip chamber.

Stand to hold the fixative flask upside down.

4.2.3 Experimental procedure

4.2.3.1 Perfusion fixation through the heart

1. Place the closed flask containing the fixative upside down about 150 cm above the animal.
2. Connect the perfusion needle via the infusion set into the flask containing the fixative and ventilate the flask. Check that there is no air bubble in the tubing of the infusion set.
3. Fix the anesthetized animal onto the operating table with its back down.
4. Open the thoracic cavity of the animal.

5. Grasp the heart close to its apex with a pair of forceps. Cut a small hole in the wall of the left ventricle close to the apex with a pair of fine scissors. Rapidly insert a blunt syringe needle into the ventricle and move it to the ascending aorta. Place a clamp on the aorta to hold the needle.
6. Cut a hole in the right atrium of the heart and start perfusion immediately.
7. Check the flow rate in the drip chamber. The flow rate is set at 10 mL/min for an adult mouse. Perfuse for 3 minutes. Stop the perfusion and remove the tissues.
8. Excise and trim the tissues with a razor blade. Immersion fix the tissues in the same fixative for 24 hours at 4°C.
9. Rinse the tissues two times with 1 × PBS.
10. Store the tissues in vials filled with 1 × PBS.

4.2.3.2 Perfusion fixation through the abdominal aorta

1. Place the closed flask containing the fixative upside down about 150 cm above the animal.
2. Connect the perfusion needle via the infusion set into the flask containing the fixative and ventilate the flask. Check that there is no air bubble in the tubing of the infusion set.
3. Fix the anesthetized animal onto the operating table with its back down.
4. Open the abdominal cavity by a long midline incision with lateral extension and move the intestines gently to the left side of the animal.
5. Carefully expose the aorta below the origin of the renal arteries and gently free the aorta from overlaying adipose and connective tissues.
6. Hold the wall of the aorta firmly with a pair of small forceps at 0.5–1.0 cm from the distal bifurcation. Insert the short-beveled syringe needle toward the heart into the lumen of the aorta (with the beveled side facing down). Place a clamp on the aorta to hold the needle.
7. Rapidly cut a hole in the inferior caval vein with a pair of fine scissors and start the perfusion. Clamp the aorta below the diaphragm but above the origin of the renal arteries. Clamping is done by pressing the aorta toward the posterior wall of the peritoneal cavity with a finger, which is then replaced by a clamp. Finally, cut the aorta above the compression.
8. The kidney surface starts to blanch immediately. Check the flow rate in the drip chamber. The flow rate is set at 5 mL/min for an adult mouse. Perfuse for 3 minutes. Stop the perfusion and remove the tissues.
9. Excise and trim the tissues with a razor blade. Immersion fix the tissues in the same fixative for 24 hours at 4°C.
10. Rinse the tissues two times with 1 × PBS.
11. Store the tissues in vials filled with 1 × PBS.

4.2.3.3 Immersion fixation

1. Hold the tissue gently with a pair of forceps and cut thin slices with a razor blade. The slices should not exceed 0.5 mm in thickness.
2. Trim the slices into $5 \times 5 \times 5\ mm^3$ blocks. Immerse the tissue blocks into the fixative solution.
3. Fix the tissue slices for 24 hours at 4°C.
4. Rinse the tissues two times with 1 × PBS.
5. Store the tissues in vials filled with 1 × PBS.

4.2.4 Data analysis

Perfusion through the heart provides efficient fixation of most organs including the brain, the heart, the lung, the liver, the pancreas, and the gastrointestinal tract. Fixation of the kidney, however, is sensitive to the route of perfusion. Perfusion through the abdominal aorta better preserves the kidney tubules and the normal relationships between the tubular cells in the kidney (Maunsbach, 1966a, 1966b). Perfusion through the abdominal aorta also results in excellent fixation of the liver, the pancreas, and the small intestine. Immersion fixation is recommended for small tissues with or without prior perfusion, such as the inner ear, the retina, the epididymis, and the nerves.

4.2.5 Troubleshooting

If the fixative flow is reduced or stopped during perfusion fixation, sufficient concentration of fixative is not obtained throughout the tissue and cells may undergo abnormal alterations, such as apoptosis or necrosis, before they are fixed. In a successful perfusion, the surface of an organ blanches quickly and the organ hardens rapidly. Tissues tend to swell if the osmolality of the fixative solution is low, whereas they shrink if the osmolality is high (Bohman & Maunsbach, 1970). For this reason, the osmotic composition of the fixative solution has to be adjusted to match that in the extracellular fluid. The pH of the fixative solution is normally between 7.0 and 7.5 and fine adjustment of the pH value is not crucial in most immunohistochemical or ultrastructural studies. In highly vascularized organs, such as the liver, the kidney, and the pancreas, perfusion fixation tends to expand the extravascular space. Such expansion can be prevented by adding 2% dextran to the fixative solution.

4.2.6 Concluding remarks

Fixation is the most effective method to preserve the structure of cells with minimal alteration from the living state with regard to volume, morphology, and spatial relationship of organelles and macromolecules. Fixation is vital to the preservation of tight junction ultrastructure and the immunolabeling of tight junction protein. Fixatives differ in their effects on cells and tissues. To achieve a good fixing result, the fixation protocol must be determined independently for each type of cell or tissue by experimentation.

References

Bohman, S. O., & Maunsbach, A. B. (1970). Effects on tissue fine structure of variations in colloid osmotic pressure of glutaraldehyde fixatives. *Journal of Ultrastructure Research, 30*, 195–208.

Eltoum, I., Fredenburgh, J., Myers, R. B., & Grizzle, W. E. (2001). Introduction to the theory and practice of fixation of tissues. *Journal of Histotechnology, 24*, 173–190.

Maunsbach, A. B. (1966a). The influence of different fixatives and fixation methods on the ultrastructure of rat kidney proximal tubule cells. I. Comparison of different perfusion fixation methods and of glutaraldehyde, formaldehyde and osmium tetroxide fixatives. *Journal of Ultrastructure Research, 15*, 242–282.

Maunsbach, A. B. (1966b). The influence of different fixatives and fixation methods on the ultrastructure of rat kidney proximal tubule cells. II. Effects of varying osmolality, ionic strength, buffer system and fixative concentration of glutaraldehyde solutions. *Journal of Ultrastructure Research, 15*, 283–309.

Chapter 4.3

Tight junction atlas

4.3.1 Introduction

Tight junctions are abundant in epithelia (Hou, 2018). Epithelia are traditionally classified according to the number of cell layers and the shape of component cells in an organ. A single layer of epithelial cells is termed simple epithelium, whereas epithelia composed of several layers are termed stratified epithelia. Simple epithelia are found at interfaces involved in selective diffusion, absorption and secretion from the lungs, the kidney, the pancreas, the salivary glands, the stomach, and the intestines. Stratified epithelia have mainly a protective function and are found in the skin, the pharynx, the esophagus, the bladder, and the uterine cervix. The shape of epithelial cells is based upon the appearance in sections taken at right angles to the epithelial surface, which varies from squamous, cuboidal to columnar. Epithelia may derive from ectoderm, mesoderm or endoderm, although in the past it was thought that true epithelia were only of ectodermal or endodermal origin. Two types of epithelia derived from mesoderm, that is, the cells lining blood vessels and the cells lining serous cavities, are given special terms—endothelium and mesothelium, respectively. Besides epithelia, tight junctions are found in the glial cells of the nervous tissues, in the Sertoli cells of the testis, and in the hair cells of the cochlea (Hou, 2018).

4.3.2 Survey of tight junction in organ systems

4.3.2.1 Cardiovascular system

The cardiovascular system comprises a circuit of vessels through which blood flow is maintained by continuous pumping of the heart. The arterial system provides a distribution network to the capillaries which are the main sites of interchange of gases and metabolites between blood and tissues. The venous system returns blood from the capillaries to the heart. The epithelial cells lining the blood vessels are known as the endothelium. There are three types of endothelium: continuous endothelium, discontinuous endothelium, and fenestrated endothelium. Continuous endothelium forms an uninterrupted lining and is found in large vessels and in capillaries from the muscle and the brain. Discontinuous endothelium, only existing in the sinusoids of the liver, does not form a continuous interface between blood and tissue. Fenestrated endothelium contains numerous large pores or fenestrations and is found in capillaries from the intestines, the endocrine glands and the glomerulus of the kidney. Tight junctions are made only by the continuous endothelium (Dejana, 2004). Illustrated are examples of the mesothelium from the heart (Fig. 4.3.1), the endothelium from the heart (Fig. 4.3.2), the endothelium from the retina (Fig. 4.3.3), and the endothelium from the brain (Fig. 5.5.3).

4.3.2.2 Skin

The skin has three major layers—the outer keratinizing stratified squamous epithelium, known as the epidermis, the underlying supporting layer made of

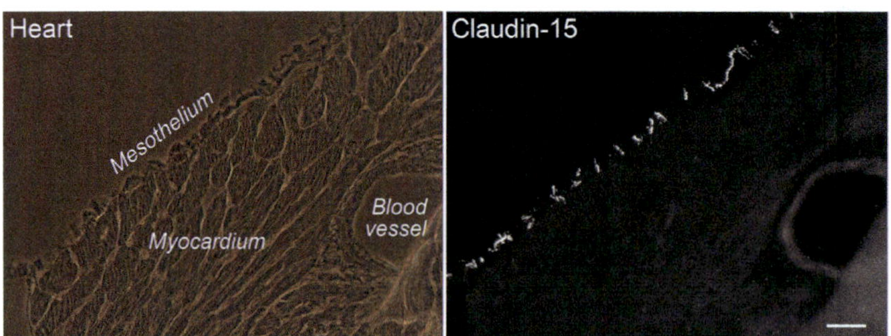

FIGURE 4.3.1 Tight junction in the heart mesothelium. The ventricular wall from the mouse heart was stained with anti-claudin-15 antibody to label the tight junction. Note that claudin-15 positive tight junctions are only found in the mesothelium. Bar: 20 μm.

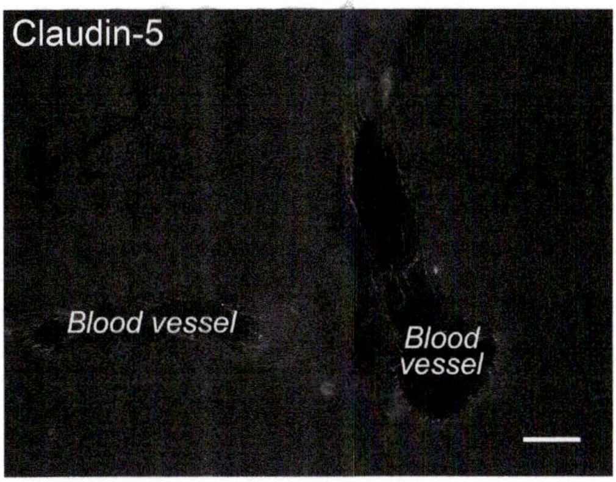

FIGURE 4.3.2 Tight junction in heart blood vessels. The ventricular wall from the mouse heart was stained with anti-claudin-5 antibody to label the tight junction. Note that claudin-5 positive tight junctions are only found in the blood vessels. Bar: 20 μm.

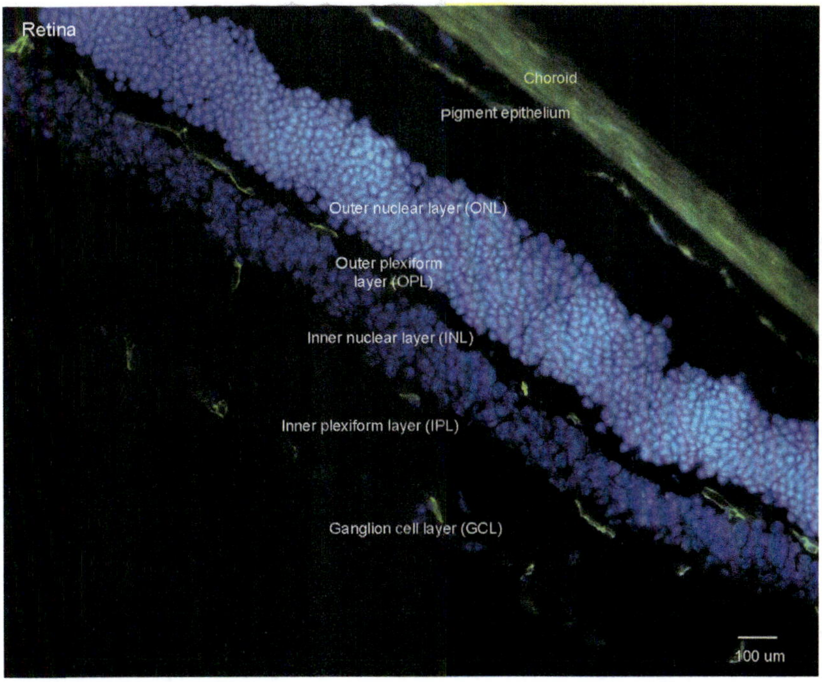

FIGURE 4.3.3 Tight junction in blood–retina barrier. The mouse retina was stained with anti-claudin-5 antibody to label the tight junction. Note that claudin-5 positive tight junctions are only found in the blood vessels. Bar: 100 μm.

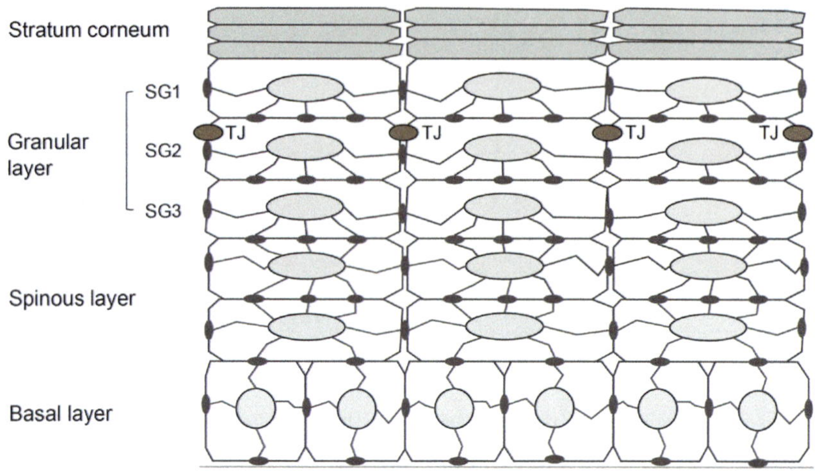

FIGURE 4.3.4 Schematic diagram of the epidermis. The different strata of the skin epidermis are indicated on the left. The granular layer is composed of three epithelial cell layers (SG1–SG3). TJs are found in the second layer (SG2). The outer layer of the skin, termed the stratum corneum, consists in dead cells and intercellular lipids. *Reproduced with permission from Hou, J. (2018).* The paracellular channel – biology, physiology and disease. *Academic Press.*

fibroelastic tissue, known as the dermis, and the deep layer made of adipose tissue, known as subcutis. The epidermis itself can be divided into four layers (Fig. 4.3.4). The innermost basal layer—stratum basale—consists in undifferentiated keratinocytes, stem cells, melanocytes, and Merkel cells. On top of this layer resides the spinous layer—stratum spinosum. The subsequent granular layer—stratum granulosum—consists of three to five cell layers. Continuous tight junctions are found in the second layer of the stratum granulosum (Fig. 4.3.5) (Yoshida et al., 2013). The uppermost layer—stratum corneum, consists in corneocytes, that is, dead cells, and intercellular lipids. The skin has four major functions—protection, sensation, thermoregulation, and metabolism.

4.3.2.3 Lung

The lungs have two functional components: the alveoli for transport of inspired and expired gases into and out of the lungs and the capillaries for passive exchange of gases between air and blood. Epithelium provides a continuous lining to each alveolus and consists of two types of cells (Fig. 4.3.6). Most of the alveolar surface area is covered by large squamous cells, known as the type I pneumocytes. The type II pneumocytes are cuboidal in shape and occupy a small proportion of the alveolar surface area. The type I pneumocytes perform gaseous exchange, whereas the type II pneumocytes secrete

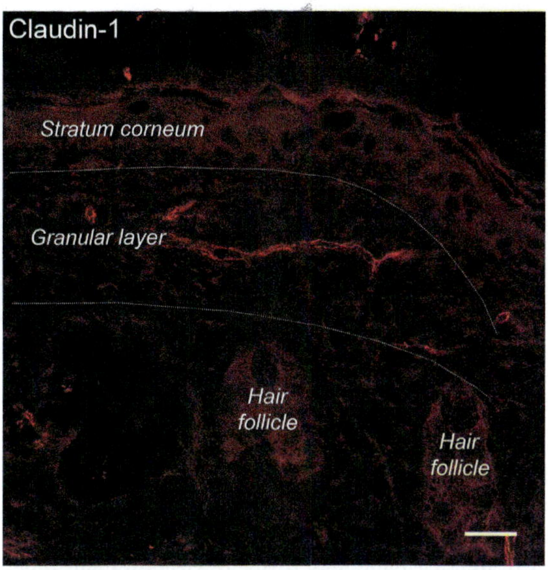

FIGURE 4.3.5 Tight junction in the skin. The mouse skin was stained with anti-claudin-1 antibody to label the tight junction. Note that claudin-1 positive tight junctions are only found in the granular layer. Bar: 20 μm.

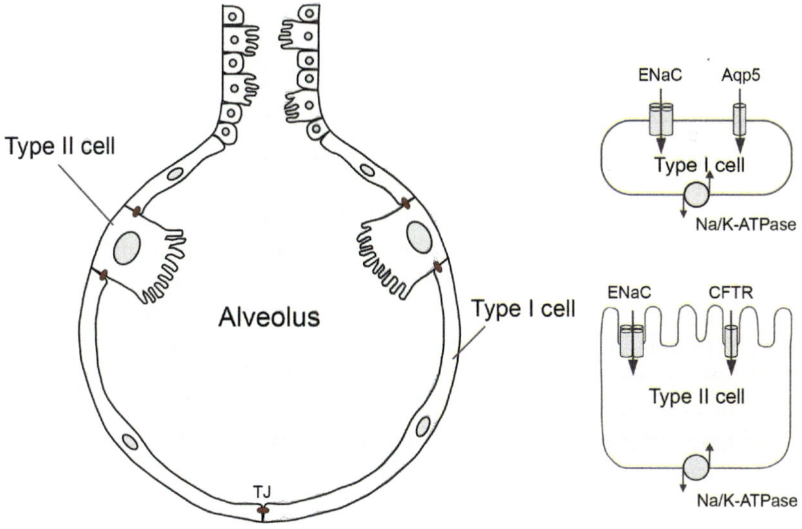

FIGURE 4.3.6 Schematic diagram of the pulmonary alveolus. The alveolar epithelium consists in two types of epithelial cells. The type I pneumocytes express epithelial Na^+ channel (ENaC) and aquaporin-5 (Aqp5), whereas the type II pneumocytes express ENaC and cystic fibrosis transmembrane conductance regulator (CFTR). Tight junctions are made by both type I and type II pneumocytes. *Reproduced with permission from Hou, J. (2018).* The paracellular channel – biology, physiology and disease. *Academic Press.*

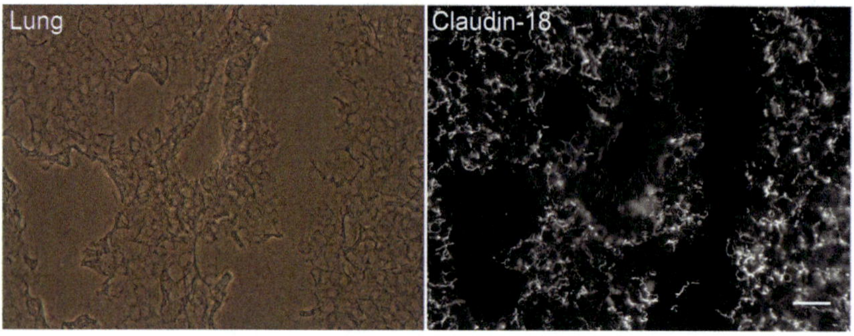

FIGURE 4.3.7 Tight junction in the pulmonary alveolus. The mouse lung was stained with anti-claudin-18 antibody to label the tight junction. Note that claudin-18 positive tight junctions are abundant in the alveolar epithelium. Bar: 20 μm.

a surfactant that reduces surface tension within the alveoli. Tight junctions are found in both type I and type II pneumocytes taking homotypic or heterotypic form (Fig. 4.3.7). The heterotypic tight junctions made between type I and type II alveolar cells may differ in structure and function from the homotypic tight junctions in type I or type II cells. Even though these heterotypic tight junctions are rare, they may possess unique permeability characteristics important for alveolar fluid reabsorption (LaFemina et al., 2010).

4.3.2.4 Gastrointestinal tract

The gastrointestinal tract consists in the pharynx, the esophagus, the stomach, the small intestines, the colon and the anal canal. The gastrointestinal tract has four distinct functional layers: the inner secretory or absorptive tissue known as the mucosa, the underlying collagenous tissue known as the submucosa, the muscular wall known as the muscularis propria, and the outer supporting adipose tissue known as the adventitia. The mucosa is made up of three components—an epithelium, a matrix layer of the lamina propria, and a smooth muscle layer of the muscularis mucosae. The epithelial cells lining the mucosa are mostly of the columnar shape. The three-dimensional structure of mucosal epithelium exhibits significant variation along the gastrointestinal tract. For example, the stomach is organized into pit, neck, and base; the small intestine into villus and crypt; the colon into surface and crypt (Fig. 4.3.8). Tight junctions are abundant throughout the gastrointestinal epithelium, but the composition of tight junction proteins may vary along the crypt-villus axis, as a part of the sophisticated program of epithelial differentiation (Noah, Donahue, & Shroyer, 2011). Illustrated are examples of the epithelium from the small intestines, for example, the duodenum, the jejunum, the ileum (Fig. 4.3.9), and the epithelium from the colon (Fig. 4.3.10).

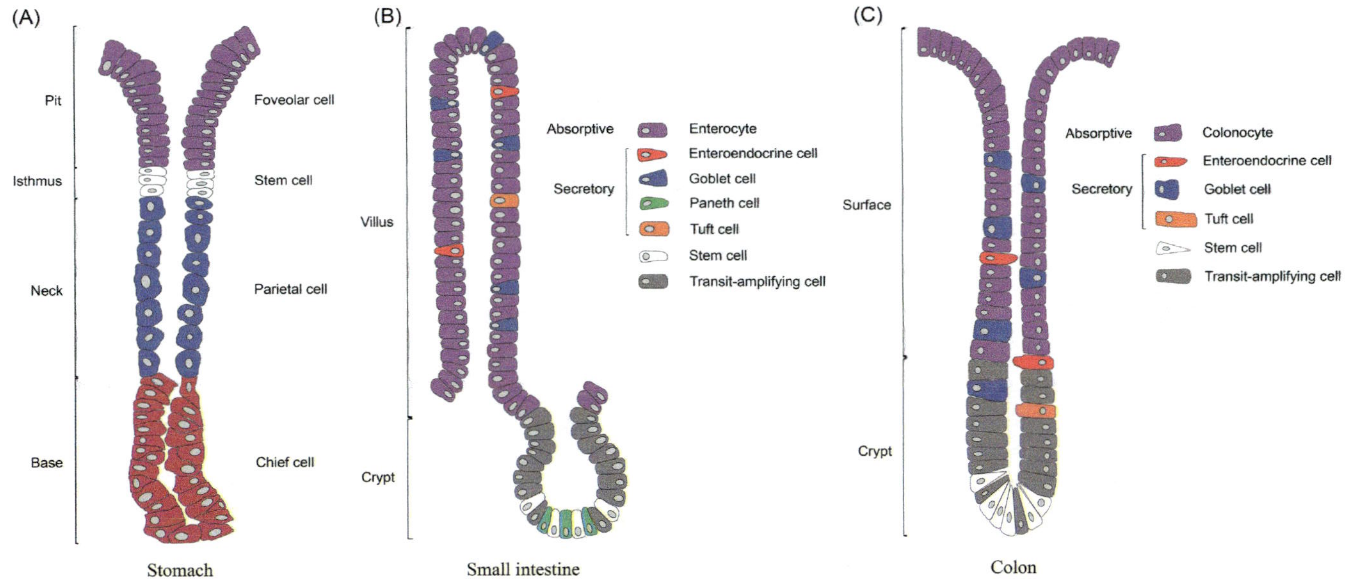

FIGURE 4.3.8 Schematic diagram of the gastrointestinal epithelium. (A) The gastric epithelium is organized into pit, neck, and base regions, and composed of foveolar cell, parietal cell, chief cell, and stem cell. (B) The small intestinal epithelium is organized into villus and crypt regions, and composed of enterocyte, enteroendocrine cell, goblet cell, Paneth cell, tuft cell, transit-amplifying cell, and stem cell. (C) The colonic epithelium is organized into surface and crypt regions, and composed of colonocyte, enteroendocrine cell, goblet cell, tuft cell, transit-amplifying cell, and stem cell. *Reproduced with permission from Hou, J. (2018).* The paracellular channel – biology, physiology and disease. *Academic Press.*

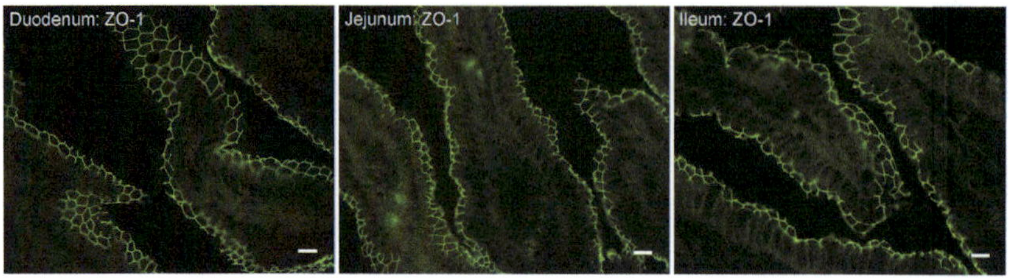

FIGURE 4.3.9 Tight junction in the small intestines. The mouse small intestines, including the duodenum, the jejunum, and the ileum, were stained with anti-ZO-1 antibody to label the tight junction. Note that ZO-1 positive tight junctions are abundant in the mucosal layer. Bar: 20 μm. *Reproduced with permission from Gong, Y, Himmerkus, N, Sunq, A, Milatz, S, Merkel, C, Bleich, M, & Hou, J. (2017). ILDR1 is important for paracellular water transport and urine concentration mechanism.* Proceedings of the National Academy of Sciences of the United States of America, 114*(20), 5271–5276.*

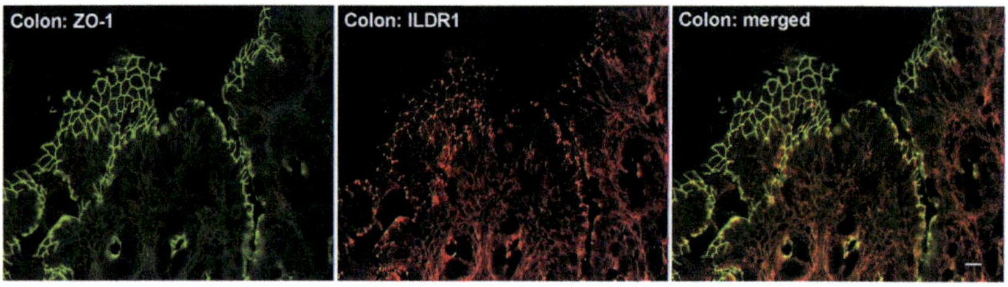

FIGURE 4.3.10 Tight junction in the colon. The mouse colon was stained with anti-ZO-1 and anti-ILDR1 antibodies to label the bicellular tight junction and the tricellular tight junction, respectively. Bar: 20 μm. *Reproduced with permission from Gong, Y, Himmerkus, N, Sunq, A, Milatz, S, Merkel, C, Bleich, M, & Hou, J. (2017). ILDR1 is important for paracellular water transport and urine concentration mechanism.* Proceedings of the National Academy of Sciences of the United States of America, 114*(20), 5271–5276.*

4.3.2.5 Liver

The structural components of the liver include plates of liver cells, known as hepatocytes, separated by wide vascular channels known as sinusoids. Hepatocytes are large polyhedral cells with round nuclei. Unlike the epithelial cells in other tissues, which are polarized into two membrane domains: apical and basolateral domains along the plane of the tissue, hepatocytes display a unique polarization arrangement in which two tight junctions are made between adjacent cells to partition the plasma membrane into three domains: a luminal domain delimiting the bile canaliculus, and two basolateral domains surrounded by the perisinusoidal space (also known as the space of Disse) (Fig. 4.3.11) (Treyer & Musch, 2013). The liver tight junctions appear as continuous lines running along the lateral hepatocyte surface and straddling the bile canaliculus (Fig. 4.3.12).

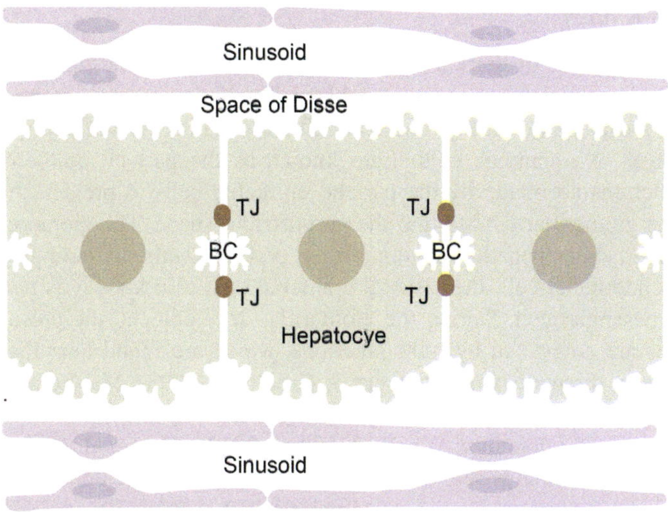

FIGURE 4.3.11 Schematic diagram of the hepatic lobule. Hepatocytes are polarized in a different way from squamous or cuboidal epithelia. Adjacent hepatocytes make two tight junctions (TJs) to create a tubular lumen in the lateral extracellular space known as the bile canaliculus (BC). The TJs can regulate electrolyte and water balances between the BCs and the sinusoids via the space of Disse. *Reproduced with permission from Hou, J. (2018).* The paracellular channel – biology, physiology and disease. *Academic Press.*

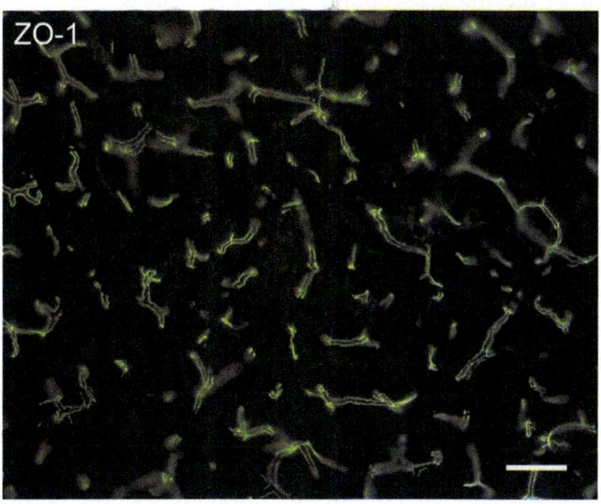

FIGURE 4.3.12 Tight junction in the liver. The mouse liver was stained with anti-ZO-1 antibody to label the tight junction. Note that ZO-1 positive tight junctions are continuous along the luminal surface of the bile canaliculus. Bar: 20 μm.

4.3.2.6 Kidney

The functional unit of the kidney, the nephron, consists in a glomerulus and a long folded renal tubule (Fig. 4.3.13A). The glomerulus is composed of Bowman's capsule and the glomerular tuft. Bowman's capsule is covered by a thin layer of squamous epithelium known as the parietal epithelial cell. Tight junctions are made by the parietal epithelial cells to prevent the leakage of the glomerular filtrate into the perivascular space. The glomerular tuft is a microvascular bed that contains three types of cells, including the glomerular endothelial cell, the visceral epithelial cell, also known as podocyte, and the mesangial cell. During the glomerular development, the presumptive podocytes are connected by tight junctions, which are found near the apical membrane of the podocytes (Quaggin & Kreidberg, 2008). Mature podocytes are devoid of the tight junction but form a special cell junction known as the slit diaphragm (Grahammer, Schell, & Huber, 2013). The renal tubule can be divided into three major segments: the proximal tubule, including the proximal convoluted tubule and the proximal straight tubule (Fig. 4.3.13B), the loop of Henle, including the thin descending and ascending limb and the thick ascending limb (Fig. 4.3.13C), and the distal tubule, including the distal convoluted tubule and the collecting duct (Fig. 4.3.13D). Each tubular segment is responsible for reabsorbing a fraction of the glomerular filtrate. The epithelium lining the renal tubule is of simple cuboidal shape. Tight junctions are abundant in epithelia from all segments of the renal tubule (Hou, Rajagopal, & Yu, 2013). Illustrated are examples of the parietal epithelium from Bowman's capsule (Fig. 4.3.14), the epithelium of the proximal tubule, the epithelium of the thick ascending limb of Henle's loop, and the epithelium of the collecting duct (Fig. 4.3.15).

4.3.2.7 Nerve

The nervous system is composed of an intercommunicating network of specialized cells known as neurons, which perform the sensory roles, constitute the conducting pathways, and analyze the electrical signals. The ensheathment of neurons with the myelin enables rapid saltatory conduction of action potentials in the nervous system. To facilitate this process, the intramyelinic space must be sealed by tight junctions to prevent electric current leakage (Fig. 4.3.16) (Dermietzel, Leibstein, & Schunke, 1980; Reale, Luciano, & Spitznas, 1975). This type of tight junction is formed by one cell, for example, the oligodendrocyte in the central nervous system or the Schwann cell in the peripheral nervous system, and termed as autotypic tight junction. Autotypic tight junctions can be found between the paranodal loops, in the inner and outer mesaxons, and in the Schmidt-Lanterman incisures of the myelin (Fig. 4.3.17).

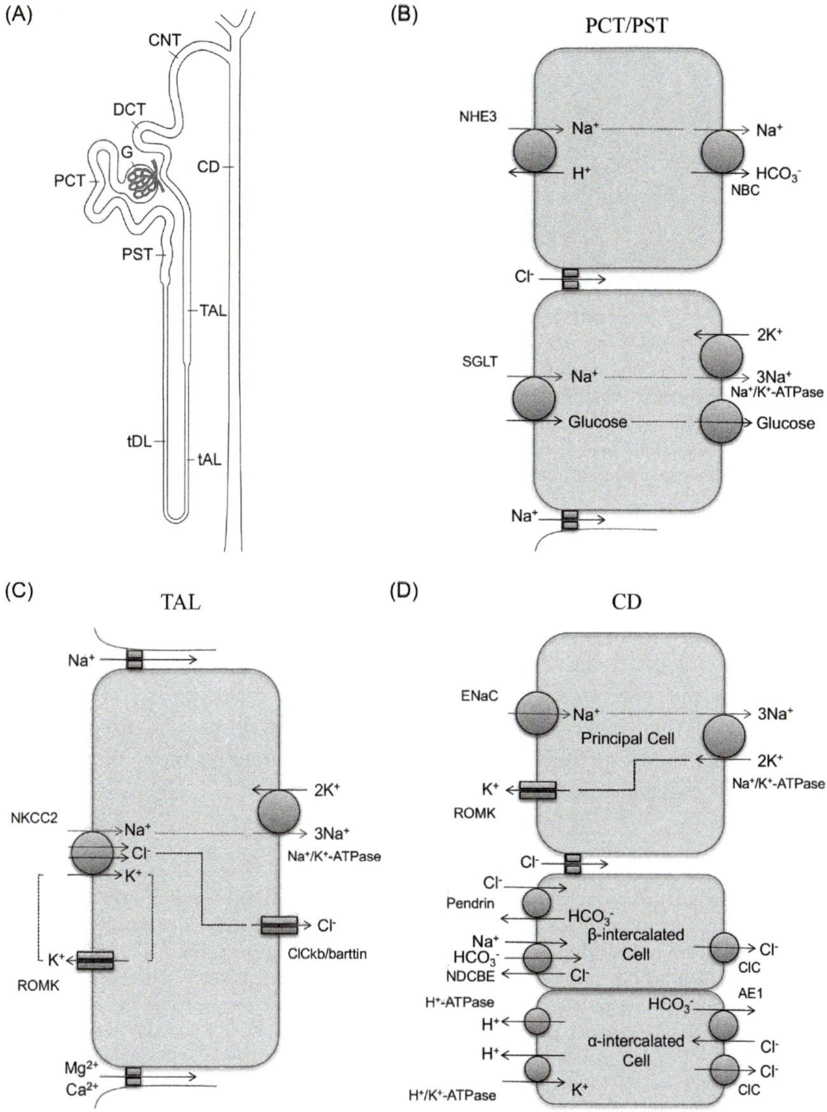

FIGURE 4.3.13 Schematic diagram of the renal epithelium. (A) The nephron consists in the glomerulus (G) and the renal tubule, which includes the proximal convoluted tubule (PCT), the proximal straight tubule (PST), the thin descending limb (tDL), the thin ascending limb (tAL), the thick ascending limb (TAL), the distal convoluted tubule (DCT), the connecting tubule (CNT), and the collecting duct (CD). (B) In the proximal tubule, both Na^+ and HCO_3^- are reabsorbed via the Na^+/H^+ exchanger 3 (NHE3) on the apical membrane and the $Na^+-HCO_3^-$ cotransporter (NBC) on the basolateral membrane. This process increases the Cl^- concentration in the proximal tubule luminal fluid, driving paracellular Cl^- reabsorption down the chemical gradient of Cl^-. Paracellular Cl^- reabsorption creates a lumen-positive voltage, which then *(Continued)*

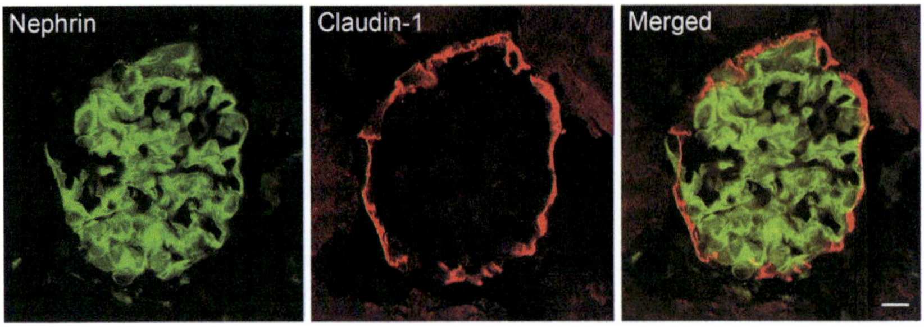

FIGURE 4.3.14 Tight junction in glomerular parietal epithelium. The mouse glomerulus was stained with anti-nephrin and anti-claudin-1 antibodies to label the cell junctions in the glomerular podocytes and parietal epithelial cells, respectively. Note that claudin-1 positive tight junctions are made by the parietal epithelial cells but not by the podocytes. Bar: 10 μm. *Reproduced with permission from Gong, Y., Sunq, A., Roth, R. A., & Hou, J. (2017). Inducible expression of claudin-1 in glomerular podocytes generates aberrant tight junctions and proteinuria through slit diaphragm destabilization.* Journal of the American Society of Nephrology, 28*(1), 106–117.*

4.3.3 Concluding remarks

The histological foundation of tight junction biology is established on the information of tight junction localization in organ systems. The epithelia from every organ make tight junctions. The appearance of tight junction under light microscopy may differ among epithelia in organs. Some nonepithelial cells can make tight junctions too. The structure and function of nonepithelial tight junctions are less understood but may become a new frontier of tight junction research.

◀ drives paracellular reabsorption of Na^+. The reabsorption of Na^+ and glucose via the Na^+-glucose cotransporter (SGLT) in the proximal tubule generates a significant osmotic gradient that drives the reabsorption of water. (C) In thick ascending limb, Na^+, K^+, and Cl^- are absorbed through the luminal membrane $Na^+-K^+-2Cl^-$ cotransporter (NKCC2). Na^+ is secreted into the basolateral side via the Na^+/K^+-ATPase. Cl^- is secreted into the basolateral side via the chloride channel (ClCkb/barttin). K^+ is recycled into the luminal side through the renal outer medullary potassium channel (ROMK). Due to continuous reabsorption of NaCl, a NaCl gradient develops from interstitial to luminal side, which generates a lumen-positive diffusion voltage to drive paracellular Mg^{++} and Ca^{++} reabsorption. (D) In the collecting duct, Na^+ is absorbed through the epithelial sodium channel (ENaC). Na^+ is secreted into the basolateral side via the Na^+/K^+-ATPase. K^+ is secreted into the luminal side via the renal outer medullary potassium channel (ROMK). Because of the unilateral Na^+ absorption, a lumen-negative voltage develops, which drives paracellular Cl^- absorption. A parallel electroneutral transport pathway exists for Cl^- using the Cl^-/HCO_3^- exchanger (pendrin) and the Na^+-driven Cl^-/HCO_3^- exchanger (NDCBE) on the apical membrane and the chloride channel (ClC) on the basolateral membrane of the β-intercalated cell. The α-intercalated cell primarily handles H^+ secretion via the H^+-ATPase and the H^+/K^+-ATPase in the luminal membrane. *Reproduced with permission from Hou, J. (2018).* The paracellular channel – biology, physiology and disease. *Academic Press.*

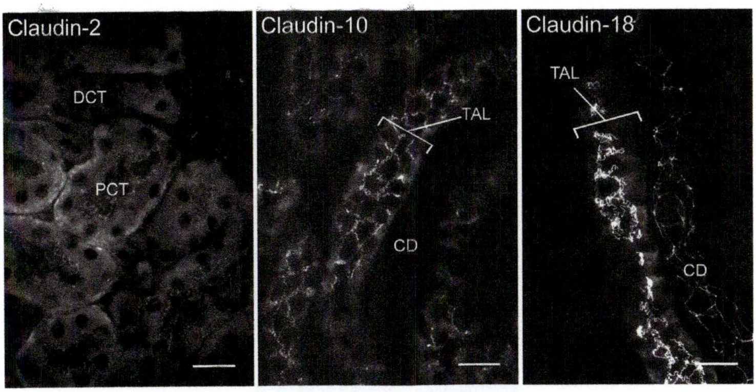

FIGURE 4.3.15 Tight junction in kidney tubular epithelia. The mouse kidney tubules were stained with anti-claudin-2, anti-claudin-10, or anti-claudin-18 antibody to label the tight junction in the epithelium of the proximal tubule, the thick ascending limb, or the collecting duct, respectively. Note that the tight junction strand appears to differ among these tubular segments with regard to its shape and structure. Bar: 10 μm. *Reproduced with permission from Hou, J., Renigunta, A., Gomes, A. S., Hou, M., Paul, D. L., Waldegger, S., & Goodenough, D. A. (2009). Claudin-16 and claudin-19 interaction is required for their assembly into tight junctions and for renal reabsorption of magnesium.* Proceedings of the National Academy of Sciences of the United States of America, 106*(36),15350–15355.*

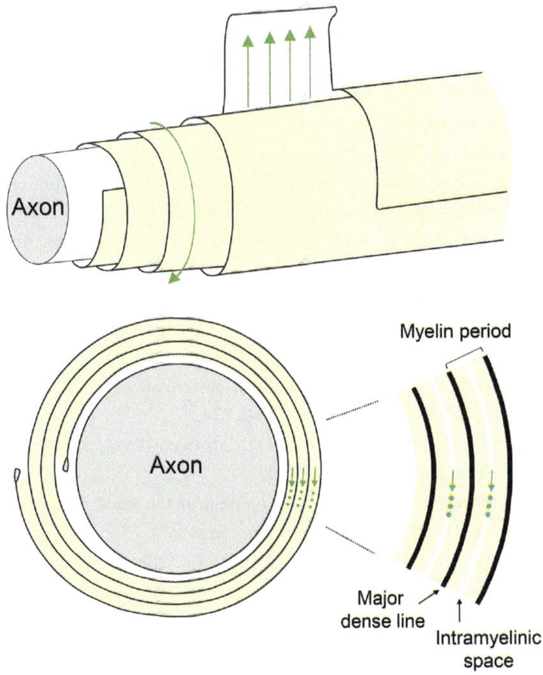

FIGURE 4.3.16 Schematic diagram of an axon ensheathed by the myelin lamellae. A simplified model unravels the myelin as membrane sheets spiraling around the axon (*top*). The cross-sectional view reveals the key features of the myelin including the major dense line and the intramyelinic space, which are organized into the myelin period (*bottom*). Current flow (*green arrow*) into the intramyelinic space is normally blocked by autotypic tight junctions (*green dot*). *Reproduced with permission from Hou, J. (2018).* The paracellular channel – biology, physiology and disease. *Academic Press.*

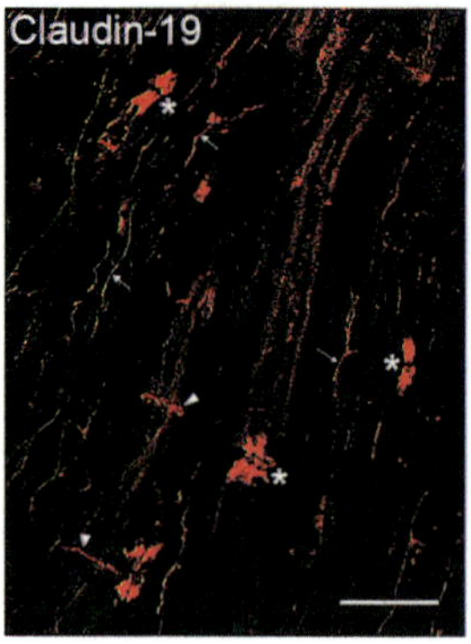

FIGURE 4.3.17 Tight junction in the myelin. The mouse sciatic nerve was stained with anti-claudin-19 antibody to label the autotypic tight junction in the myelin. Note that claudin-19 positive tight junctions are found in the paranodes (*asterisk*), the inner and outer mesaxons (*arrow*), and the Schmidt-Lanterman incisures (*arrowhead*). Bar: 10 μm.

References

Dejana, E. (2004). Endothelial cell–cell junctions: Happy together. *Nature Reviews Molecular Cell Biology*, *5*, 261.

Dermietzel, R., Leibstein, A. G., & Schunke, D. (1980). Interlamellar tight junctions of central myelin. II. A freeze fracture and cytochemical study on their arrangement and composition. *Cell and Tissue Research*, *213*, 95–108.

Grahammer, F., Schell, C., & Huber, T. B. (2013). The podocyte slit diaphragm—From a thin grey line to a complex signalling hub. *Nature Reviews Nephrology*, *9*, 587–598.

Hou, J. (2018). *The paracellular channel – Biology, physiology and disease*. Academic Press.

Hou, J., Rajagopal, M., & Yu, A. S. (2013). Claudins and the kidney. *Annual Review of Physiology*, *75*, 479–501.

LaFemina, M. J., Rokkam, D., Chandrasena, A., Pan, J., Bajaj, A., Johnson, M., & Frank, J. A. (2010). Keratinocyte growth factor enhances barrier function without altering claudin expression in primary alveolar epithelial cells. *American Journal of Physiology Lung Cellular and Molecular Physiology*, *299*, L724–L734.

Noah, T. K., Donahue, B., & Shroyer, N. F. (2011). Intestinal development and differentiation. *Experimental Cell Research*, *317*, 2702–2710.

Quaggin, S. E., & Kreidberg, J. A. (2008). Development of the renal glomerulus: Good neighbors and good fences. *Development (Cambridge, England)*, *135*, 609–620.

Reale, E., Luciano, L., & Spitznas, M. (1975). Zonulae occludentes of the myelin lamellae in the nerve fibre layer of the retina and in the optic nerve of the rabbit: A demonstration by the freeze-fracture method. *Journal of Neurocytology*, *4*, 131–140.

Treyer, A., & Musch, A. (2013). Hepatocyte polarity. *Comprehensive Physiology*, *3*, 243–287.

Yoshida, K., Yokouchi, M., Nagao, K., Ishii, K., Amagai, M., & Kubo, A. (2013). Functional tight junction barrier localizes in the second layer of the stratum granulosum of human epidermis. *Journal of Dermatological Science*, *71*, 89–99.

Chapter 5

Light microscopy for tight junction

Chapter 5.1

Theory of light microscopy

5.1.1 Lateral resolution in light microscopy

In microscopy, the term "resolution" is used to describe the minimal distance (d) at which two distinct points in a focal plane can be distinguished by an observer or a camera—as separate entities. The resolution of a microscope is related to the numerical aperture (NA) of the lens as well as the wavelength of light (λ) that is used to examine a specimen according to Abbe's diffraction limit equation (Inoue & Oldenbourg, 1995).

$$d = \frac{0.61\lambda}{NA} \tag{5.1.1}$$

The numerical aperture is related to the refractive index (n) of a medium through which light passes as well as the angular aperture (α) of an objective according to the following equation.

$$NA = n \times \sin(\alpha) \tag{5.1.2}$$

The Airy disc is the diffraction pattern that a perfect lens with a circular aperture can make. The Airy disc appears as a bright spot surrounded by a series of concentric rings, also known as the Airy pattern (Fig. 5.1.1). The central point in the Airy disc contains approximately 84% of the luminous intensity according to the Airy function, also known as the point spread function (Yoo, Song, & Gweon, 2006). If the center (or the diffraction maximum) of an Airy disc overlaps with the first minimum of another Airy disc, then they can be considered to be two separate points of light or "resolved" according the Rayleigh criterion (Fig. 5.1.2).

A Laboratory Guide to the Tight Junction. DOI: https://doi.org/10.1016/B978-0-12-818647-3.00005-2

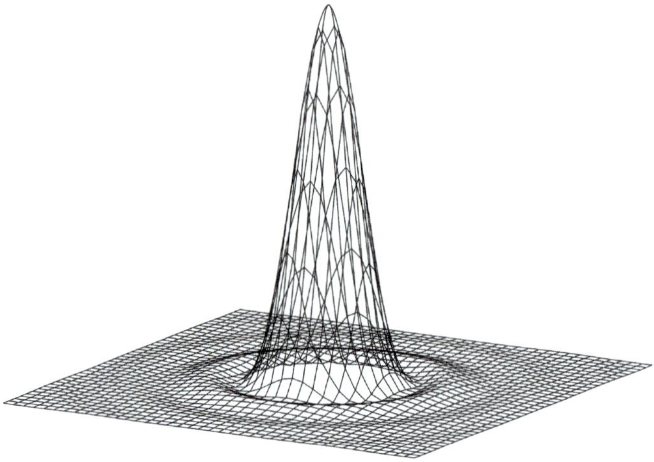

FIGURE 5.1.1 Airy disc. An illustration of airy disc or airy pattern as a central point of light surrounded by the concentric diffractive rings.

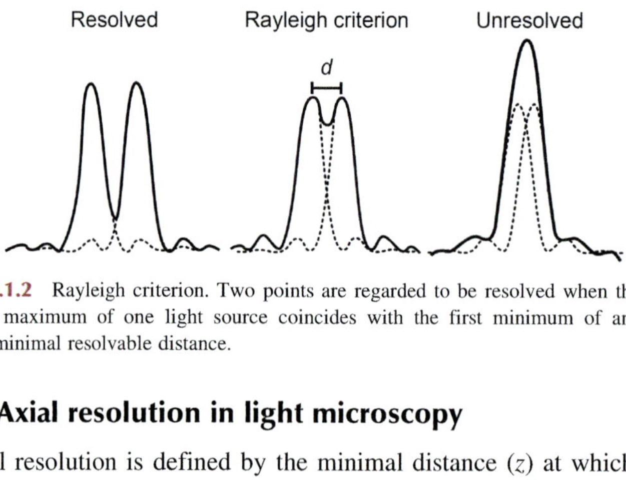

FIGURE 5.1.2 Rayleigh criterion. Two points are regarded to be resolved when the principal diffraction maximum of one light source coincides with the first minimum of another light source. *d*, minimal resolvable distance.

5.1.2 Axial resolution in light microscopy

The axial resolution is defined by the minimal distance (z) at which the diffraction images of two points can be separated along the axis of a microscope. Studies of wave optics have led to the discovery that the diffraction image of a point light source is not only periodic around the point of focus in a focal plane, but is also periodic above and below the focal plane along the axis of the microscope (Fig. 5.1.3) (Linfoot & Wolf, 1956). The axial resolution is given by the following equation.

$$z = \frac{1.67\lambda}{(NA)^2} \tag{5.1.3}$$

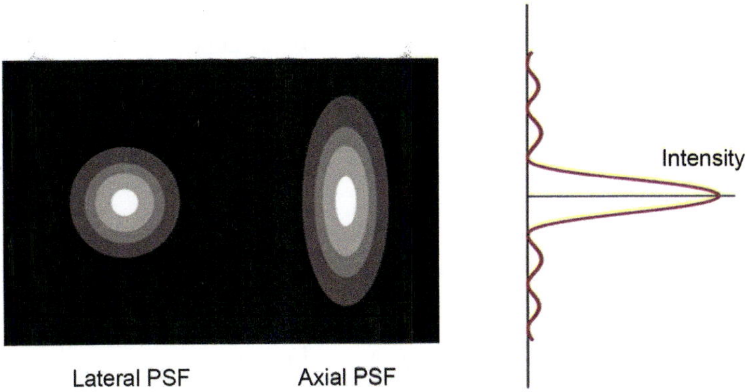

FIGURE 5.1.3 Axial diffraction pattern. The diffraction image of a point light source is periodic above and below the focal plane along the axis of a lens, which can be described by the axial point spread function (PSF).

5.1.3 Depth of field in light microscopy

The depth of field of a microscope refers to the distance from the nearest focal plane to the farthest focal plane and is measured in microns. In the theory, the depth of field equals to the axial resolution. The actual depth of field can be affected by several factors, which includes (1) the geometric shape of diffraction pattern above and below the focal plane, (2) the accommodation of the observer's eye, and (3) the final magnification of the image (Berek, 1927). When an ultrathin detector is used and the total magnification is raised sufficiently, the depth of field (δ) is given by

$$\delta = \frac{1}{4}\left(z_{\text{min}}^{+} - z_{\text{min}}^{-}\right) \tag{5.1.4}$$

that is, one-quarter of the distance between the first axial minima above ($z_{\text{min}}{}^{+}$) and below ($z_{\text{min}}{}^{-}$) the central maximum in the axial point spread function.

5.1.4 Fluorescence microscopy

Fluorescence is the emission of light by a substance that has absorbed light or other forms of electromagnetic radiation. In most cases, the emitted light has a longer wavelength and therefore lower energy than the absorbed light (Fig. 5.1.4). A fluorescence microscope is an optical microscope that uses fluorescence to study the properties of cellular substances. The majority of fluorescence microscopes used in life sciences are the epifluorescence microscopes. In this type of microscopy, the light of an excitation wavelength illuminates the specimen through the objective lens. The fluorescence emitted by the specimen is focused to the detector by the same objective

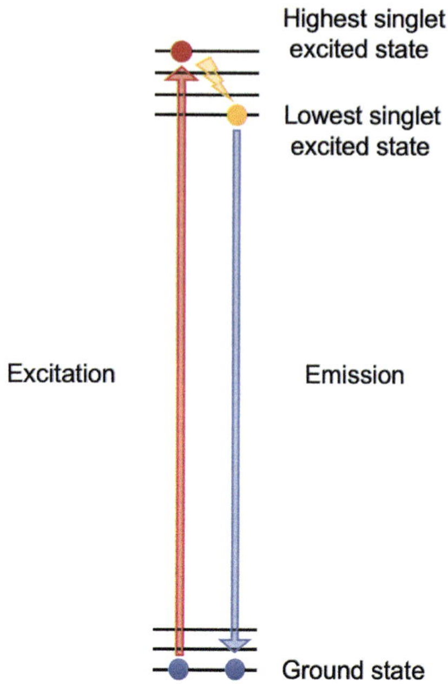

FIGURE 5.1.4 Jablonski energy diagram of fluorescence. Before excitation, the electronic configuration of the molecule is described as being in the ground state. Upon absorbing a photon of excitation light, usually of short wavelengths, electrons are raised to a higher energy and vibrational excited state. The excited electrons then lose some vibrational energy to the surrounding environment and return to the ground state with simultaneous emission of the fluorescent light of longer wavelength than the excitation light.

lens that is used for the excitation purpose. As most of the excitation light is transmitted through the specimen, only reflected excitatory light reaches the objective lens together with the emitted light, which, referred to as "the epifluorescence design," gives a high signal-to-noise ratio (Stockert & Blázquez-Castro, 2017).

5.1.5 Fluorescent labels

Antibodies labeled with fluorophores are among the most commonly used agents in fluorescence microscopy. Reactive fluorophores have chemical groups that can form covalent bonds with proteins including antibodies. Examples of reactive fluorophores are fluorescein isothiocyanate (FITC), tetramethyl-rhodamine isothiocyanate (TRITC), Cy3, Cy5, Cy7, and Alexa 488 and Alexa 546-tyramide. In addition to antibodies, biomolecules with high affinity to nucleic acids, lipids, and polysaccharides can be labeled with fluorophores, resulting in selective "immuno-like" fluorescence reactions. Some proteins are naturally fluorescent. Among them, the green fluorescent

TABLE 5.1.1 Natural fluorescent proteins and their excitation and emission spectra.

Protein	λ_{exc} (nm)	λ_{em} (nm)
GFP	475	509
RGFP	466	505
EGFP	488	507
ECFP	434	474
EYFP	514	527
DsRed	558	583
HcRed	590	615
mCherry	585	615

E, Enhanced; *DsRed*, from *Discosoma* sp.; *HcRed*, from the coral *Heteractis crispa*; *RGFP*, from *Rhacostoma atlantica*.

protein (GFP, 27 kDa) from the light-emitting cells (photocytes) of the jellyfish *Aequorea victoria* is most widely used in life sciences. When excited by violet or blue light (395 or 475 nm), GFP emits green light (peak: 509 nm, shoulder: 540 nm) with high efficiency. The GFP fluorophore is formed by spontaneous cyclizing and oxidizing reactions involving three key amino acids (aa. 65–67) in the GFP polypeptide (Tsien, 1998). Several GFP variants have been made by modifying the GFP fluorophore to confer different excitation and emission spectra (Table 5.1.1).

5.1.6 Autofluorescence

The occurrence of autofluorescence is a very important subject in fluorescence microscopy. Examples of fresh biological structures with autofluorescence due to endogenous fluorophores include elastic fibers, mitochondria, lysosomes, eosinophil granules, amyloid, and ceroid. Fixation can exacerbate autofluorescence. Formaldehyde and glutaraldehyde are known to generate pyridinium products from lysine residues in proteins, which contribute to autofluorescence (Hayat, 2012). Ageing cells are another major source of autofluorescence due to the presence of ageing pigments known as lipofuscin granules (Colcolough, Helmy, & Hack, 1970). Lipofuscin contains *N*-retinylidene-*N*-retinyl-ethanolamine (A2E), a yellow-emitting fluorophore with emission wavelength (λ_{em}) at 565–570 nm.

5.1.7 Photobleaching

Photobleaching (also termed fading) occurs when a fluorophore permanently loses the ability to fluoresce due to photon-induced chemical damage. One of the main environmental agents that induce photobleaching is oxygen (O_2). When fluorophores are in the excited state, they transfer the excitation energy to O_2. This transfer leads to the generation of the first excited state of O_2 known as the singlet oxygen ($^1O_2$). The $^1O_2$ is very reactive and can photobleach the fluorophores (Turro, Ramamurthy, & Scaiano, 2017). For this reason, O_2 must be excluded whenever possible from samples subjected to fluorescence analysis. Quenching solutions containing sacrificial reductants such as *n*-propyl gallate are commonly used to delay the production of $^1O_2$.

References

Berek, M. (1927). Grundlagen der Tiefenwahrnehmung im Mikroskop: mit einem Anhang über die Bestimmung der obersten Grenze des unvermeidlichen Fehlers einer Messung aus der Häufigkeitsverteilung der zufälligen Maximalfehler.

Colcolough, H. L., Helmy, F. M., & Hack, M. H. (1970). Some histochemical observations on the lipofuscin of vertebrate liver, kidney and cardiac muscle. *Acta Histochemica*, *35*, 343–356.

Hayat, M. E. (2012). *Fixation for electron microscopy*. Elsevier.

Inoue, S., & Oldenbourg, R. (1995). *Handbook of optics*. New York: McGrawHill.

Linfoot, E., & Wolf, E. (1956). Phase distribution near focus in an aberration-free diffraction image. *Proceedings of the Physical Society Section B*, *69*, 823.

Stockert, J. C., & Blázquez-Castro, A. (2017). *Fluorescence microscopy in life sciences*. Bentham Science Publishers.

Tsien, R. Y. (1998). The green fluorescent protein. *Annual Review of Biochemistry*, *67*, 509–544.

Turro, N. J., Ramamurthy, V., & Scaiano, J. C. (2017). *Modern molecular photochemistry of organic molecules*. Viva Books, published by arrangement with University Science Books ....

Yoo, H., Song, I., & Gweon, D. G. (2006). Measurement and restoration of the point spread function of fluorescence confocal microscopy. *Journal of Microscopy*, *221*, 172–176.

Chapter 5.2

Wide-field fluorescence microscopy for cells on cover glass

5.2.1 Background knowledge

5.2.1.1 Immunofluorescence labeling

Immunofluorescence as a method to study protein localization to tight junction has gained prominence with the demonstration that antibodies can be produced to tight junction components such as ZO-1, occludin, and claudins (Furuse et al., 1993; Furuse, Fujita, Hiiragi, Fujimoto, & Tsukita, 1998; Stevenson, Siliciano, Mooseker, & Goodenough, 1986). Tight junction proteins in cells grown on cover glass can be visualized by wide-field fluorescence microscopy. Cover glass, however, does not support cells to establish apicobasal polarity or develop mature tight junctions. Tight junction proteins are instead found in the plasma membrane or intracellular vesicles in cells grown on cover glass (Hou, Renigunta, Yang, & Waldegger, 2010).

5.2.1.2 Fixation and permeabilization

The fixation and permeabilization conditions depend upon the specimen, the antigen, and the localization of the antigen within the cell. Three requirements have to be met to facilitate the immunofluorescence labeling of membrane proteins including tight junction (TJ) proteins in the plasma membrane. First, the fixation protocol must retain the antigen within the membrane. Second, the membrane structure must be preserved by fixation without destroying the antigenic determinant recognized by the antibody. Third, the antibody must be able to reach the antigen. The permeabilization step must extract sufficient lipidic components from the membrane so that the antibody can penetrate into the cell. This step is particularly important if the antibody is raised against the cytoplasmic domain in a membrane protein. There are three commonly used fixation and permeabilization protocols for immunofluorescence labeling of membrane proteins.

1. Formaldehyde-methanol: 3.7% formaldehyde in PBS at 4°C for 10 minutes (to fix the cell) and then methanol at −20°C for 10 minutes (to permeabilize the cell).
2. Formaldehyde-Triton: 3.7% formaldehyde in PBS at 4°C for 10 minutes (to fix the cell) and then PBS with 0.2% Triton X-100 at 4°C for 1 minute (to permeabilize the cell).

3. Methanol: methanol at −20°C for 20 minutes (to fix and permeabilize the cell).

5.2.2 Materials and reagents

5.2.2.1 Equipment

NIKON 80i fluorescence microscopy equipped with CCD camera.

Thermo Fisher Scientific round (12 mm diameter) cover glass (thickness no. 1.5).

Thermo Fisher Scientific 6-well tissue culture plate.

5.2.2.2 Cell model

MDCK type II kidney epithelial cells; available from ATCC.

5.2.2.3 Cell culture medium

Complete medium	
500 mL	Dulbecco's Modified Eagle's medium (DMEM) (high glucose/ + sodium pyruvate/ + glutamine); available from Gibco
50 mL	Fetal bovine serum (FBS); available from Gibco
5 mL	Penicillin/streptomycin; available from Gibco

5.2.2.4 Buffers

3.7% Paraformaldehyde in PBS	
NaCl	8 g
KCl	0.2 g
Na_2HPO_4	1.44 g
KH_2PO_4	0.24 g

Bring to 1 L with dH_2O and heat to 60°C–80°C. Add 37 g paraformaldehyde and stir until it dissolves to give a clear solution. Adjust pH to 7.4. Filter, sterilize, and store @4°C.

0.2% Triton X-100 in PBS	
NaCl	8 g
KCl	0.2 g
Na_2HPO_4	1.44 g
KH_2PO_4	0.24 g
Triton X-100	2 mL

Bring to 1 L with dH_2O, stir to dissolve and adjust pH to 7.4.
Filter, sterilize, and store @4°C.

BLOCK solution	
NaCl	8 g
KCl	0.2 g
Na_2HPO_4	1.44 g
KH_2PO_4	0.24 g
Tween 20	2 mL
Fetal calf serum	50 mL

Bring to 1 L with dH_2O, stir to dissolve, and adjust pH to 7.4.
Filter, sterilize, and store @4°C.

0.2 M Tris buffer	
Tris base	24.2 g

Bring to 1 L with dH_2O, stir to dissolve, and adjust pH to 8.0.
Filter, sterilize, and store @4°C.

Mowiol solution	
Mowiol 4–88	2.4 g
Glycerol	6 g
dH_2O	6 mL
Stir for 2 hours at room temperature. Add,	
0.2 M Tris	12 mL

Heat to 50°C for 10 minutes and stir overnight at room temperature.

Centrifuge the solution to remove precipitates at 5,000 × *g* for 15 minutes; aliquot and freeze the supernatant at −20°C.

5.2.3 Experimental procedure

1. Clean cover glass with ethanol. Let air-dry and place cover glass in a 6-well plate.
2. Trypsinize MDCK type II cells and load into the 6-well plate at 1×10^4 per well.
3. When cells reach two-thirds confluent, aspirate medium and add 5 mL of 3.7% paraformaldehyde in PBS to each well in the 6-well plate.
4. Fix at 4°C for 10 minutes. Wash the cover glass with 1 × PBS once.
5. Add 5 mL of 0.2% Triton X-100 in PBS to each well in the 6-well plate. Fix at 4°C for 1 minute.
6. Wash the cover glass with 1 x PBS once.
7. Add 5 mL of BLOCK solution to each well in the 6-well plate. Incubate the plate at 4°C for 1 hour.
8. Aspirate the BLOCK solution. Dilute primary antibodies (anti-TJ proteins) at 1:100–1:200 into the BLOCK solution. Add the primary antibody solution to each well in the 6-well plate.
9. Incubate the plate at 4°C for overnight.

10. Aspirate the primary antibody solution. Wash cover glass with the BLOCK solution for three times.
11. Dilute secondary antibodies (antimouse, rat, or rabbit conjugated to fluorescein isothiocyanate (FITC), tetramethyl-rhodamine isothiocyanate (TRITC), or other fluorophores) at 1:200–1:500 in BLOCK solution. Add the secondary antibody solution to each well in the 6-well plate.
12. Incubate the plate at room temperature for 2 hours.
13. Aspirate the secondary antibody solution. Wash cover glass with the BLOCK solution for three times.
14. Aspirate the BLOCK solution.
15. Remove the cover glass from the 6-well plate with a pair of forceps. Add 20 μL of Mowiol solution to the cover glass (cell side facing up). Flip the cover glass over and place the cover glass (cell side facing down) onto a microscope glass slide. Apply a small pressure onto the cover glass to remove any trapped air bubble.
16. Let Mowiol solution dry at room temperature for 2 hours. Store glass slides at 4°C for no longer than 1 month.

5.2.4 Data analysis

When epithelial cells are cultured on impermeable supports such as cover glass, they spontaneously form fluid-filled, multicellular hemicysts, known as domes (Fig. 5.2.1) (Auersperg, 1969; Handler et al., 1979; Leighton, Brada, Estes, & Justh, 1969; Lever, 1979; Stoos, Naray-Fejes-Toth, Carretero, Ito, & Fejes-Toth, 1991; Sugahara, Caldwell, & Mason, 1984). Domes result from fluid accumulation between the cell layer and the impermeable support. The formation of domes indicates the need of cells to establish the apicobasal polarity and a functional tight junction barrier. Because the domes are randomly formed and scattered over the cover glass, they are not considered as a robust research model for tight junction. A proper way to allow cells to form tight junction is described in Chapter 5.4, Confocal microscopy for cells on Transwell, using permeable supports such as Transwell. Cover glass can be used instead to study the subcellular localization of tight junction (TJ) proteins in unpolarized epithelial cells prior to the formation of tight junction (Fig. 5.2.2).

5.2.5 Troubleshooting

5.2.5.1 Fixation artifact

Formaldehyde and methanol are two commonly used fixatives for membrane proteins. They act by different mechanisms. Formaldehyde crosslinks the proteins in the membrane to form a single interlocking meshwork. Methanol replaces free water in the cell and precipitates the proteins in situ in the

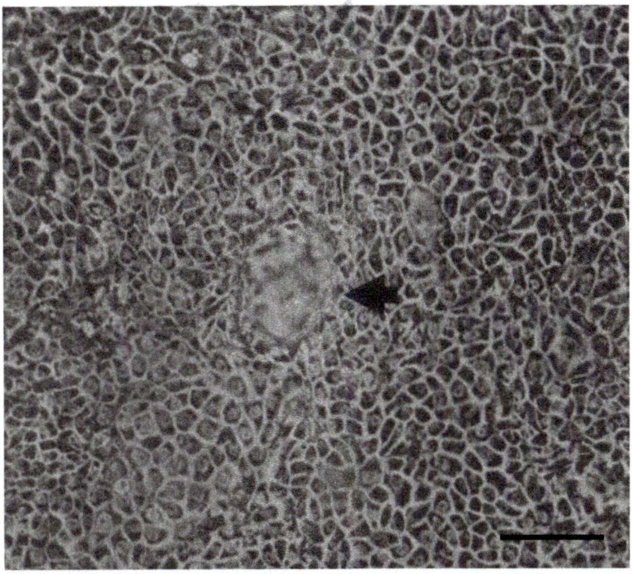

FIGURE 5.2.1 Dome formation in MDCK cells. Domes are spontaneously formed in cells cultured on impermeable supports. Bar: 100 μm. *Reproduced with permission from Lever, J. E. (1979). Inducers of mammalian cell differentiation stimulate dome formation in a differentiated kidney epithelial cell line (MDCK).* Proceedings of the National Academy of Sciences of the United States of America, 76, *1323–1327.*

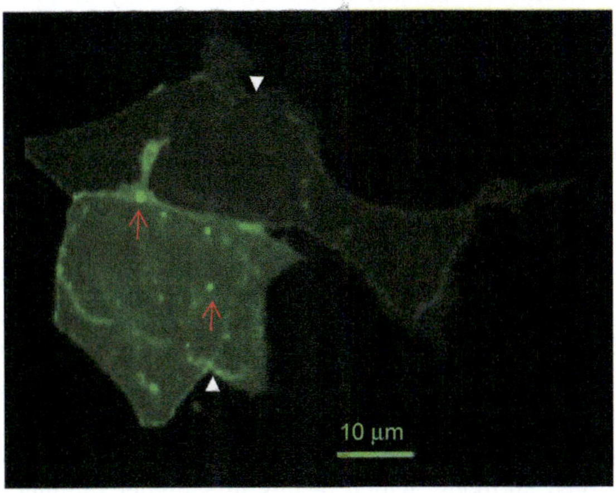

FIGURE 5.2.2 Claudin-4 protein subcellular localization in unpolarized M-1 cells. In subconfluent, not fully polarized, M-1 cells, claudin-4 is localized predominantly at the plasma membrane (*white arrowhead*) and occasionally in traveling intracellular vesicles (*red arrow*). Bar: 10 μm. *Reproduced with permission from Hou, J., Renigunta, A., Yang, J., & Waldegger, S. (2010). Claudin-4 forms paracellular chloride channel in the kidney and requires claudin-8 for tight junction localization.* Proceedings of the National Academy of Sciences of the United States of America, 107, *18010–18015.*

membrane. Both fixatives may generate artifacts. For example, formaldehyde treatment can block the access of the antibody into the meshwork or inactivate the antigen if the antigenic region is crosslinked. Methanol treatment can cause conformational changes in the protein molecules by destabilizing the hydrophilic domains. In general, both methods work for TJ proteins in unpolarized cells grown on cover glass. Methanol has a lower background level of autofluorescence than formaldehyde but does not preserve intracellular structures as well as formaldehyde. Formaldehyde-induced autofluorescence can be reduced by treating samples in 50 mM Glycine after the fixation step.

5.2.5.2 Antibody specificity

The primary antibody used for immunofluorescence must be well characterized. There are several ways to define the specificity of an antibody to TJ protein. (1) The antibody can generate immunofluorescence signals higher than background. (2) The antibody signals are selectively lost in cells that lack the target protein (the KO cell line). (3) Different antibodies directed against different epitopes on the target protein show the same localization pattern. False positive results may arise from nonspecific antibody binding. A negative control antibody must be included into the experimental design, which is closely matched to the specific antibody in host species and immunoglobulin isotype (Lane & Harlow, 1999). For a mouse monoclonal antibody, the proper control is another monoclonal antibody of the same isotype; for a rabbit serum it is the preimmune serum from the same rabbit, and for a rabbit polyclonal antibody it is an irrelevant rabbit polyclonal antibody.

5.2.5.3 Antibody avidity

Polyclonal antibodies generally have higher avidity, a measure of the stability of antigen–antibody interaction, than monoclonal antibodies. When an antigen with more than one epitope is mixed with polyclonal antibodies, the antigen–antibody complex can be stabilized by multiple intermolecular interactions. Even though the rate of dissociation of any single antigen binding to a polyclonal antibody is the same as to a monoclonal antibody, because the antigen is held by other interactions in the polyclonal antibody lattice, the overall rate of dissociation of polyclonal antibody is lower than that of monoclonal antibody, which results in a higher immunofluorescence signal.

5.2.5.4 Double or triple immunofluorescence labeling

It is often advantageous to visualize two or three antigens in the TJ at the same time. To perform double or triple immunofluorescence labeling, the fluorophores conjugated to the secondary antibody must be carefully chosen to give good color separation. For example, the FITC/Texas Red combination is better separated than FITC/TRITC. The host animal species from which the secondary antibody is raised has to be paid attention to. The rule of thumb is that the primary and the secondary antibodies cannot come from the same animal species. Hypothetically, if the primary antibodies, antiprotein A and antiprotein B are raised in rabbit and goat respectively, then an antirabbit secondary antibody from goat will be cross-bound by any antigoat secondary antibody in a tertiary reaction, resulting in overlapping signals.

5.2.5.5 Limit of resolution

Theoretically, the lateral resolution of a fluorescence microscope is 200 nm when the emission wavelength is set at 515 nm and the numerical aperture (*NA*) of 1.40 is used. Objects separated by a distance above 200 nm can be resolved. Objects within a 200 nm radius will be seen as a single signal by wide-field fluorescence microscopy. New forms of microscopy, known as superresolution microscopy, can break the diffraction limit and improve the lateral resolution to as low as 20 nm (Huang, Bates, & Zhuang, 2009).

5.2.6 Concluding remarks

Immunofluorescence microscopy is an important technique not only for fixed cells immunolabeled with fluorophore conjugated antibodies, but also for live cells expressing naturally fluorescent proteins such as green fluorescent protein. Although the cells grown on cover glass do not form mature tight junctions, they are still widely used to study the cell biology of TJ proteins such as endocytosis, trafficking, mobility, and interaction. More advanced imaging techniques, such as fluorescence recovery after photobleaching and fluorescence resonance energy transfer, are established upon the premise of immunofluorescence microscopy. The basic form of immunofluorescence microscopy allows rapid screening of TJ protein antibodies and preliminary visualization of TJ protein localization in various models of cultured cells.

References

Auersperg, N. (1969). Histogenetic behavior of tumors. I. Morphologic variation in vitro and in vivo of two related human carcinoma cell lines. *Journal of the National Cancer Institute, 43*, 151–173.

Furuse, M., Fujita, K., Hiiragi, T., Fujimoto, K., & Tsukita, S. (1998). Claudin-1 and −2: Novel integral membrane proteins localizing at tight junctions with no sequence similarity to occludin. *The Journal of Cell Biology, 141*, 1539–1550.

Furuse, M., Hirase, T., Itoh, M., Nagafuchi, A., Yonemura, S., Tsukita, S., & Tsukita, S. (1993). Occludin: A novel integral membrane protein localizing at tight junctions. *The Journal of Cell Biology, 123*, 1777–1788.

Handler, J. S., Steele, R. E., Sahib, M. K., Wade, J. B., Preston, A. S., Lawson, N. L., & Johnson, J. P. (1979). Toad urinary bladder epithelial cells in culture: Maintenance of epithelial structure, sodium transport, and response to hormones. *Proceedings of the National Academy of Sciences of the United States of America*, *76*, 4151–4155.

Hou, J., Renigunta, A., Yang, J., & Waldegger, S. (2010). Claudin-4 forms paracellular chloride channel in the kidney and requires claudin-8 for tight junction localization. *Proceedings of the National Academy of Sciences of the United States of America*, *107*, 18010–18015.

Huang, B., Bates, M., & Zhuang, X. (2009). Super-resolution fluorescence microscopy. *Annual Review of Biochemistry*, *78*, 993–1016.

Lane, D., & Harlow, E. (1999). *Using antibodies: A laboratory manual*. New York: Ed Harlow.

Leighton, J., Brada, Z., Estes, L. W., & Justh, G. (1969). Secretory activity and oncogenicity of a cell line (MDCK) derived from canine kidney. *Science (New York, NY)*, *163*, 472–473.

Lever, J. E. (1979). Inducers of mammalian cell differentiation stimulate dome formation in a differentiated kidney epithelial cell line (MDCK). *Proceedings of the National Academy of Sciences of the United States of America*, *76*, 1323–1327.

Stevenson, B. R., Siliciano, J. D., Mooseker, M. S., & Goodenough, D. A. (1986). Identification of ZO-1: A high molecular weight polypeptide associated with the tight junction (zonula occludens) in a variety of epithelia. *The Journal of Cell Biology*, *103*, 755–766.

Stoos, B. A., Naray-Fejes-Toth, A., Carretero, O. A., Ito, S., & Fejes-Toth, G. (1991). Characterization of a mouse cortical collecting duct cell line. *Kidney International*, *39*, 1168–1175.

Sugahara, K., Caldwell, J. H., & Mason, R. J. (1984). Electrical currents flow out of domes formed by cultured epithelial cells. *The Journal of Cell Biology*, *99*, 1541–1544.

Chapter 5.3

Wide-field fluorescence microscopy for thin tissue section

5.3.1 Background knowledge

5.3.1.1 Cryostat section versus paraffin section

Cryostat sectioning and paraffin sectioning are two common methods of tissue processing. Both have advantages and disadvantages. Paraffin sections provide better tissue morphology than cryostat sections. Cryostat sections are more suited to immunolabeling than paraffin sections. Many TJ protein antibodies that react well on cryostat sections fail to react on the same tissue after it has been fixed in formaldehyde and embedded in paraffin wax. It is largely because the heat and chemicals used to prepare the paraffin sections have destroyed the antigens in the tissue.

5.3.1.2 Cryostat sectioning

5.3.1.2.1 Freezing of fresh unfixed tissue

The principle of cutting frozen sections is simple—when the tissue is frozen the water within the tissue turns to ice, and in this state the tissue is firm and the ice acts as the embedding medium. Tissue for freezing should be fresh, and frozen as rapidly as possible. Slow freezing can cause distortion of tissues due to ice crystal artifact (Bancroft & Gamble, 2008). A number of suitable freezing techniques, also known as cooling baths, are listed below.

- Liquid nitrogen (−196°C)
- Liquid nitrogen in isopentane (−160°C)
- Dry ice (−78°C)
- Dry ice in ethanol (−78°C)
- Dry ice in acetone (−78°C)

One problem with the use of liquid nitrogen or dry ice is that a vapor phase is formed around the tissue, acting as an insulator to prevent the rapid cooling of tissue. This problem can be overcome by freezing the tissue in a solvent with high thermal conductivity which has been cooled by immersion in liquid nitrogen or dry ice. Cryostat sectioning of unfixed tissues may cause a loss of labile substances such as cytosolic water-soluble proteins. To study water-soluble proteins, the tissue has to be fixed by formaldehyde before sectioning in a cryostat. Membrane proteins, cytoskeletal proteins, and nuclear proteins are well preserved in cryostat sections of unfixed tissues.

5.3.1.2.2 Microtome temperature

When the tissue block is ready for sectioning, the temperature of microtome must be set at a suitable level for the tissue. Reducing the temperature produces a harder block and by raising the temperature the tissue is made softer. Most unfixed tissues section well between −15°C and −23°C (Table 5.3.1) (Liyanage et al., 2017). Tissues containing large amounts of water section best at warmer temperatures. As there is more water, the tissue will have a harder consistency, requiring a higher temperature to obtain the ideal morphology in sections. Tissues containing large amounts of fat require colder temperatures to reach optimal hardness. Most fixed tissues section best at warmer temperatures within the range of −7°C to −12°C, because fixation increases water content in the tissue (Sternberger, 1979).

5.3.1.2.3 Sectioning technique

The disposable blade must be sharp on the edge and precooled before sectioning. The antiroll plate must be carefully adjusted against the blade to stop the frozen section from curling upward during sectioning. The success of sectioning is often determined by the microadjusting of antiroll plate, which includes the following:

- correct height to blade edge,
- correct angle to blade,
- top edge not damaged, and
- at cryostat chamber temperature.

The speed of sectioning and the temperature of microtome are also critical. Soft tissues cut best at a slow rate while hard tissues cut best at a fast speed. If the section is shattered with chatter lines, it is an indication that the temperature is set too low.

5.3.1.2.4 Postsectioning fixation

Postsectioning fixation is carried out in dehydrants such as alcohols and acetone for two purposes. First, dehydrants remove and replace free water in the tissue section. Second, dehydrants precipitate the proteins in situ in the cell membrane. Because cryostat sections are <10 μm in thickness, most cells are cut open, so the permeabilization step is not necessary. There are three commonly used postsectioning fixation protocols for immunofluorescence labeling of tight junction.

1. Methanol: methanol at −20°C for 20 minutes.
2. Ethanol: ethanol at −20°C for 20 minutes.
3. Acetone: acetone at −20°C for 20 minutes.

TABLE 5.3.1 Optimal cryostat sectioning temperatures for unfixed tissues.

Tissue	−10–15°C	−15–25°C	−25–50°C
Adrenal gland	*		
Bone marrow		*	
Brain		*	
Bladder			*
Breast—fatty			*
Breast—little fat		*	
Cartilage	*		
Cervix		*	
Fat			*
Heart and blood vessel		*	
Intestine		*	
Kidney			*
Larynx		*	
Liver			*
Lung			*
Lymphoid		*	
Muscle		*	
Nose			*
Pancreas		*	
Prostate		*	
Ovary		*	
Rectum		*	
Skin with fat			*
Skin without fat		*	
Spleen	*		
Testis	*		
Thyroid		*	
Tongue			*
Uterus	*		

5.3.2 Materials and reagents

5.3.2.1 Equipment

Leica CM1950 automatic sectioning cryostat.

NIKON 80i wide-field fluorescence microscopy equipped with CCD camera.

Thermo Fisher Scientific round (12 mm diameter) cover glass (thickness no. 1.5).

Thermo Fisher Scientific ColorFrost Plus slides made of positively charge uncoated glass.

5.3.2.2 Tissue-tek

Tissue-tek OCT 4583 compound from VWR.

Tissue-tek cryomolds from VWR.

5.3.2.3 Animals

Strain of mouse: C57BL/6; age of mouse: 8-week old; gender of mouse: male.

5.3.2.4 Buffers

BLOCK solution	
NaCl	8 g
KCl	0.2 g
Na_2HPO_4	1.44 g
KH_2PO_4	0.24 g
Tween 20	2 mL
Fetal calf serum	50 mL

Bring to 1 L with dH_2O, stir to dissolve, and adjust pH to 7.4. Filter, sterilize, and store @4°C.

0.2 M Tris buffer	
Tris base	24.2 g

Bring to 1 L with dH_2O, stir to dissolve, and adjust pH to 8.0. Filter, sterilize, and store @4°C.

Mowiol solution	
Mowiol 4–88	2.4 g
Glycerol	6 g
dH_2O	6 mL
Stir for 2 hours at room temperature. Add,	
0.2 M Tris	12 mL

Heat to 50°C for 10 minutes and stir overnight at room temperature.

Centrifuge the solution to remove precipitates at 5,000 × *g* for 15 minutes; aliquot and freeze the supernatant at −20°C.

5.3.3 Experimental procedure

5.3.3.1 Freezing tissues

1. Dissect tissues. Cut into small blocks (~5 mm cubic shape) using a scalpel.
2. Place tissue blocks into Tissue-tek cryomolds and fill the cryomolds with Tissue-tek OCT 4583 compound.
3. Prepare a cooling bath by immersing dry ice in acetone. Float the cryomolds in the cooling bath.
4. Leave for at least 1 minute. Remove the cryomolds from the cooling bath with plastic tweezers and transfer directly to a cryostat.

5.3.3.2 Cutting cryostat sections

1. Wash ColorFrost Plus slides (made of positively charged glass) by dipping in acetone, air-dry, and store at room temperature. Store plastic tweezers and brush in the cryostat. Precool cryostat chamber, microtome, and a specimen holder (chuck).

   (The use of positively charged glass ensures that sections stick firmly to the slides during the immunolabeling and washing steps.)
2. Remove a tissue block from the cryomolds. Place Tissue-Tek OCT 4583 compound on precooled cryostat chuck and mount the tissue block onto the chuck so that the side of the cut section is at 90° to the blade.
3. The optimal cutting temperature varies for different tissues (Table 5.3.1). Most tissues cut well at −15°C to −20°C. If the sections wrinkle as they are cut and look mushy, decrease the temperature. If the sections shatter and look brittle, increase the temperature.

   (Some cryostats allow independent control of chamber and microtome temperatures. In these cases, the chamber temperature is often set at 5°C higher than the microtome temperature—the specimen temperature.)
4. Start to cut sections. First trim the block to get a good cutting surface. Then, adjust the section thickness to 5 μm and cut three sections so that the preset section thickness is achieved. Now bring the antiroll plate on the blade to stop the section from rolling up. The antiroll plate must be parallel to the blade edge and only protrude slightly over the edge (Fig. 5.3.1).
5. Remove the antiroll plate and hold a glass slide over but not touching the cut section. The section will spring onto the slide because of the difference in temperature between section and slide.

6. Place the slide in a precooled slide box and store in a −20°C freezer. Cut sections are stable for up to 1 month at −20°C.

   (Do not deep freeze the slides at − 80°C because deep freezing may damage the morphology of the section.)

5.3.3.3 Immunolabeling cryostat sections

1. Take out the slides from the −20°C freezer, and dip into a jar filled with precooled methanol (−20°C). Put the jar into the −20°C freezer and incubate for 20 minutes to postfix the sections.
2. Take the jar out of the freezer and the slides out of methanol. Air-dry the slides.
3. Make a wet chamber by lining a 24 × 24 cm petri dish with two or three sheets of filter paper and add water to moisten the filter paper.
4. Wash slides once with BLOCK solution. Drain BLOCK solution off against a towel and put slides into the wet chamber.
5. Dilute primary antibody (anti-TJ proteins) at 1:100–1:200 in BLOCK solution. Add the primary antibody solution to each slide (300–500 μL is sufficient to cover the section on each slide).
6. Incubate the wet chamber at 4°C for overnight.
7. Take slides out of the wet chamber, drain off antibody, and wash three times with the BLOCK solution.
8. Dilute secondary antibodies (antimouse, rat, or rabbit conjugated to fluorescein isothiocyanate (FITC), tetramethyl-rhodamine isothiocyanate (TRITC), or other fluorophores) at 1:200–1:500 in BLOCK solution. Add the secondary antibody solution to each slide and put slides back into the wet chamber.
9. Incubate the wet chamber at room temperature for 2 hours.
10. Take slides out of the wet chamber, drain off antibody, and wash three times with the BLOCK solution.
11. Drain BLOCK solution off against a towel. Add 20 μL of Mowiol solution to the section on each slide. Place a cover glass onto the section and apply a small pressure to remove any trapped air bubble.
12. Let Mowiol solution dry at room temperature for 2 hours. Store glass slides at 4°C for no longer than 1 month.

5.3.4 Data analysis

The morphology of cryostat sections can be examined under a light microscope with no histological staining (Fig. 5.3.2). After immunofluorescence labeling, tight junctions appear as fine, sharp, and interdigitating strands on the luminal surface of an epithelial structure, that is, the kidney tubule (Fig. 5.3.2) (Gong et al., 2014; Hou et al., 2009; Kiuchi-Saishin et al., 2002).

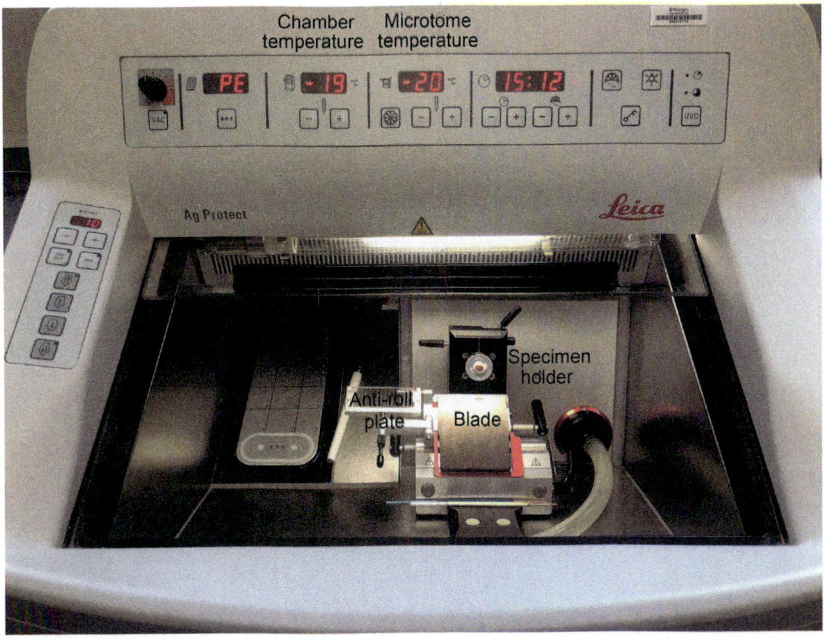

FIGURE 5.3.1 Cryostat sectioning. The sectioning of mouse kidney tissue is demonstrated in a Leica cryostat. The cryostat chamber temperature and the microtome temperature are set at −19°C and −20°C, respectively. Note that the antiroll plate must be parallel to the blade edge and only protrude slightly over the edge.

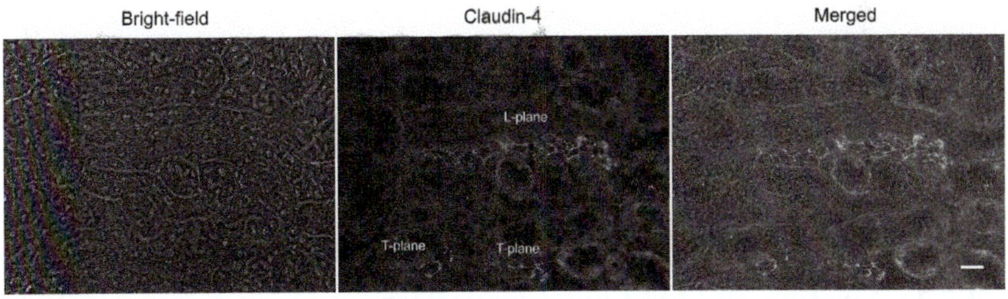

FIGURE 5.3.2 Mouse kidney section for immunofluorescence labeling. The kidney section consists in epithelial tubular structures that are cleaved along the transverse axis or the longitudinal (sagittal axis) (left panel). Tight junctions are labeled with an anticlaudin-4 antibody and appear as short circular strands on the transverse plane (middle panel: T-plane) and long straight strands on the longitudinal plane (middle panel: L-plane). The merged image shows that tight junctions are on the luminal surface of the epithelial tubules (right panel). Scale bar: 20 μm.

The epithelial structure can be counter-stained with cell-type specific protein markers to reveal the cellular origin of a tight junction (TJ) protein (Fig. 5.3.3).

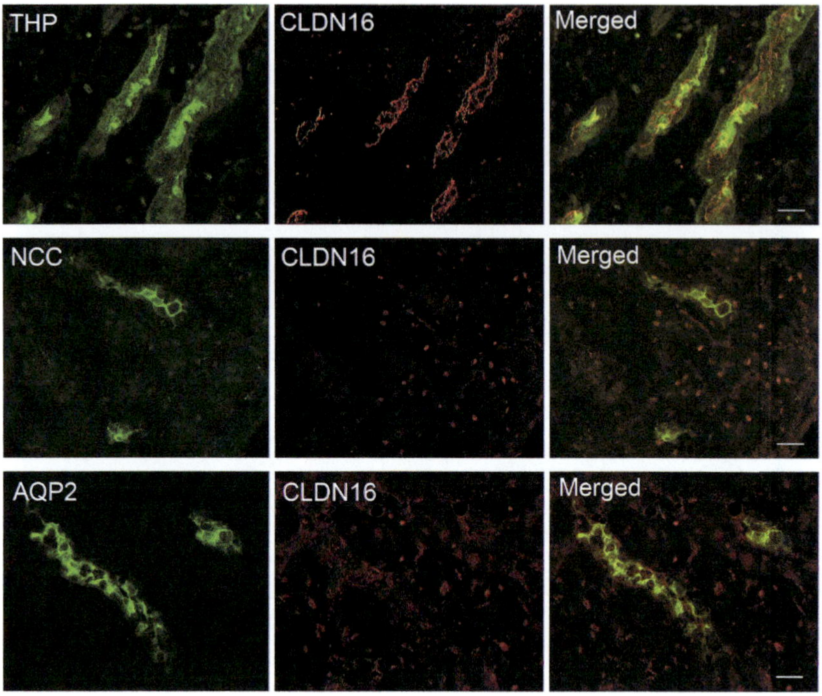

FIGURE 5.3.3 Dual immunofluorescence labeling of mouse kidney sections. The cellular origin of claudin-16 protein localization in the mouse kidney can be revealed by dual immunofluorescence labeling of claudin-16 protein (CLDN16) with a thick ascending limb marker (THP), with a distal convoluted tubule marker (NCC), and with a collecting duct marker (AQP2). Scale bar: 20 μm. *Reproduced with permission from Hou, J., Renigunta, V., Nie, M., Sunq, A., Himmerkus, N., Quintanova, C., Bleich, M., Renigunta, A., & Wolf, M. T. F. (2019). Phosphorylated claudin-16 interacts with Trpv5 and regulates transcellular calcium transport in the kidney.* Proceedings of the National Academy of Sciences of the United States of America, *116(38):19176–19186.*

5.3.5 Troubleshooting

5.3.5.1 Fixation and tight junction pattern

The characteristic pattern of tight junction is best revealed in freshly cut frozen cryostat sections postfixed with alcohols or acetone. Formaldehyde fixed tissue samples can give positive TJ protein signals but not the pattern of TJ strands. The signal decreases when the concentration of formaldehyde increases, which suggests that crosslinked tight junctions lose either the antigenicity or the accessibility to antibodies. Freshly cut sections without crosslinking are, however, unable to preserve cytosolic labile substances including proteins. Alcohol or acetone fixation often shrinks tissue sections, leaving large gaps in the interstitia. Extracellular matrix, resident macrophages, or capillaries are not amenable to alcohol or acetone fixation (van der Loos, 2007).

5.3.5.2 Nonspecific antibody binding

Nonspecific antibodies bind to a target protein with lower affinity. In some applications such as immunoprecipitation from cell lysates (Chapter 2.3) and immunofluorescence labeling for unpolarized cells (see Chapter 5.2: Wide-field fluorescence microscopy for cells on cover glass), nonspecific antibody binding can be easily removed by increasing the ionic strength or the detergent content in the washing buffer. Tissue sections, however, may retain nonspecific antibody binding to a level similar to specific antibody binding. There are several potential causes. (1) Antibody precipitation due to air drying can generate signals resistant to any washing buffer. (2) Elastin fibers are a notorious binder to many irrelevant antibodies. (3) Uneven sections contain lumps that may trap antibodies nonspecifically. Beyond these common considerations, the labeling of tight junction itself can cause signal reinforcement, because tight junction is a compacted macromolecular complex. For this reason, the antibody against claudin-3 cross-binds to claudin-4 in tissue sections, and vice versa, as claudin-3 and claudin-4 proteins are similar in amino-acid sequences and colocalize in many cell types (Li et al., 2014).

5.3.6 Concluding remarks

Wide-field fluorescence microscopy for tissue sections can yield valuable information as to the location of TJ proteins. It determines whether a TJ protein is ubiquitous or is present only in certain tissues. It further determines whether all cells in a given tissue express the TJ protein or whether the TJ protein is restricted to one or a few cell types in the tissue. Its use is not limited to normal tissues. When applied to pathology, wide-field fluorescence microscopy can provide a semiquantitative tool to measure the abundance levels of TJ proteins in various diseases. TJ protein antibodies may be tested on both frozen sections and paraffin sections. In general, more antibodies will react on frozen sections than on paraffin sections. Frozen sections can give the characteristic pattern of tight junction, which paraffin sections are unable to replicate.

References

Bancroft, J. D., & Gamble, M. (2008). *Theory and practice of histological techniques*. Elsevier Health Sciences.

Gong, Y., Yu, M., Yang, J., Gonzales, E., Perez, R., Hou, M., ... Hou, J. (2014). The Cap1-claudin-4 regulatory pathway is important for renal chloride reabsorption and blood pressure regulation. *Proceedings of the National Academy of Sciences of the United States of America, 111*, E3766–E3774.

Hou, J., Renigunta, A., Gomes, A. S., Hou, M., Paul, D. L., Waldegger, S., & Goodenough, D. A. (2009). Claudin-16 and claudin-19 interaction is required for their assembly into tight junctions and for renal reabsorption of magnesium. *Proceedings of the National Academy of Sciences of the United States of America, 106*, 15350–15355.

Kiuchi-Saishin, Y., Gotoh, S., Furuse, M., Takasuga, A., Tano, Y., & Tsukita, S. (2002). Differential expression patterns of claudins, tight junction membrane proteins, in mouse nephron segments. *Journal of the American Society of Nephrology: JASN, 13*, 875–886.

Li, X., Iida, M., Tada, M., Watari, A., Kawahigashi, Y., Kimura, Y., ... Fukasawa, M. (2014). Development of an anti-claudin-3 and −4 bispecific monoclonal antibody for cancer diagnosis and therapy. *The Journal of Pharmacology and Experimental Therapeutics, 351*, 206–213.

Liyanage, S., Dassanayake, R. S., Bouyanfif, A., Rajakaruna, E., Ramalingam, L., Moustaid-Moussa, N., & Abidi, N. (2017). Optimization and validation of cryostat temperature conditions for trans-reflectance mode FTIR microspectroscopic imaging of biological tissues. *MethodsX, 4*, 118–127.

Sternberger, L. A. (1979). *Immunohistochemistry*. Wiley.

van der Loos, C. M. (2007). A focus on fixation. *Biotechnic & histochemistry: Official Publication of the Biological Stain Commission, 82*, 141–154.

Chapter 5.4

Confocal microscopy for cells on Transwell

5.4.1 Background knowledge

5.4.1.1 Confocal microscopy

In conventional (i.e., wide-field) fluorescence microscopy, the entire specimen is illuminated evenly by a light source. All layers of the sample are excited at the same time and the resultant fluorescence includes a large amount of unfocused background signals. Confocal microscopy is an optical imaging technique that uses a spatial pinhole to block out-of-focus light as a means to increase the contrast and resolution in a micrograph (Pawley, 2010). Capturing multiple two-dimensional images at different depths enables the reconstruction of the three-dimensional structure (a process known as optical sectioning) of the specimen.

Resolution is inseparable from contrast. Resolution is defined as the minimal separation distance at which two objects can be distinguished. The Rayleigh criterion for resolution states that two points are resolved when the first minimum of one Airy disc is aligned with the central maximum of the other Airy disc (Fig. 5.1.2). Contrast is defined as the difference between the maximal intensity of each object and the minimal intensity in the area between two objects (Fig. 5.4.1). The relationship between contrast and separation distance for two objects is referred to as the contrast transfer function, which estimates that, under optimal imaging conditions, the Rayleigh criterion separation distance corresponds to a contrast value of 26.4% (Oldenbourg, Terada, Tiberio, & Inoue, 1993). Obviously, contrast is related to the width of the intensity peak described by the point spread function, more precisely, the full width at half maximum (FWHM). In confocal microscopy, illumination and detection are both limited by a spatial pinhole, which reduces the FWHM by approximately 30% compared to that in wide-field fluorescence microscopy (Sheppard, Gu, & Roy, 1992). Because of the narrower intensity width in point spread function, the resolution required to produce acceptable contrast in confocal microscopy is reduced to a distance approximated by the following equation.

$$d = \frac{0.4\lambda}{NA} \tag{5.4.1}$$

in which d is the minimal distance at which two distinct points in a focal plane can be distinguished, numerical aperture (NA) is the numerical

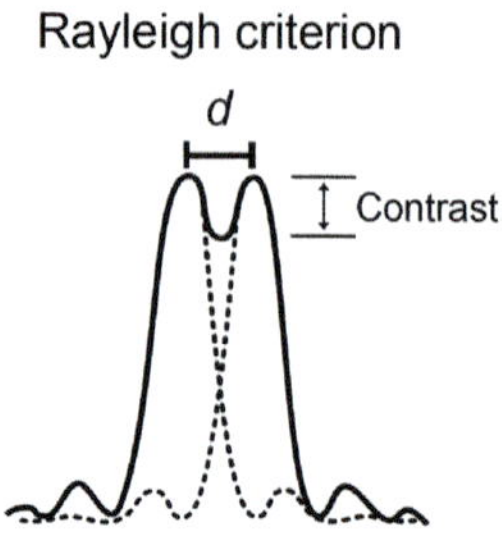

FIGURE 5.4.1 Contrast in microscopy. Contrast is the difference between the maximal intensity in point spread function and the minimal intensity in overlapped area. Contrast is expressed as percentage of the maximal intensity. *d*, Minimal resolvable distance.

aperture of the lens, and λ is the wavelength of light that is used to examine the specimen. Note that Eq. (5.4.1) has broken Abbe's diffraction limit (compared to Eq. 5.1.1).

The axial resolution in confocal microscopy is given by the following equation.

$$z = \frac{1.4\lambda}{(NA)^2} \tag{5.4.2}$$

Although confocal microscopy only modestly improved the axial resolution over that in wide-field fluorescence microscopy (compared to Eq. 5.1.3), the true advantage of confocal microscopy is the optical sectioning capability to examine thick specimens, which often cause signal saturation due to out-of-focus fluorescence in wide-field fluorescence microscopy (Carlsson et al., 1985).

5.4.1.2 Apicobasal polarity

Tight junction formation is intrinsically connected to the establishment of apicobasal polarity in epithelial cells. How epithelial cells polarize is not fully understood, but thought to involve several key processes, such as interaction among polarity proteins, adhesion via cadherins, directed exocytosis, and partitioning of membrane lipids (Shin, Fogg, & Margolis, 2006). When epithelial cells are cultured on impermeable supports such as cover glass, they are either unpolarized or partially polarized into fluid-filled, multicellular hemicysts, known as domes (Fig. 5.2.1). Permeable supports such as Transwell allow epithelial cells to grow on microporous filters (pore size: 0.4–3.0 μm) that permit fluid exchange from the basal side of the cells. Within a few days of culture on the filters, epithelial cells spontaneously polarize and form tight junctions (Cereijido, Robbins, Dolan, Rotunno, & Sabatini, 1978; Misfeldt, Hamamoto, & Pitelka, 1976).

5.4.1.3 Fixation of tight junction

Polarized cells grown on permeable supports are fixed in dehydrants such as alcohols and acetone to preserve tight junction structure. Formaldehyde fixation often gives erroneous signals of tight junction (TJ) protein localization. It is believed that crosslinking prevents antibodies from accessing the tight junction but promote antibody binding to secondary targets. Because alcohols and acetone also extract lipids from the plasma membrane, the permeabilization step is not necessary. There are three commonly used fixation protocols for immunofluorescence labeling of tight junction in polarized cells. Because polarized cells in culture are thicker than cryostat sections, the fixation duration is longer for cells than for sections (see Chapter 5.3.1.2.4).

1. Methanol: methanol at −20°C for 1 hour.
2. Ethanol: ethanol at −20°C for 1 hour.
3. Acetone: acetone at −20°C for 1 hour.

5.4.2 Materials and reagents

5.4.2.1 Equipment

NIKON A1Rsi laser scanning confocal microscope.

Thermo Fisher Scientific round (12 mm diameter) cover glass (thickness no. 1.5).

Thermo Fisher Scientific glass slides.

Corning Transwell (Table 3.2.1).

5.4.2.2 Cell model

MDCK type II epithelial cells; available from ATCC.

5.4.2.3 Cell culture medium

Complete medium	
500 mL	Dulbecco's Modified Eagle's medium (DMEM) (high glucose/+ sodium pyruvate/+ glutamine); available from Gibco
50 mL	Fetal bovine serum (FBS); available from Gibco
5 mL	Penicillin/streptomycin; available from Gibco

5.4.2.4 Buffers

BLOCK solution	
NaCl	8 g
KCl	0.2 g
Na_2HPO_4	1.44 g
KH_2PO_4	0.24 g
Tween 20	2 mL
Fetal calf serum	50 mL

Bring to 1 L with dH_2O, stir to dissolve, and adjust pH to 7.4. Filter, sterilize, and store @4°C.

0.2 M Tris buffer	
Tris base	24.2 g

Bring to 1 L with dH_2O, stir to dissolve, and adjust pH to 8.0. Filter, sterilize, and store @4°C.

Mowiol solution	
Mowiol 4–88	2.4 g
Glycerol	6 g
dH_2O	6 mL
Stir for 2 hours at room temperature. Add,	
0.2 M Tris	12 mL

Heat to 50°C for 10 minutes and stir overnight at room temperature. Centrifuge the solution to remove precipitates at 5,000 × *g* for 15 minutes; aliquot and freeze the supernatant at −20°C.

5.4.3 Experimental procedure

1. Trypsinize MDCK type II cells and load into the Transwell insert (12-mm diameter) at 1×10^6 per insert. Add culture medium to the interior and the exterior compartments of the Transwell.
2. Culture for nine consecutive days and change medium every 3 days.
3. Record transepithelial resistance (R_{te}) with an epithelial ohmmeter (see Chapter 3.4). The R_{te} value of fully polarized MDCK type II cells is normally over 90 Ω cm^2.
4. Aspirate medium and fill the Transwell (both interior and the exterior compartments) with precooled methanol (−20°C). Put the Transwell into a −20°C freezer and incubate for 1 hour to fix the cells.
5. Aspirate methanol and air-dry the Transwell.
6. Add 1 mL of BLOCK solution to the interior and the exterior compartments of Transwell, respectively. Incubate the Transwell at 4°C for 1 hour.
7. Aspirate the BLOCK solution. Dilute primary antibodies (anti-TJ proteins) at 1:100–1:200 into the BLOCK solution. Add the primary antibody solution to the interior compartment of Transwell.
8. Incubate the Transwell at 4°C for overnight.
9. Aspirate the primary antibody solution. Wash the interior compartment of Transwell with the BLOCK solution for three times.
10. Dilute secondary antibodies (antimouse, rat, or rabbit conjugated to fluorescein isothiocyanate (FITC), tetramethyl-rhodamine isothiocyanate (TRITC), or other fluorophores) at 1:200–1:500 in BLOCK solution. Add the secondary antibody solution to the interior compartment of Transwell.

11. Incubate the Transwell at room temperature for 2 hours.
12. Aspirate the secondary antibody solution. Wash the interior compartment of Transwell with the BLOCK solution for three times.
13. Aspirate the BLOCK solution.
14. Remove the insert from the Transwell and cut the filter support (with cells grown on top) from the insert with a blade. Be careful not to damage the cell monolayer. Use a pair of forceps to place the filter on top of a microscope glass slide (cell side facing up). Add 20 μL of Mowiol solution to the filter and place a cover glass on top of the filter. Apply a small pressure onto the cover glass to remove any trapped air bubble.
15. Let Mowiol solution dry at room temperature for 2 hours. Store glass slides at 4°C for no longer than 1 month.

5.4.4 Data analysis

Polarized epithelial cells grown on permeable supports exhibit many characteristics of epithelial cells in tissue, including assuming a cuboidal or columnar shape, partitioning the plasma membrane into apical and basolateral domains, forming circumferential intercellular junctions including the tight junction, and performing directional ion transport (Cereijido et al., 1978; Misfeldt et al., 1976; Richardson & Simmons, 1979). Polarized epithelial cells grown on permeable supports are thicker than unpolarized cells grown on cover glass. Molecules are distributed in a 3D space rather than on a 2D plane. A z stack of confocal images is first taken for an area of polarized epithelial cells, which constitute blur-free optical sections. Among these sections, tight junctions appear near the apical membrane (Fig. 5.4.2). An algorithm for 3D reconstruction can then be applied to these optical sections, which provides a sagittal view of tight junction localization with a reasonable axial resolution (Fig. 5.4.3).

5.4.5 Troubleshooting

5.4.5.1 Signal sensitivity

Four relevant imaging parameters can be adjusted to enhance signal sensitivity. (1) The excitation intensity (or the laser power) defines the number of photons that reach the cells and excite the fluorophores. (2) The exposure time defines the amount of emission light being collected by the camera. (3) The bin number of the camera defines how many photons are sampled into a single image point (or a pixel). (4) The gain in the camera defines the overall sensitivity of the camera.

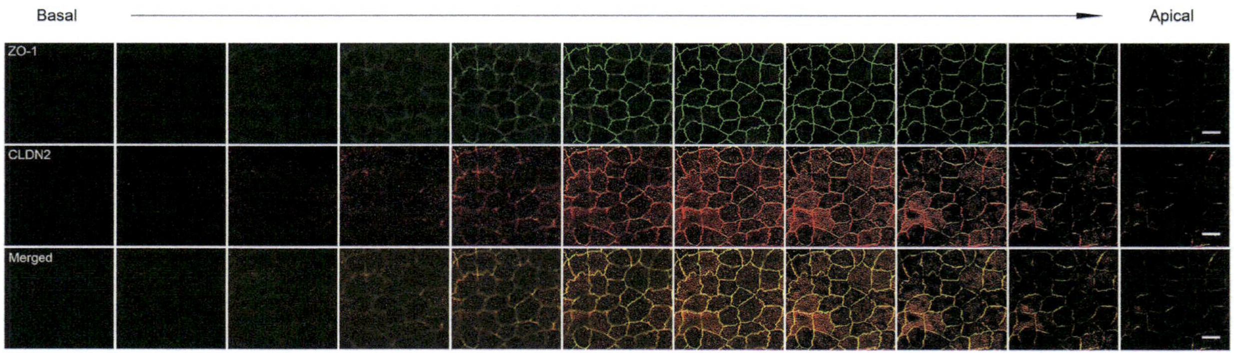

FIGURE 5.4.2 Collage of confocal micrographs showing optical sections of polarized epithelial cells. MDCK type II cells were polarized on Transwell, immunolabeled with antibodies against ZO-1 and claudin-2 (CLDN2), and visualized by confocal microscopy. Note that tight junctions appear near the apical surface of the epithelial cells. Bar: 10 μm.

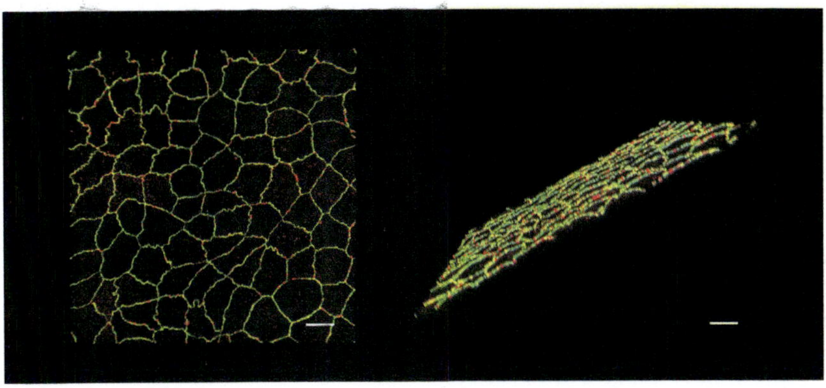

FIGURE 5.4.3 Three-dimensional reconstruction of tight junction. Optical sections of polarized MDCK type II cells from Fig. 5.4.2 were combined to reconstruct the three-dimensional structure of tight junction. An *en face* view of tight junction is shown at left; a sagittal view of tight junction is shown at right. The axial length of tight junction is less than 200 nm. Bar: 10 μm.

5.4.5.2 Axial resolution

Adjusting the diameter of the spatial pinhole varies the thickness of an optical section in confocal microscopy, and hence the axial resolution. The *NA* of the objective also defines the thickness of the optical section. With higher NA objectives, the optical section is thinner and the axial resolution is higher. Thinner optical sections require more time to take and longer exposure to laser can significantly photobleach the specimen. The number of optical sections should be set according to the required axial resolution, which, to avoid unnecessary oversampling, should not be set smaller than the axial resolution defined by the size of the spatial pinhole and the *NA* of the objective.

5.4.5.3 Live-cell imaging

Live-cell imaging for polarized epithelial cells grown on filter supports has been a daunting task largely because the translucent filter obstructs the light path in a microscope. Alternatively, the filter may be excised but this eliminates selective access to apical and basolateral surfaces. To overcome the difficulty of traditional Transwell in live-cell imaging, an inverted type of Transwell is invented, which allows the epithelial cells to grow on the underside of filter supports (Wakabayashi, Chua, Larkin, Lippincott-Schwartz, & Arias, 2007). Now that tight junctions in the epithelial cells are facing downward, they can be easily visualized in a glass chamber with an inverted confocal microscope.

5.4.6 Concluding remarks

Confocal microscopy is essential to the study of tight junction biology in polarized epithelial cells. It offers the ability to visualize macromolecular complexes such as tight junction in three-dimensional datasets. Different TJ proteins can be labeled and presented with pseudocolors representing monochrome images collected from different fluorophores. The structural integrity of tight junction is revealed as sharp and interdigitating strands by confocal microscopy. Pharmacological compounds can be added to the apical or basolateral side of polarized epithelial cells to elicit structural changes in tight junction. When combined with electrophysiological recording, confocal microscopy can establish a structure–function relationship for tight junction.

References

Carlsson, K., Danielsson, P.-E., Lenz, R., Liljeborg, A., Majlöf, L., & Åslund, N. (1985). Three-dimensional microscopy using a confocal laser scanning microscope. *Optics Letters*, *10*, 53–55.

Cereijido, M., Robbins, E. S., Dolan, W. J., Rotunno, C. A., & Sabatini, D. D. (1978). Polarized monolayers formed by epithelial cells on a permeable and translucent support. *The Journal of Cell Biology*, *77*, 853–880.

Misfeldt, D. S., Hamamoto, S. T., & Pitelka, D. R. (1976). Transepithelial transport in cell culture. *Proceedings of the National Academy of Sciences of the United States of America*, *73*, 1212–1216.

Oldenbourg, R., Terada, H., Tiberio, R., & Inoue, S. (1993). Image sharpness and contrast transfer in coherent confocal microscopy. *Journal of Microscopy*, *172*, 31–39.

Pawley, J. (2010). *Handbook of biological confocal microscopy*. Springer Science & Business Media.

Richardson, J. C., & Simmons, N. L. (1979). Demonstration of protein asymmetries in the plasma membrane of cultured renal (MDCK) epithelial cells by lactoperoxidase-mediated iodination. *FEBS Letters*, *105*, 201–204.

Sheppard, C. J., Gu, M., & Roy, M. (1992). Signal-to-noise ratio in confocal microscope systems. *Journal of Microscopy*, *168*, 209–218.

Shin, K., Fogg, V. C., & Margolis, B. (2006). Tight junctions and cell polarity. *Annual Review of Cell and Developmental Biology*, *22*, 207–235.

Wakabayashi, Y., Chua, J., Larkin, J. M., Lippincott-Schwartz, J., & Arias, I. M. (2007). Four-dimensional imaging of filter-grown polarized epithelial cells. *Histochemistry and Cell Biology*, *127*, 463–472.

Chapter 5.5

Confocal microscopy for thick tissue sections

5.5.1 Background knowledge

Tight junctions are found in complex three-dimensional organ structures, such as blood vessel, gastrointestinal tract, pulmonary alveolus, renal tubule, and so on (Hou, 2018). Although thin sections allow locating tight junction (TJ) proteins in specific cell types in a tissue, the morphology of tight junction, in particular, the continuity of TJ strands, cannot be preserved in these tissue sections. Thick tissue sections offer several advantages over thin tissue sections, including better morphology, stronger fluorescence signals, and more importantly, allowing three-dimensional arrangement of tight junction to be visualized by confocal microscopy.

5.5.2 Materials and reagents

5.5.2.1 Equipment

Leica CM1950 automatic sectioning cryostat.

NIKON A1Rsi laser scanning confocal microscope.

Thermo Fisher Scientific round (12 mm diameter) cover glass (thickness no. 1.5).

Thermo Fisher Scientific ColorFrost Plus slides made of positively charge uncoated glass.

5.5.2.2 Tissue-tek

Tissue-tek OCT 4583 compound from VWR.

Tissue-tek cryomolds from VWR.

5.5.2.3 Animals

Strain of mouse: C57BL/6; age of mouse: 8-week old; gender of mouse: male.

5.5.2.4 Buffers

BLOCK solution	
NaCl	8 g
KCl	0.2 g
Na_2HPO_4	1.44 g
KH_2PO_4	0.24 g
Tween 20	2 mL
Fetal calf serum	50 mL

Bring to 1 L with dH_2O, stir to dissolve, and adjust pH to 7.4. Filter, sterilize, and store @4°C.

0.2 M Tris buffer	
Tris base	24.2 g

Bring to 1 L with dH_2O, stir to dissolve, and adjust pH to 8.0. Filter, sterilize, and store @4°C.

Mowiol solution	
Mowiol 4–88	2.4 g
Glycerol	6 g
dH_2O	6 mL
Stir for 2 hours at room temperature. Add,	
0.2 M Tris	12 mL

Heat to 50°C for 10 minutes and stir overnight at room temperature.

Centrifuge the solution to remove precipitates at 5,000 × *g* for 15 minutes; aliquot and freeze the supernatant at −20°C.

5.5.3 Experimental procedure

The tissue preparation, cryostat sectioning, postsectioning fixation, and immunolabeling protocols for thick tissue sections are identical to those described for thin tissue sections (see Chapter 5.3.3). In brief, mouse tissues are rapidly frozen in a cooling bath (−78°C). Tissue sections of 20 μm in thickness are cut with a cryostat and postfixed in precooled methanol (−20°C). Immunolabeling of tissue sections is carried out in a wet chamber by sequential binding to primary and secondary antibodies.

5.5.4 Data analysis

Three-dimensional datasets of confocal micrographs can reveal the pattern of TJ structural organization in thick tissue sections unseen by conventional (i.e., wide-field) fluorescence microscopy. For example, confocal microscopy uncovers a spatial heterogeneity in tight junctions of kidney thick ascending limb cells (Milatz et al., 2017). The localization of claudin-10 and claudin-16 is mutually exclusive in the thick ascending limb tubule (Fig. 5.5.1). The TJ strands appear to be discontinuous at the transition from claudin-10 to claudin-16 localization. Three-dimensional reconstruction of thick ascending limb tubules reveal that the TJ strands made of claudin-10 are in fact either above or below the TJ strands made of claudin-16 along the *z* axis (Fig. 5.5.2). The TJ structures are intact at locations where claudin-10 or claudin-16 each resides.

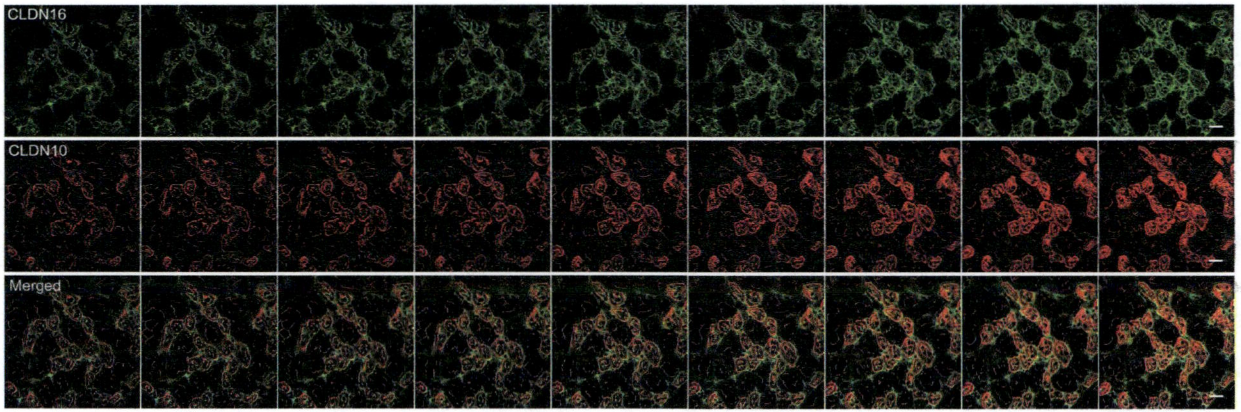

FIGURE 5.5.1 Collage of confocal micrographs showing optical sections of kidney tubules. Mouse kidney sections of 20 μm in thickness were cut with a cryostat, immunolabeled with antibodies against claudin-10 (CLDN10) and claudin-16 (CLDN16), and visualized by confocal microscopy. Tight junctions appear as short circular strands in the lumen of kidney thick ascending limbs on each transverse plane. Note that claudin-10 and claudin-16 never colocalize in the same TJ strand. Scale bar: 20 μm.

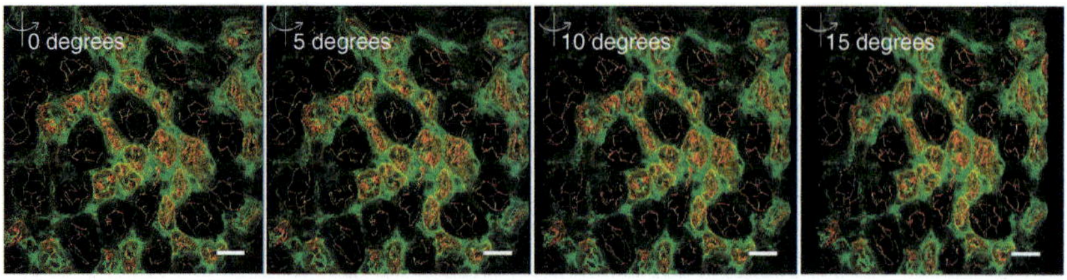

FIGURE 5.5.2 Three-dimensional reconstruction of tight junction in kidney tubules. Optical sections of kidney thick ascending limbs from Fig. 5.5.1 were combined to reconstruct the three-dimensional structure of tight junction. An *en face* view of tight junction in kidney tubules is shown at left, which is then turned counterclockwise by 5, 10, and 15 degrees, respectively. Note that the TJ strands made of claudin-10 (*red*) are not in the same plane as those of claudin-16 (*green*). Scale bar: 20 μm.

5.5.5 Troubleshooting

Nonspecific antibody binding is a major problem for immunolabeling in thick tissue sections (Sternberger, 1979). When antibodies permeate deep into tissues, they can be trapped nonspecifically. Extensive washing with high salt or detergent is not effective against antibody trapping. For example, in Fig. 5.5.1, claudins show significant binding to the basolateral membrane in kidney tubules, which is absent in thin tissue sections from the same animal (Fig. 5.3.3). Undoubtedly, when tissue sections become thicker, fluorescence signals are higher. However, there is a limit to the thickness of tissue section, beyond which an abrupt loss of signals will occur. The limit of tissue section thickness is 25 μm for immunolabeling experiments. Hypothetically, when tissue section thickness exceeds the limit, antibody penetration and diffusion into the tissue are made impossible. Tissues often consist in many types of cells, which are scattered in different areas of the tissues. Although high-magnification micrographs allow examining the subcellular structure of tight junction, the cellular localization pattern of TJ proteins can only be captured at low magnification in a large field of view (Fig. 5.5.3).

5.5.6 Concluding remarks

Confocal microscopy for thick tissue sections can reveal important information not only about the cellular localization of TJ proteins, but also of the subcellular structure of tight junction. Compared with wide-field fluorescence microscopy, the optical sectioning capacity of confocal microscopy allows three-dimensional reconstruction of the entire tissue section and interrogation of the TJ structural alterations in a 3D space. The spatial relationship of tight junction to various cellular or subcellular structures in a tissue section is critical for the understanding of tight junction physiology, and might be a key factor that varies to cause diseases.

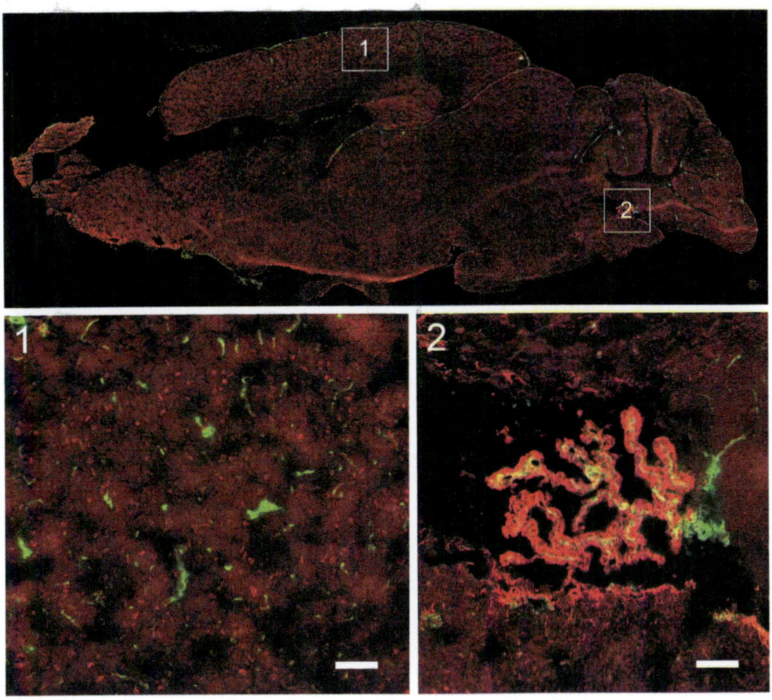

FIGURE 5.5.3 Low-magnification confocal micrograph of tissue section. Mouse brain section of 20 μm in thickness was cut with a cryostat, immunolabeled with antibodies against claudin-1 (*red*) and claudin-5 (*green*), and visualized by confocal microscopy. Claudin-1 and claudin-5 are expressed by different cells in the brain. Claudin-5 proteins are found in the tight junction of brain capillary making the blood–brain barrier (Box 1). Claudin-1 proteins are found in the tight junction of choroid plexus making the blood–cerebrospinal fluid barrier (Box 2). Scale bar: 20 μm.

References

Hou, J. (2018). *The paracellular channel - biology, physiology and disease*. Academic Press.

Milatz, S., Himmerkus, N., Wulfmeyer, V. C., Drewell, H., Mutig, K., Hou, J., ... Bleich, M. (2017). Mosaic expression of claudins in thick ascending limbs of Henle results in spatial separation of paracellular Na + and Mg2 + transport. *Proceedings of the National Academy of Sciences of the United States of America, 114*, E219–e227.

Sternberger, L. A. (1979). *Immunohistochemistry*. Wiley.

Chapter 6

Electron microscopy for tight junction

Chapter 6.1

Theory of electron microscopy

6.1.1 Wave-particle duality of electron

Every quantum entity including electron exhibits properties of both particle and wave. The wavelength of an electron is expressed by the equation of de Broglie:

$$\lambda = \frac{h}{mv} \tag{6.1.1}$$

in which, h = Planck's constant, m = mass of the electron, and v = electron velocity (Dirac, 1981).

After substituting kinetic energy (mv) with accelerating voltage (V), de Broglie principle can be written into the following equation.

$$\lambda = \frac{1.23}{\sqrt{V}} \tag{6.1.2}$$

If an electron microscope is operated at an accelerating voltage of 60 kV, the wavelength of the electron is 0.005 nm. The resolution of the electron microscope, according to Abbe's diffraction limit equation (see Section 5.1.1), is approximately 0.1 nm, which is 1000 times greater than a light microscope.

6.1.2 Electromagnetic lens

As electrons are particles with such a small mass that they will even be stopped by gas molecules in the air, glass lenses are of no value to an electron

A Laboratory Guide to the Tight Junction. DOI: https://doi.org/10.1016/B978-0-12-818647-3.00006-4

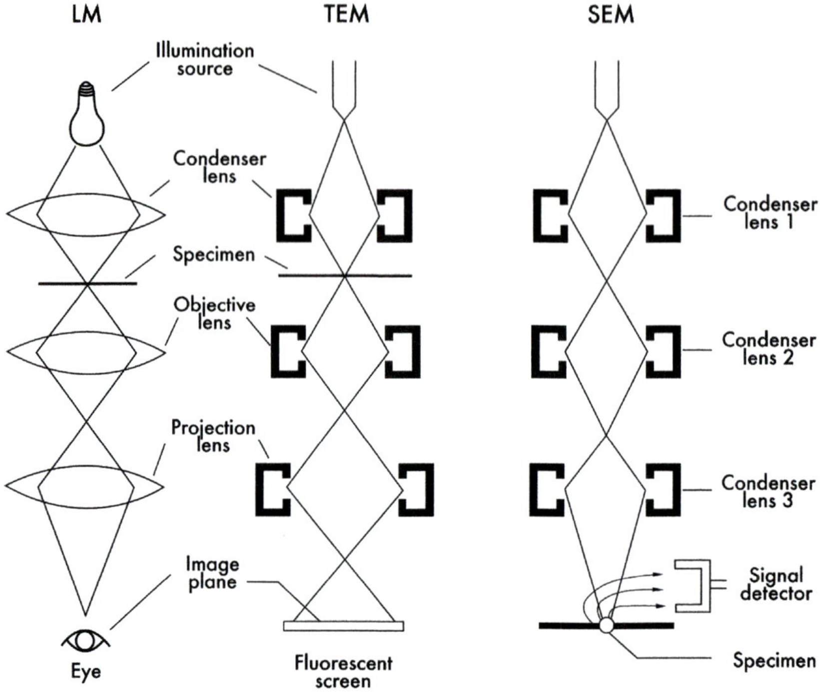

FIGURE 6.1.1 Comparison of light microscopy (LM) to transmission electron microscopy (TEM) and scanning electron microscopy (SEM). In LM and TEM, specimen is illuminated by light or electron beam, and magnified by a series of glass lenses or electromagnetic lenses, respectively. *Reproduced with permission from Płaczek, M., & Kosela, M. (2016). Microscopic methods in analysis of submicron phospholipid dispersions.* Acta Pharmaceutica, 66, *1–22.*

microscope. On the other hand, because electrons have a charge, they can be controlled by an electromagnetic field. The electromagnetic field is generated by a coil of wire with a direct current running through it, known as the solenoid, or the electromagnetic lens. The transmission electron microscope is similar in many ways to the compound light microscope (Bozzola & Russell, 1999). In both microscopes, an electromagnetic radiation (light or electron beam) is converged onto a thin specimen by the means of a condenser lens. The illumination transmitted through the specimen is focused and magnified by a series of lenses including objective lens and projector lens (Fig. 6.1.1). Both types of electromagnetic radiation may be recorded by a silver-based photographic emulsion that is sensitive to electromagnetic radiant energy.

6.1.3 Specimen preparation

The purpose of tissue fixation for electron microscopy is by twofold. First, fixation preserves the structure of tissues with minimal alteration from the living state with regard to morphology, volume, and spatial localization of organelles and macromolecules. Second, fixation protects tissues from disruption by embedding, sectioning, and bombarding of electron beams. The principles of chemical fixation are described in Chapter 4.1, Fixation and fixatives. For examination under transmission electron microscopy, tissue specimens must be sectioned into thin sections, usually thinner than the cells being studied. To achieve this, tissues are embedded into a suitable medium made of resins. Both water-immiscible and water-miscible resins are in use. Water-immiscible resins such as epoxy resins are polymerized by heat and found to preserve tissue morphology better than water-miscible resins. Water-miscible resins such as acrylic resins are cured by UV irradiation at low temperatures, which allows immunolabeling of proteins in tissue sections (Hayat, 1981).

6.1.4 Ultramicrotomy

Transmission electron microscopy for thin tissue sections provides in situ information on the morphology and structure of subcellular components and their spatial localization. The major factor that determines the quality of sections is the ultramicrotome. The ultramicrotome is an instrument for producing extremely thin tissue sections, the thickness of which ranges from 10 to 100 nm, depending upon the type of specimen (Fig. 6.1.2). Sections must be

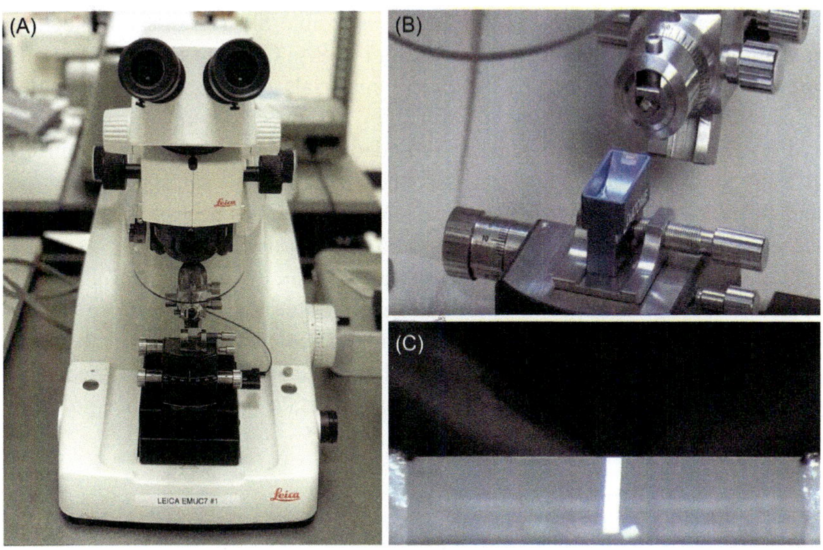

FIGURE 6.1.2 Ultramicrotome sectioning. (A) The sectioning of a tissue block is demonstrated in a Leica ultramicrotome. (B) Tissue sections are cut with a diamond knife. (C) Cut sections float on the surface of water in the knife trough. The interference color of the section may indicate its thickness. Silver: 70–90 nm.

TABLE 6.1.1 Relation of interference color to thickness of ultramicrotome-cut biological section.

Color	Approximate thickness (nm)
Gray	60
Silver	90
Gold	150
Purple	190
Blue	240

of known and uniform thickness and free from chatter, wrinkle, break, or fold. The thickness of a section is usually estimated by observing the interference color in light reflected from the section while it is floating on the surface of water in the knife trough (Table 6.1.1) (Sakai, 1980). These colors are produced by interference between the wave of light reflected by the top surface of the section and the wave of light reflected by the section–water interface.

6.1.5 Positive staining

Different parts of a tissue section are recognizable by transmission electron microscopy because they differ in electron opacity. Contrast is a result of differential electron opacity. As biological materials consist largely of molecules containing light elements, such as carbon, oxygen, nitrogen, and hydrogen, the contrast information provided by endogenous atoms is insignificant for electron microscopy. Introduction of heavy metals to the tissue section by a process known as positive staining becomes necessary to acquire satisfactory contrast information (Hayat, 1975). Osmium, uranium, and lead are commonly used to increase the contrast in tissue sections. Osmium stains lipids and proteins via redox reactions with unsaturated fatty acids and sulfhydryl groups; uranium stains nucleic acids via ionic interaction with phosphate groups; and lead stains carbohydrates via chelation with hydroxyl groups. Admittedly, the deposition of heavy metals is nonspecific, but it is reproducible for many types of tissue.

6.1.6 Negative staining

Negative staining is based upon the principle that there is no reaction between the stain and the specimen. This is accomplished when the stain is used at a pH at which the interaction between stain and biological materials is negligible. The specimen is embedded in electron-dense metal atoms such as tungsten and uranium. The difference between the specimen and the surrounding heavy metal atoms with respect to their density grants the necessary contrast. The electron beam passes through the specimen due to its low electron density, but not through the metallic background, producing an image of light specimen surrounded by dark background. Negative staining allows the determination of morphology and structure of subcellular components (e.g., nuclei, ribosomes, and membranes) and isolated macromolecules (e.g., DNA and protein molecules), based upon the penetration of stain into the holes and crevices of biological molecules that imposes the structural information onto the stain (Kuo, 2007).

6.1.7 Low temperature methods

Cryofixation is a physical approach used to preserve the localization and morphology of subcellular components in living tissues. Cryofixation is achieved by vitrification, a process that allows the solidification of water into amorphous ice (glass) without any formation of ice crystal. Vitrification avoids phase separation, and the vitrified state is believed to be the most faithful representation of the liquid state. The vitrification temperature of water (i.e., the temperature above which vitrified water transforms into the crystalline state), is approximately $-143°C$. Complete vitrification of water occurs at a cooling rate of $\sim 3 \times 10^{6}$°C/s (Brüggeller & Mayer, 1980). Among the rapid freezing methods, high-pressure freezing has the potential to freeze, near vitrification, thick tissue specimens (up to 0.6 mm thick) (Graber & Studer, 2007). The frozen specimens can then be transferred to organic solvents such as acetone, ethanol, and methanol to replace water, in a process known as freeze substitution. The solvent, in turn, is replaced with a resin. Freeze fracturing is a method used to understand the membrane structure. Freeze fracturing involves splitting the frozen cell membrane along its central hydrophobic plane and replicating the fractured surface with platinum and carbon. The fracture plane follows the contour of the membrane and leaves bumps or depressions on the replica where it passes around protein particles or nonbilayer lipid conformations present in the interior of the membrane (Severs, 2007). Freeze etching is a modification of freeze fracturing in that water is allowed to sublimate (etching) to expose macromolecules and intracellular organelles just below the fracture plane before replication. The details of membrane protein, intracellular vesicle, and cytoskeleton, which would otherwise be masked from view, can be visualized with this method (Heuser & Salpeter, 1979).

6.1.8 Immunolabeling techniques

A major stumbling block for the localization of proteins using immunoelectron microscopy might be the harsh conditions behind the fixation and embedding steps that mask or denature the antigenic determinants in proteins. As it turns out, for most antigens, glutaraldehyde must be limited to less than 1% because it is a strong denaturing agent. Osmium is rarely used because of its oxidizing effect. The next decision to make is whether to perform the immunolabeling applications before or after the embedding step. There are distinct advantages and disadvantages to both types of application. Preembedding labeling is advantageous for preserving the antibody binding site. Because antibodies do not penetrate cell membranes, detergents are often used to facilitate the access of antibodies to the interior of cell body, which may severely damage the ultrastructure of tissue specimens (Humbel, de Jong, Müller, & Verkleij, 1998). Postembedding labeling is advantageous for preserving tissue morphology and contrast. When using the postembedding labeling technique, however, it is always presumed that antibody binding affinity has been partially compromised by the embedding protocol. Dehydration, infiltration and embedding steps must be carried out under conditions that optimally preserve antibody binding affinity (Hirst, Johnson, Li, & Raisman, 2000). Because immunoglobulins or immunoglobulin fragments are not directly visible by electron microscopy, they must be tagged with electron-dense substances to fulfill the localization purpose (Bullock & Petrusz, 2012). The types of tag fall into three categories based upon the nature of the tag. The first category includes organic molecules that possess electron opacity, the commonly used one of which is ferritin, a 450-kDa protein rich in iron. The second category includes enzymes such as horseradish peroxidase whose reaction generates water-insoluble and electron-dense products. The last category includes heavy metals such as colloidal gold that can be visualized directly under an electron microscope.

References

Bozzola, J. J., & Russell, L. D. (1999). *Electron microscopy: Principles and techniques for biologists*. Jones & Bartlett Learning.

Brüggeller, P., & Mayer, E. (1980). Complete vitrification in pure liquid water and dilute aqueous solutions. *Nature*, *288*, 569.

Bullock, G. R., & Petrusz, P. (2012). *Techniques in immunocytochemistry*. Academic Press.

Dirac, P. A. M. (1981). *The principles of quantum mechanics*. Oxford university press.

Graber, W., & Studer, D. (2007). High-pressure freezing for electron microscopy of biological specimens. *Microscopy and Microanalysis*, *13*, 232–233.

Hayat, M. A. (1975). *Positive staining for electron microscopy*. Van Nostrand Reinhold New York.

Hayat, M. A. (1981). *Principles and techniques of electron microscopy. Biological applications*. Edward Arnold.

Heuser, J. E., & Salpeter, S. R. (1979). Organization of acetylcholine receptors in quick-frozen, deep-etched, and rotary-replicated Torpedo postsynaptic membrane. *The Journal of Cell Biology*, *82*, 150–173.

Hirst, E. M., Johnson, T. C., Li, Y., & Raisman, G. (2000). Improved post-embedding immunocytochemistry of myelinated nervous tissue for electron microscopy. *Journal of Neuroscience Methods*, *95*, 151–158.

Humbel, B. M., de Jong, M. D., Müller, W. H., & Verkleij, A. J. (1998). Pre-embedding immunolabeling for electron microscopy: An evaluation of permeabilization methods and markers. *Microscopy Research and Technique*, *42*, 43–58.

Kuo, J. (2007). *Electron microscopy: Methods and protocols* (Vol. 369). Springer Science & Business Media.

Sakai, T. (1980). Relation between thickness and interference colors of biological ultrathin section. *Journal of Electron Microscopy*, *29*, 369–375.

Severs, N. J. (2007). Freeze-fracture electron microscopy. *Nature Protocols*, *2*, 547–576.

Chapter 6.2

Transmission electron microscopy for cell culture

6.2.1 Background knowledge

6.2.1.1 Electron microscopy for tight junction

Our understanding of the structure of tight junction and its associated proteins has been greatly advanced by the development in fluorescence imaging techniques for light microscopy. Light microscopy offers a simple and robust solution to the needs of protein localization in various subcellular organelles including the tight junction, but its spatial resolution is too low to produce any meaningful result of the macromolecular architecture of the organelles in study. Electron microscopy has the capacity to provide information on TJ architecture and protein localization at the nanometer level. Because TJ formation is intrinsically connected to the establishment of apicobasal polarity, permeable supports such as Transwell are essential for epithelial or endothelial cells to grow into confluence, establish polarized membrane domains, and develop mature cell junctions including the tight junction (see Chapter 5.4: Confocal microscopy for cells on Transwell).

6.2.1.2 Fixation for electron microscopy

Owing to the superior resolution of electron microscopy, specimens must be optimally fixed. Many fixatives commonly used for light microscopy, such as methanol and acetone, are not suitable for electron microscopy, because they result in structural damage that is detectable at the ultrastructural level. Fixation for electron microscopy serves two purposes: (1) to immobilize cellular components with minimal changes in morphology; (2) to maintain the maximal degree of morphological preservation during subsequent procedures of dehydration, embedding, and bombardment by electron beam (Bullock & Petrusz, 2012). The most widely used fixative for electron microscopy is glutaraldehyde, a type of aldehyde-based chemical crosslinker. For biological specimens, glutaraldehyde alone or in combination with formaldehyde makes the most reliable regimen of fixation. Fixation of cultured cells is generally more straightforward than fixation of tissues. Before fixation, it is necessary to wash the cells with physiological buffers to remove the culture medium that contains the protein components that might be cross-linked by glutaraldehyde. Fixation is often carried out at low temperatures, that is, 4°C for a long period, that is, 16–20 hours, because of the slow diffusion rate of glutaraldehyde (Hayat, 2012).

6.2.2 Materials and reagents

6.2.2.1 Equipment

JEOL JEM-1400 120 kV TEM
Leica UC7 ultramicrotome
Corning Transwell (Table 3.2.1)
Embedding mold from Polyscience Inc.
Formvar-coated grids from Polyscience Inc.

6.2.2.2 Cell model

βEnd3 endothelial cells; available from ATCC.

6.2.2.3 Cell culture medium

Complete medium	
500 mL	EBM-2 basal medium; available from Lonza
50 mL	EGM-2 supplements; available from Lonza
5 mL	Penicillin/streptomycin; available from Gibco

6.2.2.4 Buffers

EM fixation buffer (2% paraformaldehyde, 2% glutaraldehyde in 0.1 M cacodylate buffer)	
16% Paraformaldehyde	12.5 mL
25% Glutaraldehyde	8 mL
0.2 M cacodylate buffer	50 mL
dH_2O	29.5 mL

Adjust pH to 7.4. Filter, sterilize, and store at 4°C.

0.2 M cacodylate buffer	
Sodium cacodylate (trihydrate)	42.8 g

Bring to 1 L with dH_2O, stir to dissolve, and adjust pH to 7.4. Filter, sterilize, and store at 4°C.

1% OsO_4 in 0.1 M cacodylate buffer	
OsO_4	1 g
0.2 M cacodylate buffer	50 mL
dH_2O	50 mL

Adjust pH to 7.4. Filter, sterilize, and store at 4°C.

1% Uranyl acetate buffer

Uranyl acetate	1 g
dH_2O	100 mL

Adjust pH to 7.4. Filter, sterilize, and store at 4°C.

Araldite resin mix

Araldite resin (Polyscience Inc.; Cat No.: 00552)	50 g
DDSA (Polyscience Inc.; Cat No.: 00563)	55 g
DMP-30 (Polyscience Inc.; Cat No.: 00553)	2.1 g

Mix and centrifuge at 500 × *g* for 5 minutes at room temperature to remove air bubble; store at −20°C.

1/3 (v/v) Araldite resin mix with 2/3 (v/v) propylene oxide

Araldite resin mix	100 mL
Propylene oxide	200 mL

Mix and centrifuge at 500 × *g* for 5 minutes at room temperature to remove air bubble; store at −20°C.

1/2 (v/v) Araldite resin mix with 1/2 (v/v) propylene oxide

Araldite resin mix	100 mL
Propylene oxide	100 mL

Mix and centrifuge at 500 × *g* for 5 minutes at room temperature to remove air bubble; store at −20°C.

3/4 (v/v) Araldite resin mix with 1/4 (v/v) propylene oxide

Araldite resin mix	300 mL
Propylene oxide	100 mL

Mix and centrifuge at 500 × *g* for 5 minutes at room temperature to remove air bubble; store at −20°C.

Lead citrate buffer

Lead citrate	0.01 g
NaOH (10 M)	0.1 mL
dH_2O	10 mL

Mix in a 10-mL centrifuge tube and shake vigorously until lead dissolves.

6.2.3 Experimental procedure

6.2.3.1 Cell culture and fixation

1. Trypsinize βEnd3 endothelial cells and load into the Transwell insert (12-mm diameter) at 1×10^6 per insert. Add culture medium to the interior and the exterior compartments of the Transwell.
2. Culture for 9 consecutive days and change medium every 3 days.
3. Aspirate medium and wash cells with 1 × PBS.

4. Aspirate 1× PBS and fill the Transwell (both interior and the exterior compartments) with EM fixation buffer. Fix the cells at 4°C for overnight.

6.2.3.2 Embedding

1. Aspirate fixative, cut Transwell membrane (with cells grown on top) into strips and place into Eppendorf tubes. Add 0.1 M sodium cacodylate (pH 7.2–7.4) at room temperature with rotation for 10 minutes; repeat three times.
2. Aspirate the sodium cacodylate buffer; add 1% OsO_4 (osmium tetroxide) in 0.1 M sodium cacodylate (pH 7.2–7.4); rotate for 1 hour at room temperature (wrap the vial with foil to protect against light).
3. Aspirate the OsO_4 buffer; add 0.1 M sodium cacodylate (pH 7.2–7.4); rotate for 5 minutes at room temperature; repeat three times.
4. Aspirate the sodium cacodylate buffer; add Nanopure water; rotate for 5 minutes at room temperature; repeat three times.
5. Aspirate Nanopure water; add 1% uranyl acetate in water; rotate for 1 hour at room temperature (wrap the vial with foil to protect against light).
6. Aspirate the uranyl acetate buffer; add Nanopure water; rotate for 5 minutes at room temperature; repeat three times.
7. Aspirate Nanopure water; add 20% ethanol; rotate for 10 minutes at room temperature; followed by rotation with 40% ethanol for 10 minutes; 60% ethanol for 10 minutes, 80% ethanol for 10 minutes, and 100% ethanol for overnight.
8. Wash twice with a new bottle of 100% ethanol for 10 minutes each time.
9. Aspirate ethanol; add propylene oxide; rotate for 10 minutes at room temperature; repeat three times.
10. Aspirate propylene oxide; add 1/3 (v/v) Araldite resin mix with 2/3 (v/v) propylene oxide; rotate for 1 hour at room temperature; protect against light.
11. Aspirate solution in Step 10; add ½ Araldite resin mix with ½ propylene oxide; rotate for 1 hour at room temperature; protect against light.
12. Aspirate solution in Step 11; add ¾ Araldite resin mix with ¼ propylene oxide; rotate for 1 hour at room temperature; protect against light.
13. Aspirate solution in Step 12; add 100% Araldite resin mix; rotate for overnight at room temperature; protect against light.
14. Add solution in Step 13 (containing the specimen) into a flat embedding mold. Place a small piece of paper with the identification number of the specimen upside down in the resin next to the specimen. Adjust the orientation of the specimen with a wooden stick so that the sagittal

sections of Transwell membrane (with cells grown on top) can be cut with an ultramicrotome.

15. Polymerize the specimen at 60°C for 2 days.

6.2.3.3 Sectioning and poststaining

1. Once the specimen is embedded, it is cut into 70-nm ultrathin sections by an ultramicrotome (Hayat, 1981).
2. Ultrathin sections are collected on Formvar-coated grids (200 mesh).
3. Prepare fresh uranyl acetate buffer and lead citrate buffer.
4. Prepare a clean staining surface by placing a piece of Parafilm at the bottom of a petri dish.
5. Place one drop of uranyl acetate buffer for each grid onto the Parafilm. Float the grid on top of the stain with the side containing section facing down.
6. Cover the petri dish with a lid and stain for 5 minutes at room temperature.
7. Pick the grid out of the stain droplet with fine forceps and dip into a series of glass beakers containing dH_2O. Agitate the grid several times up and down through each beaker.
8. Drain off excess dH_2O with filter paper.
9. Stain the grid with lead citrate buffer as described for uranyl acetate buffer.
10. Place the grid section-side-up on filter paper until it is fully dry.

6.2.4 Data analysis

Tight junction is located in the lateral membrane close to the apical surface of epithelial or endothelial cells. The three-dimensional arrangement of tight junction in the cell monolayer dictates that an *en face* view from above the cell provides the best angle to visualize the tight junction (Figure 5.4.2). While the *en face* view can be easily obtained for light microscopy, a similar view for electron microscopy will be extremely difficult to obtain, because it requires a transverse section to be cut right through the tight junction. On the other hand, because tight junctions form a continuous belt around each cell, sagittal sectioning will ensure tight junctions to be captured at cell–cell contact areas. On a sagittal section, the tight junction appears as a fusion of plasma membrane outer leaflets (Fig. 6.2.1). The cytoplasmic face of the junctional membranes is covered by an electron-opaque, fuzzy material, unique to this type of junction (Fig. 6.2.1). It is believed to be the plaque of tight junction, a protein network that anchors the tight junction onto the cytoskeleton, that is, actin filaments (Hou, 2018).

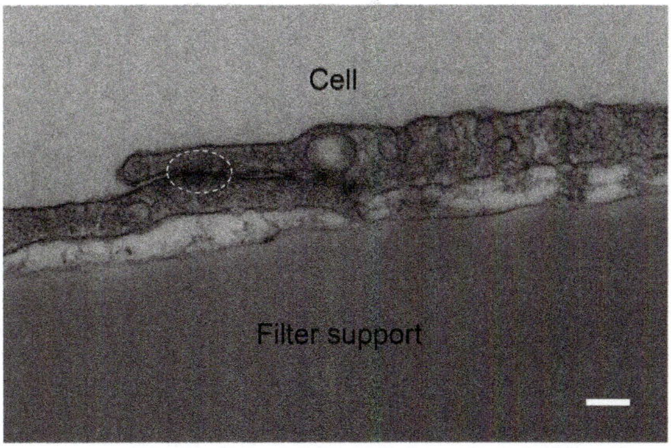

FIGURE 6.2.1 Transmission electron micrograph of endothelial tight junction. The βEnd3 endothelial cells were cultured on Transwell filter supports to allow forming tight junctions. The sagittal section of cells grown on the filter support was visualized by electron microscopy. Note the tight junction at plasma membrane fusion site (circle). Bar: 100 nm.

6.2.5 Troubleshooting

6.2.5.1 Fixation of membrane structure

Tight junction is a membrane structure consisting in both proteins and lipids. The key to preserve TJ morphology is fixation condition. Glutaraldehyde is an excellent fixative for proteins. Although it may react with phospholipids containing free amino groups, most lipids are not retained by glutaraldehyde fixation (Collins, Arborgh, & Brunk, 1977). Therefore, secondary fixation with osmium tetroxide is often needed. Osmium is the best fixative for preservation of membrane structure because it oxidizes unsaturated fatty acids in the membrane (Penttila, McDowell, & Trump, 1975). Nevertheless, osmium can destroy enzymatic activity and antigenicity in cells, which limits its use in postembedding immunolabeling or immunohistochemistry.

6.2.5.2 Poststaining of tight junction

Poststaining increases the contrast in ultrathin sections. The most common stains are uranyl acetate and lead citrate. Uranyl acetate binds to DNA and membrane whereas lead citrate stains ribosome and cytoplasmic substance. Admittedly, it is an empirical process to determine the optimal concentration and treatment duration of a stain for tight junction. Increasing the concentration of uranyl acetate while decreasing the concentration of lead citrate reinforces the signals at the TJ membrane but loses the staining of TJ plaque (Fig. 6.2.2). Reduced osmium is electron-dense and, furthermore, acts as a mordant for several stains. The contrast in TJ membrane can be improved by a protocol employing ferrocyanide-reduced osmium and tannic acid fixation

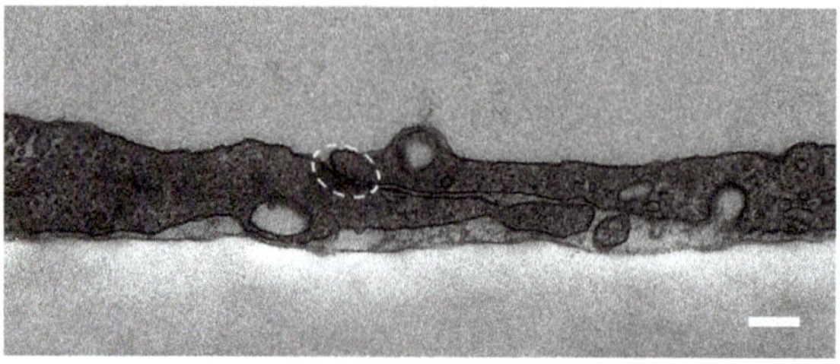

FIGURE 6.2.2 Improvement of membrane contrast in transmission electron micrograph. The βEnd3 endothelial cells were cultured on Transwell, sectioned along the sagittal plane and post-stained with 2% uranyl acetate and 0.05% lead citrate. Compared to Fig. 6.2.1, an increase in uranyl acetate concentration reinforces the membrane signals but a decrease in lead citrate reduces the cytoplasmic signals in the TJ region (circle). Bar: 100 nm.

followed by uranyl acetate and lead citrate staining (Otani et al., 2019). Introduction of heavy metal tracers such as lanthanum can also increase the membrane contrast because the tracers stain the extracellular space (Goodenough & Revel, 1970).

6.2.6 Concluding remarks

Electron microscopy has contributed immensely to the field of tight junction cell biology. The observation of fine details of tight junction has laid the foundation for experimental manipulations directed at unraveling tight junction function and understanding how tight junction ultrastructure varies in normal, experimental, and diseased states. The polarized cell culture model allows a systematic study of the structure–function relationship for tight junction. Epithelial or endothelial cells cultured on permeable supports can first be recorded for paracellular conductance, and then sectioned and examined by electron microscopy to reveal the ultrastructure of tight junction. The localization of TJ proteins visualized by light microscopy can also be related to the ultrastructural alteration in tight junction captured by electron microscopy. Some TJ proteins may even be immunolabeled and visualized directly under an electron microscope (Chapter 6.5: Transmission electron microscopy for immunolabeling application).

References

Bullock, G. R., & Petrusz, P. (2012). *Techniques in immunocytochemistry*. Academic Press.

Collins, V. P., Arborgh, B., & Brunk, U. (1977). A comparison of the effects of three widely used glutaraldehyde fixatives on cellular volume and structure. A TEM, SEM, volumetric and cytochemical study. *Acta pathologica et microbiologica Scandinavica Section A, Pathology*, *85a*, 157–168.

Goodenough, D. A., & Revel, J. P. (1970). A fine structural analysis of intercellular junctions in the mouse liver. *The Journal of Cell Biology*, *45*, 272–290.

Hayat, M. A. (1981). *Principles and techniques of electron microscopy. Biological Applications.* Edward Arnold.

Hayat, M. E. (2012). *Fixation for electron microscopy*. Elsevier.

Hou, J. (2018). *The paracellular channel – Biology, physiology and disease*. Academic Press.

Otani, T., Nguyen, T. P., Tokuda, S., Sugihara, K., Sugawara, T., Furuse, K., … Furuse, M. (2019). Claudins and JAM-A coordinately regulate tight junction formation and epithelial polarity. *The Journal of Cell Biology*.

Penttila, A., McDowell, E. M., & Trump, B. F. (1975). Effects of fixation and postfixation treatments on volume of injured cells. *The Journal of Histochemistry and Cytochemistry: Official Journal of the Histochemistry Society*, *23*, 251–270.

Chapter 6.3

Transmission electron microscopy for tissue section

6.3.1 Background knowledge

6.3.1.1 Tissue perfusion

As the fine structure of tissues is dependent on a continuous blood supply, fixation of the tissues must be carried out in vivo by vascular perfusion. Vascular perfusion allows rapid and uniform penetration of fixatives into all parts of the tissue before any dissection. It is indispensable for the fixation of highly vascular tissues such as the brain, the kidney, and the liver. Generally, the animal is first perfused with Ringer's solution to remove blood from the vasculature, followed by perfusion with the fixative. A perfusion system (gravity or mechanical pump driven) is set up whereby two input lines are available, one for Ringer's buffer and the other for the fixative. For ultrastructural studies, the use of oxygenated Ringer's solution is recommended to minimize ischemic injuries to the tissue before introduction of the fixative.

6.3.1.2 Tissue fixation

No single standard fixation method is suitable for all types of tissue. The fixative and the buffer vehicle must be adjusted for each specific tissue or cell type. The combination of fixative and buffer vehicle for each tissue has been summarized in Chapter 4, Histological approaches for tight junction. A good standard fixative is a mixture of glutaraldehyde (1%–2%) and paraformaldehyde (2%–4%) (Karnovsky, 1965). This combination exploits the advantage and minimizes the disadvantage of each fixative. Formaldehyde fixes less stably but penetrates into the tissue quicker than glutaraldehyde, whereas glutaraldehyde fixes more thoroughly but penetrates slower. The two common buffers are sodium cacodylate or sodium phosphate. Cacodylate buffer has superior buffering properties within the range of pH 5.0–7.4 and results in fewer precipitates than does phosphate buffer (Sabatini, Bensch, & Barrnett, 1963). Mitochondria and other organelles can be damaged when exposed to the high concentrations of phosphates. Phosphate buffer is frequently used when performing immunolocalization experiments. The osmolality of buffer must be isotonic or slightly hypertonic to extracellular fluids to prevent cell swelling. It has been debated whether aldehydes contribute to the final osmolality of the fixative solution. Most literatures now agree that, when used at low concentrations, aldehydes increase the osmolality of the fixative solution by less than 20 mOsm (Hayat, 1981).

6.3.2 Materials and reagents

6.3.2.1 Equipment

JEOL JEM-1400 120 kV TEM
Leica UC7 ultramicrotome
Embedding mold from Polyscience Inc.
Formvar-coated grids from Polyscience Inc.

6.3.2.2 Buffers

EM fixation buffer (2% paraformaldehyde, 2% glutaraldehyde in 0.1 M cacodylate buffer)	
16% Paraformaldehyde	12.5 mL
25% Glutaraldehyde	8 mL
0.2 M cacodylate buffer	50 mL
dH_2O	29.5 mL

Adjust pH to 7.4. Filter, sterilize, and store at 4°C.

0.2 M cacodylate buffer	
Sodium cacodylate (trihydrate)	42.8 g

Bring to 1 L with dH_2O, stir to dissolve, and adjust pH to 7.4. Filter, sterilize, and store at 4°C.

1% OsO_4 in 0.1 M cacodylate buffer	
OsO_4	1 g
0.2 M cacodylate buffer	50 mL
dH_2O	50 mL

Adjust pH to 7.4. Filter, sterilize, and store at 4°C.

1% Uranyl acetate buffer	
Uranyl acetate	1 g
dH_2O	100 mL

Adjust pH to 7.4. Filter, sterilize, and store at 4°C.

Araldite resin mix	
Araldite resin (Polyscience Inc.; Cat No.: 00552)	50 g
DDSA (Polyscience Inc.; Cat No.: 00563)	55 g
DMP-30 (Polyscience Inc.; Cat No.: 00553)	2.1 g

Mix and centrifuge at 500 × *g* for 5 minutes at room temperature to remove air bubble; store at −20°C.

1/3 (v/v) Araldite resin mix with 2/3 (v/v) propylene oxide	
Araldite resin mix	100 mL
Propylene oxide	200 mL

Mix and centrifuge at $500 \times g$ for 5 minutes at room temperature to remove air bubble; store at $-20°C$.

1/2 (v/v) Araldite resin mix with 1/2 (v/v) propylene oxide	
Araldite resin mix	100 mL
Propylene oxide	100 mL

Mix and centrifuge at $500 \times g$ for 5 minutes at room temperature to remove air bubble; store at $-20°C$.

3/4 (v/v) Araldite resin mix with 1/4 (v/v) propylene oxide	
Araldite resin mix	300 mL
Propylene oxide	100 mL

Mix and centrifuge at $500 \times g$ for 5 minutes at room temperature to remove air bubble; store at $-20°C$.

Lead citrate buffer	
Lead citrate	0.01 g
NaOH (10 M)	0.1 mL
dH_2O	10 mL

Mix in a 10-mL centrifuge tube and shake vigorously until lead dissolves.

6.3.3 Experimental procedure

6.3.3.1 Perfusion and fixation

1. Perfuse animal with EM fixation buffer by gravity flow (see details in Chapter 4: Histological approaches for tight junction).
2. Dissect tissues. Cut into small blocks (~5 mm cubic shape) using a scalpel.
3. Place tissue blocks into Eppendorf tubes filled with EM fixation buffer. Fix the tissues at 4°C for overnight.

6.3.3.2 Embedding

1. Aspirate fixative and add 0.1 M sodium cacodylate (pH 7.2–7.4) at room temperature with rotation for 10 minutes; repeat three times.
2. Aspirate the sodium cacodylate buffer; add 1% OsO_4 (osmium tetroxide) in 0.1 M sodium cacodylate (pH 7.2–7.4); rotate for 1 hour at room temperature (wrap the vial with foil to protect against light).
3. Aspirate the OsO_4 buffer; add 0.1 M sodium cacodylate (pH 7.2–7.4); rotate for 5 minutes at room temperature; repeat three times.
4. Aspirate the sodium cacodylate buffer; add Nanopure water; rotate for 5 min at room temperature; repeat three times.

5. Aspirate Nanopure water; add 1% uranyl acetate in water; rotate for 1 hour at room temperature (wrap the vial with foil to protect against light).
6. Aspirate the uranyl acetate buffer; add Nanopure water; rotate for 5 minutes at room temperature; repeat three times.
7. Aspirate nanopure water; add 20% ethanol; rotate for 10 minutes at room temperature; followed by rotation with 40% ethanol for 10 minutes; 60% ethanol for 10 minutes, 80% ethanol for 10 minutes, and 100% ethanol for overnight.
8. Wash twice with a new bottle of 100% ethanol for 10 minutes each time.
9. Aspirate ethanol; add propylene oxide; rotate for 10 minutes at room temperature; repeat three times.
10. Aspirate propylene oxide; add 1/3 (v/v) Araldite resin mix with 2/3 (v/v) propylene oxide; rotate for 1 hour at room temperature; protect against light.
11. Aspirate solution in Step 10; add ½ Araldite resin mix with ½ propylene oxide; rotate for 1 hour at room temperature; protect against light.
12. Aspirate solution in Step 11; add ¾ Araldite resin mix with ¼ propylene oxide; rotate for 1 hour at room temperature; protect against light.
13. Aspirate solution in Step 12; add 100% Araldite resin mix; rotate for overnight at room temperature; protect against light.
14. Add solution in Step 13 (containing the tissue) into a flat embedding mold. Place a small piece of paper with the identification number upside down in the resin next to the tissue. Adjust the orientation of the tissue with a wooden stick so that sections of desirable plane can be cut with an ultramicrotome.
15. Polymerize the specimen at 60°C for 2 days.

6.3.3.3 Sectioning and poststaining

1. Once the tissue is embedded, it is cut into 70 nm ultrathin sections by an ultramicrotome (Hayat, 1981).
2. Ultrathin sections are collected on Formvar-coated grids (200 mesh).
3. Prepare fresh uranyl acetate buffer and lead citrate buffer.
4. Prepare a clean staining surface by placing a piece of Parafilm at the bottom of a petri dish.
5. Place one drop of uranyl acetate buffer for each grid onto the Parafilm. Float the grid on top of the stain with the side containing section facing down.
6. Cover the petri dish with a lid and stain for 5 minutes at room temperature.
7. Pick the grid out of the stain droplet with fine forceps and dip into a series of glass beakers containing dH_2O. Agitate the grid several times up and down through each beaker.
8. Drain off excess dH_2O with filter paper.

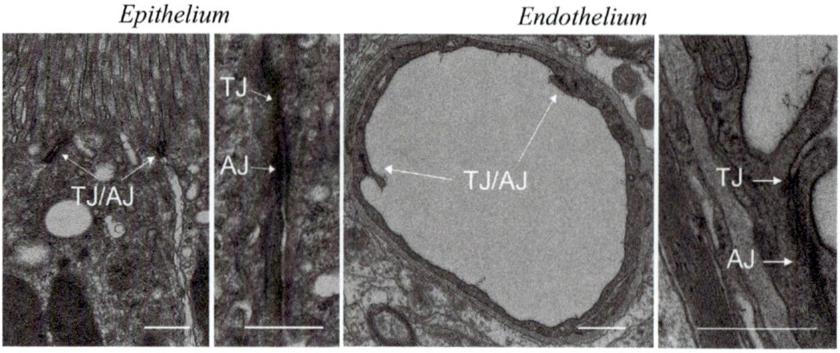

FIGURE 6.3.1 Epithelial and endothelial tight junctions. The TJ obliterates the paracellular space by fusing the plasma membranes of adjoining cells. The adherens junction (AJ) is located beneath the TJ and connects the adjoining plasma membranes with the electron-dense fibrillary molecule known as cadherin. On the cytoplasmic sides of both junctions are the electron-dense plaques that consist in the scaffold proteins such as zonula occludens-1 (ZO-1) and anchor the junctions onto the cytoskeleton. The epithelium is from the mouse kidney proximal tubule. The endothelium is from the mouse cerebral cortex. Bar: 500 nm. *Reproduced with permission from Hou, J. (2018).* The paracellular channel – Biology, physiology and disease. *Academic Press.*

9. Stain the grid with lead citrate buffer as described for uranyl acetate buffer.
10. Place the grid section-side-up on filter paper until it is fully dry.

6.3.4 Data analysis

Tight junctions are abundant in organ structures made from epithelial or endothelial cells (Hou, 2018). At low magnification, tight junctions appear where the plasma membranes of adjoining cells are in close apposition (Fig. 6.3.1). At high magnification, tight junctions are seen where the exoplasmic leaflets of the plasma membranes from adjoining cells appear to fuse (Fig. 6.3.1). Tight junctions are found where the plane of section intersects the plane of intercellular boundary at oblique or right angles. Tight junctions extend in depth for 100–500 nm depending upon the orientation of the section. At the basal end of tight junction, the membrane fusion splits again into the outer leaflets of lateral membranes. The appearance of membrane fusion at tight junction may vary from organ to organ. In most cases, the membranes remain fused throughout the tight junction. Occasionally, recurring fusions and focal splitting of the fusion line can be found in the tight junction (Fig. 6.3.2) (Farquhar & Palade, 1963).

6.3.5 Troubleshooting

6.3.5.1 Poststaining versus en bloc staining

The biggest problem associated with poststaining is the formation of electron-dense precipitates on the tight junction. To determine whether these

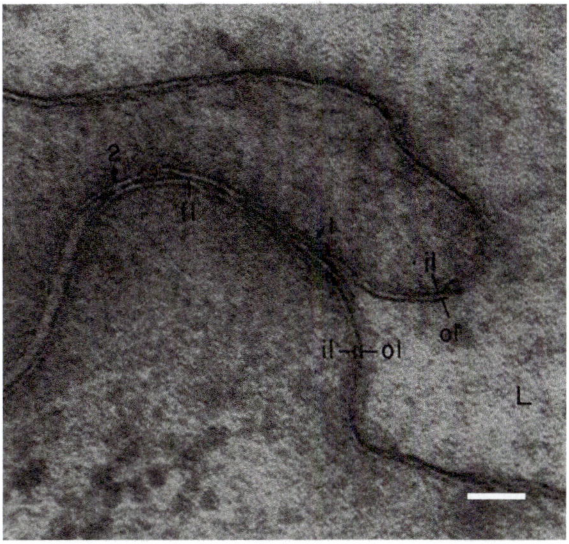

FIGURE 6.3.2 Detailed structure of tight junction. The epithelium from the rat gastric mucosa was sectioned and visualized by transmission electron microscopy at high magnification. The tight junction (between arrow 1 and 2) is recognized as an area where the lateral membranes of two adjoining cells fuse. There are recurring membrane fusions and focal splitting of the fusion line (fl) within the tight junction. L: lumen; ol: outer leaflet; il: inner leaflet. Bar: 100 nm. *Reproduced with permission from Farquhar, M. G., & Palade, G. E. (1963). Junctional complexes in various epithelia.* The Journal of Cell Biology, 17, *375–412.*

precipitates are the results of poststaining or sample preparation, an area that contains no tissue is often examined, for example, the lumen of a blood vessel or the edge of a tissue section. If precipitates are found over tissue-free areas, then it is likely introduced by poststaining. To minimize precipitate formation, all solutions must be filtered through 0.2 μm-mesh membranes. Alternatively, tissues can be stained en bloc after osmium fixation. En bloc staining with 2% uranyl acetate has been shown to increase the TJ membrane contrast in liver specimens (Goodenough & Revel, 1970). Nevertheless, en bloc staining with lead is known to decrease the quality of ultrastructural preservation so is not recommended unless necessary (Hayat, 1975).

6.3.5.2 Osmolality, electrolytes, and additives in fixation

Cells tend to swell if the osmolality of the fixative buffer is low, whereas they will shrink if the fixative buffer becomes hypertonic (Maunsbach, 1966). Cell swelling and shrinking both reduce the ultrastructural quality of tight junction in tissues. For this reason, the osmotic composition of the fixative buffer has to be adjusted for each tissue. In the renal papilla, where extracellular osmolality is high, the normal fixative buffer should be supplemented with 0.2 M sucrose to prevent cell swelling. Electrolytes, particularly Ca^{++} and Mg^{++}, can reduce the extraction of certain proteins during fixation and stabilize the membranes and cytoskeletal elements (Bullock, 1984). The working concentration is 1–3 mM. The use of divalent cations, however, is not recommended with phosphate buffers. In highly vascularized organs, such as the pancreas, the kidney, the liver and the brain, perfusion fixation may lead to a distension of the extravascular space. Such distension can be prevented by adding 2% dextran (m.w. 40 kDa) to the fixative buffer (Bohman & Maunsbach, 1970).

6.3.6 Concluding remarks

Tissue represents a three-dimensional organization of cells and contains the most intriguing information on tight junction structure with respect to how continuous tissue barrier is established to obliterate the intercellular space. Electron microscopy for tissue section can provide a direct demonstration of where tight junctions are localized in cells relative to the cell membrane and other subcellular organelles and how TJ localization varies from cell to cell as a requirement to seal the extracellular space, for example, TJ localization in bile canaliculi is different from that in renal tubules. The nanometer resolution offered by electron microscopy permits an ultrastructural understanding of TJ organization in different tissues. Such fundamental knowledge of TJ ultrastructure has laid the premise for molecular delineation of the proteins that constitute the tight junction.

References

Bohman, S. O., & Maunsbach, A. B. (1970). Effects on tissue fine structure of variations in colloid osmotic pressure of glutaraldehyde fixatives. *Journal of Ultrastructure Research*, *30*, 195–208.

Bullock, G. R. (1984). The current status of fixation for electron microscopy: A review. *Journal of Microscopy*, *133*, 1–15.

Farquhar, M. G., & Palade, G. E. (1963). Junctional complexes in various epithelia. *The Journal of Cell Biology*, *17*, 375–412.

Goodenough, D. A., & Revel, J. P. (1970). A fine structural analysis of intercellular junctions in the mouse liver. *The Journal of Cell Biology*, *45*, 272–290.

Hayat, M. A. (1975). *Positive staining for electron microscopy*. Van Nostrand Reinhold New York.

Hayat, M. A. (1981). *Principles and techniques of electron microscopy. Biological Applications*. Edward Arnold.

Hou, J. (2018). *The paracellular channel – Biology, physiology and disease*. Academic Press.

Karnovsky, M. (1965). *A formaldehyde-glutaraldehyde fixative of high osmolality for use in electron microscopy*. *Journal of Cell Biology* (27, p. 137A). .

Maunsbach, A. B. (1966). The influence of different fixatives and fixation methods on the ultrastructure of rat kidney proximal tubule cells. II. Effects of varying osmolality, ionic strength, buffer system and fixative concentration of glutaraldehyde solutions. *Journal of Ultrastructure Research*, *15*, 283–309.

Sabatini, D. D., Bensch, K., & Barrnett, R. J. (1963). Cytochemistry and electron microscopy. The preservation of cellular ultrastructure and enzymatic activity by aldehyde fixation. *The Journal of Cell Biology*, *17*, 19–58.

Chapter 6.4

Transmission electron microscopy for tracer assay

6.4.1 Background knowledge

6.4.1.1 Tissue barrier defined by tracer

Tracers are electron-dense substances that are exogenously administered to delineate the extracellular spaces and the limits to extracellular spaces, also known as tissue barriers. Tracers are particularly useful to distinguish the types of cell junctions by virtue of their ability to mark the extracellular space in the vicinity of the junctions. For instance, tight junction excludes tracer, whereas adherens junction or gap junction permits deposition of tracer within the intercellular space (Goodenough & Revel, 1970). Tracers may also be used to determine the permeability of organ systems, and specifically, the permeability of the vascular endothelium.

6.4.1.2 Lanthanum

Lanthanum is a trivalent cation and can bind to negatively charged molecules on the cell membrane, especially negatively charged glycoproteins (Doggenweiler & Frenk, 1965). When tissues are perfused with lanthanum before or during fixation, lanthanum follows an intercellular route until it is blocked by a tissue barrier made of tight junctions (Ghabriel, Jennings, & Allt, 1989). In doing so, lanthanum defines the site of tight junction, which itself appears free of lanthanum. It is also possible to trace the pathway of lanthanum as it moves through an epithelium or endothelium and thus define the permeability across the tissue barrier. Fixation with osmium is necessary to retain lanthanum in tissues because lanthanum deposits are not apparent when specimens are fixed with aldehydes alone (Bozzola & Russell, 1999). In specimens fixed with aldehydes, lanthanum is washed out during dehydration. Lanthanum must be used at room temperature because it is precipitated at low temperatures. Phosphate buffers are not compatible with lanthanum for their mixing results in precipitation.

6.4.2 Materials and reagents

6.4.2.1 Equipment

JEOL JEM-1400 120 kV TEM
Leica UC7 ultramicrotome
Embedding mold from Polyscience Inc.
Formvar-coated grids from Polyscience Inc.

6.4.2.2 Buffers

EM fixation buffer (without lanthanum) (2% paraformaldehyde, 2% glutaraldehyde in 0.1 M cacodylate buffer)

16% Paraformaldehyde	12.5 mL
25% Glutaraldehyde	8 mL
0.2 M cacodylate buffer	50 mL
dH_2O	29.5 mL

Adjust pH to 7.4. Filter, sterilize, and store at 4°C.

EM fixation buffer (with lanthanum) (1% lanthanum, 2% paraformaldehyde, 2% glutaraldehyde in 0.1 M cacodylate buffer)

Lanthanum nitrate	1 g
16% Paraformaldehyde	12.5 mL
25% Glutaraldehyde	8 mL
0.2 M cacodylate buffer	50 mL
dH_2O	29.5 mL

Slowly adjust to pH 7.7 with 1 M NaOH. Use freshly prepared solution.

0.2 M cacodylate buffer

Sodium cacodylate (trihydrate)	42.8 g

Bring to 1 L with dH_2O, stir to dissolve, and adjust pH to 7.4. Filter sterilize and store at 4°C.

1% OsO_4 in 0.1 M cacodylate buffer

OsO_4	1 g
0.2 M cacodylate buffer	50 mL
dH_2O	50 mL

Adjust pH to 7.4. Filter, sterilize, and store at 4°C.

1% Uranyl acetate buffer

Uranyl acetate	1 g
dH_2O	100 mL

Adjust pH to 7.4. Filter, sterilize, and store at 4°C.

Araldite resin mix

Araldite resin (Polyscience Inc.; Cat No.: 00552)	50 g
DDSA (Polyscience Inc.; Cat No.: 00563)	55 g
DMP-30 (Polyscience Inc.; Cat No.: 00553)	2.1 g

Mix and centrifuge at $500 \times g$ for 5 minutes at room temperature to remove air bubble; store at −20°C.

1/3 (v/v) Araldite resin mix with 2/3 (v/v) propylene oxide

Araldite resin mix	100 mL
Propylene oxide	200 mL

Mix and centrifuge at $500 \times g$ for 5 minutes at room temperature to remove air bubble; store at $-20°C$.

1/2 (v/v) Araldite resin mix with 1/2 (v/v) propylene oxide	
Araldite resin mix	100 mL
Propylene oxide	100 mL

Mix and centrifuge at $500 \times g$ for 5 minutes at room temperature to remove air bubble; store at $-20°C$.

3/4 (v/v) Araldite resin mix with 1/4 (v/v) propylene oxide	
Araldite resin mix	300 mL
Propylene oxide	100 mL

Mix and centrifuge at $500 \times g$ for 5 minutes at room temperature to remove air bubble; store at $-20°C$.

Lead citrate buffer	
Lead citrate	0.01 g
NaOH (10 M)	0.1 mL
dH_2O	10 mL

Mix in a 10-mL centrifuge tube and shake vigorously until lead dissolves.

6.4.3 Experimental procedure

1. Perfuse animal with EM fixation buffer (with lanthanum) by gravity flow (see details in Chapter 4: Histological approaches for tight junction).
2. Dissect tissues. Cut into small blocks (~5 mm cubic shape) using a scalpel.
3. Place tissue blocks into Eppendorf tubes filled with EM fixation buffer (without lanthanum). Fix the tissues at 4°C for overnight.
4. Embed tissue blocks, section tissue blocks and poststain tissue sections as described in Sections 6.3.3.2, Embedding, and 6.3.3.3, Sectioning and poststaining.

6.4.4 Data analysis

Lanthanum is unable to penetrate the tight junction and has been used as a pharmacologic agent to block the paracellular conductance of ionic currents (Yu et al., 2009). Lanthanum is generally regarded as a qualitative measure of paracellular permeability, because it can grossly define where a structural blockage exists within an epithelium or endothelium, and whether or not, the epithelium or endothelium has a major change in paracellular permeability after a treatment. When perfused into the cardiovascular system, lanthanum decorates the luminal membrane in the blood vessels and marks the site of tight junction from the luminal side (Fig. 6.4.1). Notably, the blood vessels

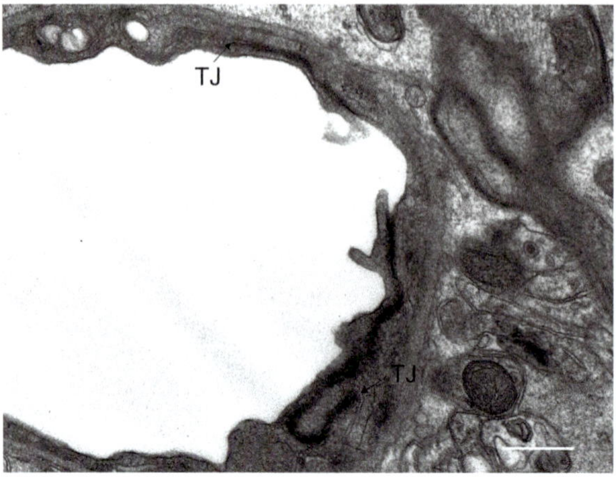

FIGURE 6.4.1 Lanthanum as a tracer to mark the blood–brain barrier. When lanthanum was introduced into the brain vascular system along with the fixative, it decorated the luminal membrane in the cerebral capillary but was stopped from passing into the brain parenchyma by tight junctions at the sites indicated by the arrows. Bar: 500 nm.

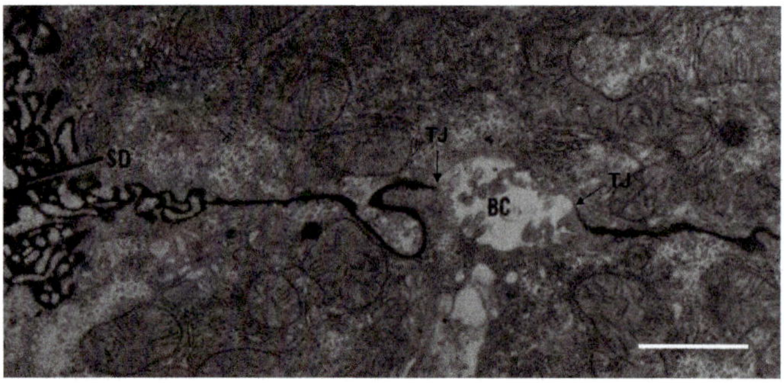

FIGURE 6.4.2 Lanthanum as a tracer to mark the tight junction in the bile canaliculus. When perfused into the liver, lanthanum accumulated in the space of Disse (SD) and the intercellular space but was stopped at the sites of tight junction (indicated by the arrows), thereby leaving the bile canaliculus (BC) free of the tracer. Bar: 1 μm. *Reproduced with permission from Goodenough, D. A., & Revel, J. P. (1970). A fine structural analysis of intercellular junctions in the mouse liver.* The Journal of Cell Biology, *45, 272–290.*

in most organs are leaky due to the lack of tight junction. Lanthanum extravasation into the parenchyma allows marking the location of tight junction in the epithelium from the stromal side (Fig. 6.4.2).

6.4.5 Troubleshooting

Tight junctions allow paracellular permeation of ion and solute on the basis of size and charge (Hou, 2018). The size selectivity of tight junction is biphasic, consisting of a high-capacity, size-restrictive pathway (for molecules of <8 Å in diameter) and a low-capacity, size-independent pathway (for molecules of >8 Å in diameter) (Watson, Rowland, & Warhurst, 2001). Lanthanum has an atomic weight of 57, and the diameter of lanthanum colloidal particle is around 2 nm. In the theory, lanthanum cannot permeate through the high-capacity, size-restrictive pathway. It may permeate through the low-capacity, size-independent pathway. The size-independent pathway is believed to either represent transient breaks in the bicellular tight junction due to dynamic TJ behavior or a novel channel of 10 nm in diameter made by the tricellular tight junction. The paracellular permeability of ionic lanthanum (La^{3+}) is higher than colloidal lanthanum [$La(OH)^{2+}$ or $La(OH)_2{}^+$]. The pH is critical for the formation of colloidal lanthanum. Experiments have shown that lanthanum at pH 7.7 is a colloid that is retarded by gel filtration on a Sephadex G-10 column (Goodenough & Revel, 1970).

6.4.6 Concluding remarks

Tracer assay used in electron microscopy has provided valuable information on the location of tight junctions in various organ systems. Lanthanum is among the most commonly used tracers for tight junction. Other tracers, such as ruthenium, ferritin, and horseradish peroxidase, can also be used to define the site of tight junction in tissues. Although not regarded as a quantitative measure of paracellular permeability, tracer assay can reveal the integrity of tight junction in vivo in physiologic systems, and more importantly, how the integrity of tight junction is compromised by pathogenic insults.

References

Bozzola, J. J., & Russell, L. D. (1999). *Electron microscopy: Principles and techniques for biologists*. Jones & Bartlett Learning.

Doggenweiler, C., & Frenk, S. (1965). Staining properties of lanthanum on cell membranes. *Proceedings of the National Academy of Sciences of the United States of America*, *53*, 425.

Ghabriel, M., Jennings, K., & Allt, G. (1989). Diffusion barrier properties of the perineurium: An in vivo ionic lanthanum tracer study. *Anatomy and Embryology*, *180*, 237–242.

Goodenough, D. A., & Revel, J. P. (1970). A fine structural analysis of intercellular junctions in the mouse liver. *The Journal of Cell Biology*, *45*, 272–290.

Hou, J. (2018). *The paracellular channel – Biology, physiology and disease*. Academic Press.

Watson, C. J., Rowland, M., & Warhurst, G. (2001). Functional modeling of tight junctions in intestinal cell monolayers using polyethylene glycol oligomers. *American Journal of Physiology Cell Physiology*, *281*, C388–C397.

Yu, A. S., Cheng, M. H., Angelow, S., Gunzel, D., Kanzawa, S. A., Schneeberger, E. E., … Coalson, R. D. (2009). Molecular basis for cation selectivity in claudin-2-based paracellular pores: Identification of an electrostatic interaction site. *The Journal of General Physiology*, *133*, 111–127.

Chapter 6.5

Transmission electron microscopy for immunolabeling application

6.5.1 Background knowledge

6.5.1.1 Immunoelectron microscopy

Immunoelectron microscopy unites two important techniques to study tight junction biology: (1) immunolabeling technique for molecular localization and (2) high-resolution imaging technique for ultrastructural determination. Immunoelectron microscopy can be divided into two classes based upon the embedding methods—preembedding and postembedding. Because detergents have to be used in the preembedding method in order for the antibody to cross the cell membrane, the ultrastructure of tight junction is poorly preserved by this method (Humbel, de Jong, Müller, & Verkleij, 1998). The postembedding method preserves tissue morphology better, yet at the cost of reduced binding affinity to antibodies. The fixation condition for immunoelectron microscopy is different from conventional electron microscopy. To retain antibody binding affinity, relatively mild fixation is carried out for immunoelectron microscopy, which limits formaldehyde concentration below 4% and glutaraldehyde concentration below 1%. Osmium is strictly excluded. As a result, the gross morphology of specimens processed for immunoelectron microscopy can hardly match that for conventional electron microscopy.

6.5.1.2 Low temperature embedding

The elevated temperatures (60°C–70°C) used to polymerize epoxy resins during tissue embedding can destroy the antigens irreversibly. A new class of acrylic resins has been developed for immunolabeling applications. The key advantage of acrylic resins over epoxy resins is that they can be polymerized at low temperatures. There are two types of acrylic resins—Lowicryl resins and London resins (LR). London resins are the simplest to use, as in the case of LR white, the resin can be polymerized by a chemical accelerator at 4°C–20°C or by UV light at −20°C.

6.5.2 Materials and reagents

6.5.2.1 Equipment

JEOL JEM-1400 120 kV TEM

Leica UC7 ultramicrotome

UV lamp from Thermo Fisher Scientific Inc.

Gelatin capsule from Polyscience Inc.
Formvar-coated grids from Polyscience Inc.

6.5.2.2 Buffers

IEM fixation buffer (4% paraformaldehyde, 0.01% glutaraldehyde in 0.1 M phosphate buffer)

16% Paraformaldehyde	25 mL
25% Glutaraldehyde	0.04 mL
0.2 M phosphate buffer	50 mL
dH_2O	25 mL

Adjust pH to 7.4. Filter, sterilize, and store at 4°C.

0.2 M phosphate buffer

Na_2HPO_4	21.8 g
NaH_2PO_4	6.2 g

Bring to 1 L with dH_2O, stir to dissolve, and adjust pH to 7.4. Filter, sterilize, and store at 4°C.

BLOCK solution

NaCl	8 g
KCl	0.2 g
Na_2HPO_4	1.44 g
KH_2PO_4	0.24 g
Fetal calf serum	50 mL

Bring to 1 L with dH_2O, stir to dissolve, and adjust pH to 7.4. Filter, sterilize, and store at 4°C.

50/50 (v/v) LR white/ethanol

LR white	50 mL
Ethanol	50 mL

Mix and centrifuge at $500 \times g$ for 5 minutes at room temperature to remove air bubble; store at -20°C.

60/40 (v/v) LR white/ethanol

LR white	60 mL
Ethanol	40 mL

Mix and centrifuge at $500 \times g$ for 5 minutes at room temperature to remove air bubble; store at -20°C.

70/30 (v/v) LR white/ethanol

LR white	70 mL
Ethanol	30 mL

Mix and centrifuge at $500 \times g$ for 5 minutes at room temperature to remove air bubble; store at -20°C.

0.1% Uranyl acetate buffer	
Uranyl acetate	0.1 g
dH_2O	100 mL

Adjust pH to 7.4. Filter, sterilize, and store at 4°C.

6.5.3 Experimental procedure

6.5.3.1 Perfusion and fixation

1. Perfuse animal with IEM fixation buffer by gravity flow (see details in Chapter 4: Histological approaches for tight junction).
2. Dissect tissues. Cut into small blocks (~5 mm cubic shape) using a scalpel.
3. Place tissue blocks into Eppendorf tubes filled with IEM fixation buffer. Fix the tissues at 4°C for overnight.

6.5.3.2 Embedding

1. Aspirate fixative and add 0.1 M phosphate buffer at room temperature with rotation for 10 minutes; repeat three times.
2. Aspirate 0.1 M phosphate buffer; add 20% ethanol; rotate for 10 minutes at room temperature; followed by rotation with 40% ethanol for 10 minutes; 60% ethanol for 10 minutes, 80% ethanol for 10 minutes, and 100% ethanol for overnight.
3. Aspirate ethanol; add 50/50 LR white/ethanol and rotate for 30 minutes at 4°C.
4. Aspirate solution in Step 3; add 60/40 LR white/ethanol and rotate for 30 minutes at 4°C.
5. Aspirate solution in Step 4; add 70/30 LR white/ethanol and rotate for 60 minutes at 4°C.
6. Aspirate solution in Step 5; add 100% LR white and rotate for overnight at 4°C.
7. Add solution in Step 6 (containing the tissue) into a gelatin capsule. Place a small piece of paper with the identification number upside down in the resin next to the tissue. Adjust the orientation of the tissue with a wooden stick so that sections of desirable plane can be cut with an ultramicrotome.
8. Place the tissue at a distance of 15 cm from a 30 W UV lamp (365 nm wavelength). Polymerize the specimen at −20°C for 24 hours.

6.5.3.3 Sectioning and immunolabeling

1. Once the tissue is embedded, it is cut into 70 nm ultrathin sections by an ultramicrotome (Hayat, 1981).
2. Ultrathin sections are collected on Formvar-coated grids (200 mesh).
3. Collect grids by floating them on BLOCK buffer.

4. Prepare a clean staining surface by placing a piece of Parafilm at the bottom of a petri dish.
5. Place one drop of BLOCK buffer for each grid onto the Parafilm. Float the grid on top of the buffer with the side containing section facing down.
6. Cover the petri dish with a lid and incubate for 30 minutes at room temperature.
7. Dilute primary antibodies (anti-TJ proteins) at 1:50–1:100 into the BLOCK solution. Place one drop of primary antibody solution for each grid onto the Parafilm. Float the grid on top of the solution with the side containing section facing down.
8. Cover the petri dish with a lid and incubate 1 hour at room temperature.
9. Pick the grid out of the antibody solution droplet with fine forceps and dip into a series of glass beakers containing BLOCK buffer. Agitate the grid several times up and down through each beaker.
10. Dilute secondary antibodies (antimouse, rat or rabbit conjugated to colloidal gold) at 1:100–1:200 in BLOCK solution. Place one drop of secondary antibody solution for each grid onto the Parafilm. Float the grid on top of the solution with the side containing section facing down.
11. Cover the petri dish with a lid and incubate 1 hour at room temperature.
12. Pick the grid out of the antibody solution droplet with fine forceps and dip into a series of glass beakers containing BLOCK buffer. Agitate the grid several times up and down through each beaker.
13. Place one drop of uranyl acetate buffer for each grid onto the Parafilm. Float the grid on top of the stain with the side containing section facing down.
14. Cover the petri dish with a lid and stain for 5 minutes at room temperature.
15. Pick the grid out of the stain droplet with fine forceps and dip into a series of glass beakers containing dH_2O. Agitate the grid several times up and down through each beaker.
16. Drain off excess dH_2O with filter paper.
17. Place the grid section-side-up on filter paper until it is fully dry.

6.5.4 Data analysis

Because claudin delocalization from tight junction is sufficient to elicit paracellular permeability change independent of its total cellular abundance (Gong et al., 2014), a high-resolution imaging approach, allowing direct visualization of claudin protein within the tight junction, is needed to address the molecular alteration in situ in the tight junction. Claudin-5 is the predominant claudin found in the tight junction of the blood–brain barrier (BBB) (Nitta et al., 2003). Immunoelectron microscopy reveals exclusive TJ localization of claudin-5 protein in the BBB (Fig. 6.5.1). Sections may be doubly labeled with primary antibodies raised in two different species by using secondary antibodies conjugated to different sizes of colloidal gold, for example, 5 nm

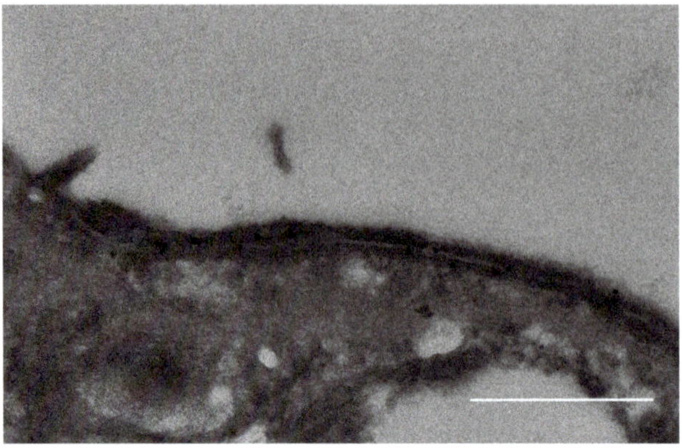

FIGURE 6.5.1 Claudin-5 localization in the tight junction of blood–brain barrier. Transmission electron micrograph shows cerebral endothelial tight junction immunolabeled for claudin-5 protein with colloidal gold particles (5 nm in diameter). Note that claudin-5 signals decorate the cell junction cleft. Bar: 500 nm.

versus 10 nm. The same procedure is followed as for single labeling, but a cocktail containing two primary antibodies is used in Step 7 and a cocktail of two secondary antibodies is used in Step 10 of Section 6.5.3.3.

6.5.5 Troubleshooting

6.5.5.1 Negative result

If there is no labeling, the first thing to consider is to try immunolabeling cryostat tissue sections with a light microscopic approach (Chapter 5.3: Wide-field fluorescence microscopy for thin tissue section). If cryostat sections of fresh specimens do not work, there is usually no point in continuing to the electron microscopic level. Provided that a strong positive signal can be seen at the light microscopic level, a negative result at the electron microscopic level might be due to two reasons. In the case of plastic section (e.g., embedded in LR white resin), a possible reason is that aldehyde fixation blocks the access of antibody into the tissue or inactivate the antigen if the antibody binding region is cross-linked. This problem can usually be solved by the cryopreparation method of cutting cryosection of vitrified specimen or plastic section of freeze-substituted specimen (Bozzola & Russell, 1999). In the latter approach, aldehyde fixation condition must be weakened significantly (e.g., in 1%–2% formaldehyde for 5–10 minutes). Although the cryopreparation method deleteriously affects the ultrastructure of tight junction, it allows determining if a negative labeling result is due to the fixation that compromises the antibody binding affinity in the tissue section. A second reason for lack of labeling is that the number of antigen molecules may be too low to detect in ultrathin sections. An ultrathin section of 70 nm is only about seven to nine times of the membrane thickness (7.5–10 nm). Conceivably, the antigen molecules available in an ultrathin section are far fewer than those in a cryostat section (5–10 μm) or in a whole cell (10–20 μm) detected by immunofluorescence microscopy (Griffiths, 2012).

6.5.5.2 Nonspecific binding

The primary antibody used for immunoelectron microscopy must first be validated for specificity using immunofluorescence labeling techniques (see Section 5.2.5.2). The secondary antibody in the case of colloidal gold is another source of nonspecific binding. The optimal concentration of secondary antibody should be titrated and defined as the highest that does not give background labeling in the absence of a primary antibody. The definition of background in immunoelectron microscopy assumes the knowledge of antigen's subcellular localization, for example, a TJ protein would not be expected in the cytoplasm.

6.5.6 Concluding remarks

Immunolabeling has long been applied to the study of protein localization in cells and tissues on the level of light microscopy. Yet, the most conclusive proof of TJ localization for a protein comes from the demonstration of gold-conjugated antibody labeling on the level of electron microscopy. As gold particles are small and dense and can specifically decorate the TJ ultrastructure, they have produced valuable information on the molecular composition of tight junction. Immunoelectron microscopy can also be used as a quantitative measurement of protein abundance in the tight junction, provided that antibody binding is linearly correlated to target protein concentration. A positive signal in immunoelectron microscopy is only significant when independent techniques such as light microscopy prove that the gold-conjugated antibody labeling is due to antibody binding with the antigen of interest in the tight junction.

References

Bozzola, J. J., & Russell, L. D. (1999). *Electron microscopy: Principles and techniques for biologists*. Jones & Bartlett Learning.

Gong, Y., Yu, M., Yang, J., Gonzales, E., Perez, R., Hou, M., ... Hou, J. (2014). The Cap1-claudin-4 regulatory pathway is important for renal chloride reabsorption and blood pressure regulation. *Proceedings of the National Academy of Sciences of the United States of America*, *111*, E3766–E3774.

Griffiths, G. (2012). *Fine structure immunocytochemistry*. Springer Science & Business Media.

Hayat, M. A. (1981). *Principles and techniques of electron microscopy. Biological Applications*. Edward Arnold.

Humbel, B. M., de Jong, M. D., Müller, W. H., & Verkleij, A. J. (1998). Pre-embedding immunolabeling for electron microscopy: An evaluation of permeabilization methods and markers. *Microscopy Research and Technique*, *42*, 43–58.

Nitta, T., Hata, M., Gotoh, S., Seo, Y., Sasaki, H., Hashimoto, N., ... sukita, S. (2003). Size-selective loosening of the blood-brain barrier in claudin-5-deficient mice. *The Journal of Cell Biology*, *161*, 653–660.

Chapter 6.6

Freeze-fracture electron microscopy

6.6.1 Background knowledge

6.6.1.1 Principle of freeze-fracture technique

Freeze-fracture electron microscopy (FF-EM) is unique among electron microscopic methods in that it gives *en face* views of the internal organization of biological membranes, allowing the study of in-plane distribution of integral proteins spanning the lipid bilayer (Bozzola & Russell, 1999). In the frozen state, the hydrophobic membrane interior is most subject to fracture when external forces are exerted on the membrane. As the energy required to split the lipid bilayer is less than what is necessary to fracture the cytoplasm, the lipid bilayer is frequently split to reveal the interior faces of the membrane. Specific names have been given to each of the fracture faces. The portion of the bilayer associated with the cytoplasm is termed the P-face or protoplasmic face. The portion associated with extracellular space is termed the E-face or exoplasmic face.

6.6.1.2 Technical consideration

6.6.1.2.1 Freezing

A small block of tissue ($\sim$2 mm cubic shape) or a cell monolayer on cover glass is first rapidly frozen by plunging it manually into nitrogen slush and then rapidly transferred into liquid nitrogen until needed for fracturing.

6.6.1.2.2 Fracturing

For fracturing, the specimen is quickly transferred to the precooled, temperature-controlled specimen table maintained at $-110°C$ within the vacuum chamber of a freeze-fracture apparatus, for example, Balzers BAF 400T (Fig. 6.6.1). Freeze fracture can be achieved by two methods—knife fracture and tensile fracture (Fig. 6.6.2) (Severs, 2007). When the fracture plane encounters a cell membrane, it passes along the center of the lipid bilayer, as it is a line of weakness at cryogenic temperatures, splitting the membrane into P- and E-faces. Integral membrane proteins may partition with one fracture face, from which they protrude to form the freeze-fracture intramembrane particles, leaving complementary pits in the other fracture face from which they are wrenched.

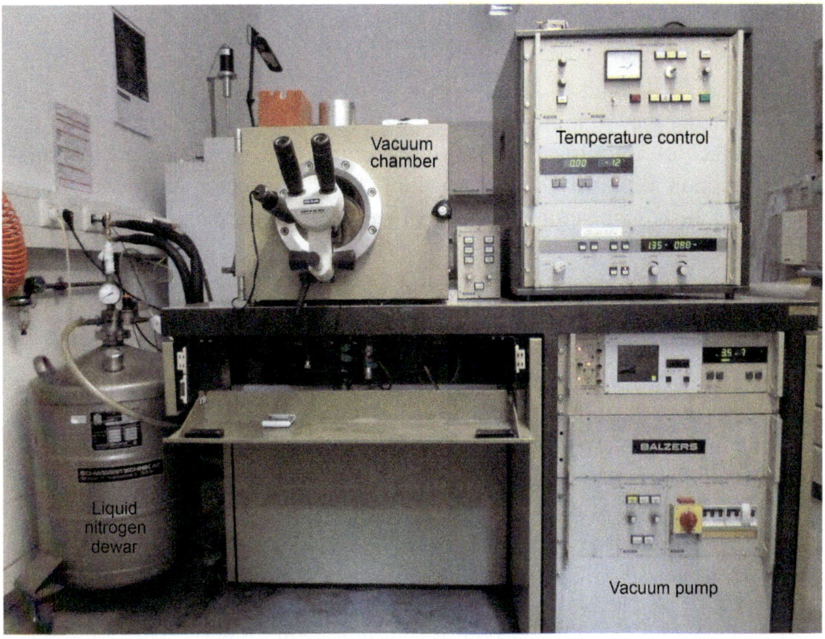

FIGURE 6.6.1 A commercial freeze-fracture apparatus. The Balzers BAF 400T device is illustrated here. The liquid nitrogen dewar provides liquid nitrogen to cool the cold stage, microtome, and cold shroud inside the vacuum chamber via the FET filling device. To fracture a specimen, the operator manually controls the microtome knife inside the vacuum chamber while simultaneously viewing the cold specimen with the binocular microscope through the window.

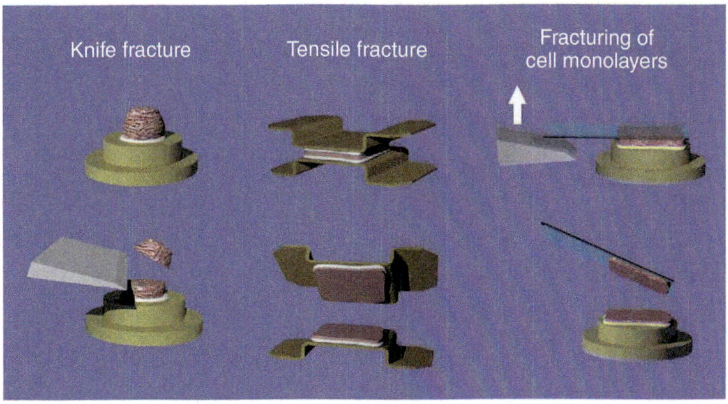

FIGURE 6.6.2 Two principal freeze-fracture methods. Knife fracture is by passing the cold microtome knife through the specimen. Tensile fracture is by breaking apart the specimen mounted between a pair of carriers. Freeze fracture of cell monolayers, which are too thin to be fractured by microtome, is accomplished using a variant of the tensile fracture method. A rectangular piece of coverslip on which the cells have been cultured is mounted, cell side down, on a gold stub and frozen. The microtome blade is maneuvered under a projecting portion of the coverslip and then raised, flicking off the coverslip and generating a fracture through the cell membrane. *Reproduced with permission from Severs, N. J. (2007). Freeze-fracture electron microscopy.* Nature Protocols, 2, *547–576.*

6.6.1.2.3 *Etching*

The specimen may optionally be etched after fracturing. During etching, water molecules are allowed to sublime from the frozen surface of the fractured specimen at −100°C in the vacuum chamber. Etching can reveal many features of the cytoskeleton and the extracellular matrix that are obscured by the overlying ice (Heuser & Salpeter, 1979).

6.6.1.2.4 *Replication*

The frozen surface of a freeze-fractured specimen is rich in topographical detail, but is labile. To view the fracture face by electron microscopy, the specimen must be obliquely shadowed with a thin layer of platinum (1.5–2 nm thick), deposited from a platinum beam gun (Fig. 6.6.3). The platinum atoms landing on the fracture surface do not form a homogeneous layer. Instead, the atoms coalesce to form small platinum grains, about 1 nm in diameter. This discontinuous platinum replica must be strengthened by a uniform layer of carbon (15–20 nm thick), deposited from a carbon beam gun (Fig. 6.6.3). The combined

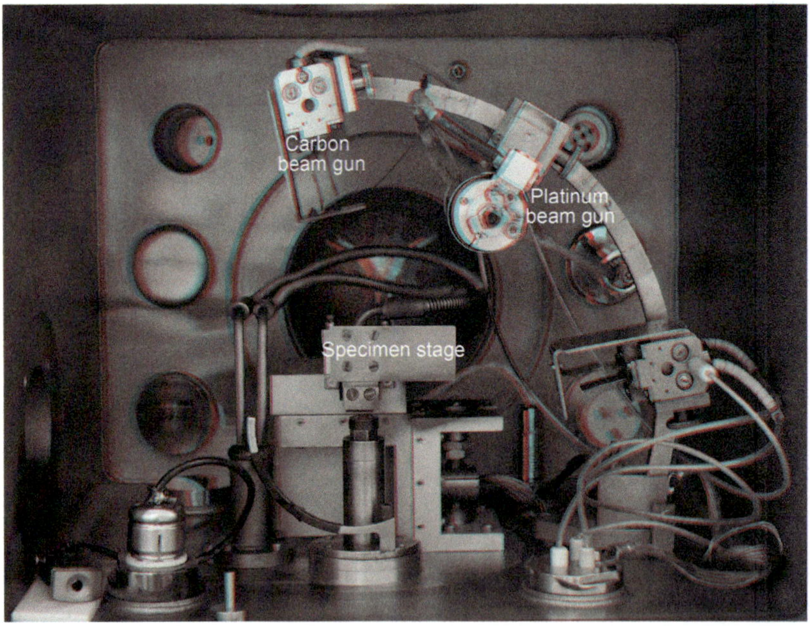

FIGURE 6.6.3 An inside view of the vacuum chamber in a freeze-fracture apparatus. The Balzers BAF 400T device is illustrated here. Replication is carried out in the vacuum chamber after freeze fracturing by a platinum beam gun and a carbon beam gun, that are obliquely arranged above the specimen.

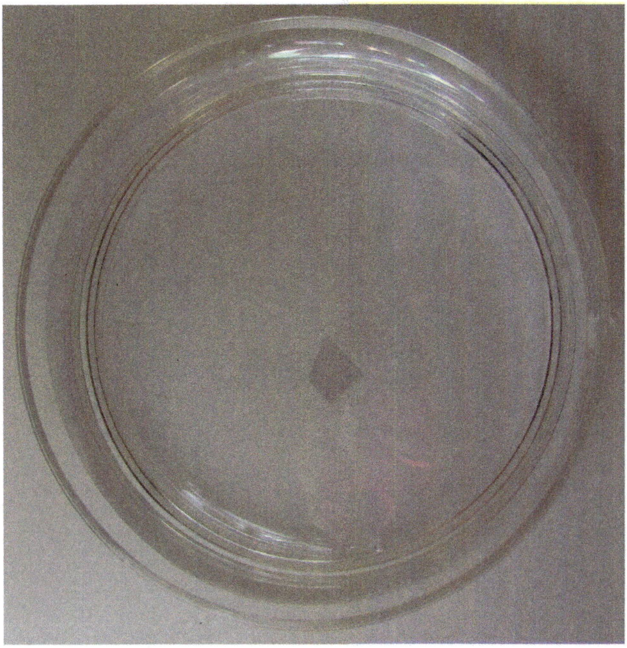

FIGURE 6.6.4 A platinum-carbon replica. The combined platinum-carbon replica is cleaned by chromic acid and floated on distilled water, ready to be mounted onto electron microscopic grids.

platinum-carbon replica is then cleaned with bleach to remove the original biological material and mounted onto the Formvar-coated grid for examination under a transmission electron microscope (Fig. 6.6.4).

6.6.2 Materials and reagents

6.6.2.1 Equipment

JEOL JEM-1400 120 kV TEM

Balzers BAF 400T freeze-fracture device

Formvar-coated grids from Polyscience Inc.

Polyvinyl alcohol adhesive (PVA) from Sigma Aldrich.

6.6.2.2 Cell model

LLC-PK1 epithelial cells; available from ATCC.

6.6.2.3 Cell culture medium

Complete medium

500 mL	Dulbecco's Modified Eagle's medium (DMEM) (high glucose/ + sodium pyruvate/ + glutamine); available from Gibco
50 mL	Fetal bovine serum (FBS); available from Gibco
5 mL	Penicillin/streptomycin; available from Gibco

6.6.2.4 Buffers

FF-EM fixation buffer (2% glutaraldehyde in 0.1 M phosphate buffer)

25% Glutaraldehyde	8 mL
0.2 M Phosphate buffer	50 mL
dH_2O	42 mL

Adjust pH to 7.4. Filter, sterilize, and store at 4°C.

0.2 M Phosphate buffer

Na_2HPO_4	21.8 g
NaH_2PO_4	6.2 g

Bring to 1 L with dH_2O, stir to dissolve, and adjust pH to 7.4. Filter, sterilize, and store at 4°C.

20% Glycerol in 0.1 M phosphate buffer

Glycerol	20 mL
0.2 M phosphate buffer	50 mL
dH_2O	30 mL

Filter, sterilize, and store at 4°C.

25% Glycerol in 0.1 M phosphate buffer

Glycerol	25 mL
0.2 M phosphate buffer	50 mL
dH_2O	25 mL

Filter, sterilize, and store at 4°C.

30% Glycerol in 0.1 M phosphate buffer

Glycerol	30 mL
0.2 M phosphate buffer	50 mL
dH_2O	20 mL

Filter, sterilize, and store at 4°C.

6.6.3 Experimental procedure

6.6.3.1a Cell culture and fixation

1. Clean cover glass with ethanol. Let air dry and place cover glass in a 6-well plate.
2. Trypsinize LLC-PK1 cells and load into the 6-well plate at 1×10^6 per well. Culture for 9 consecutive days and change medium every 3 days.
3. Aspirate medium and wash cells with 1 × PBS.
4. Aspirate 1 × PBS and fill the 6-well plate with FF-EM fixation buffer. Fix the cells at 4°C for 1 hour.

6.6.3.1b Perfusion and fixation

1. Perfuse animal with FF-EM fixation buffer by gravity flow (see details in Chapter 4: Histological approaches for tight junction).
2. Dissect tissues. Cut into small blocks (~2 mm cubic shape) using a scalpel.
3. Place tissue blocks into Eppendorf tubes filled with FF-EM fixation buffer. Fix the tissues at 4°C for 1 hour.

6.6.3.2 Freeze fracturing

1. Fold a strip of parafilm and place for viewing under a binocular microscope. Place selected specimen carriers on the parafilm base.
2. Immerse the fixed tissue pieces or cell monolayers into a series of glycerol solutions (20%, 25%, and 30% glycerol in 0.1 M phosphate buffer) for 30 minutes to 2 hours in each solution.
3. Transfer the fixed tissue pieces or cell monolayers onto the parafilm close to the specimen carriers.
4. Use an applicator stick sharpened to a point to smear PVA mounting medium on the surfaces of a pair of clean carriers with which the specimen will have contact.
5. Trim the sample to a size and shape that fits snugly into the recess of one carrier, leaving a suitably shaped portion protruding to contact the surface of apposing carrier.
6. Invert second carrier over the first to make a specimen sandwich.
7. Fill a bench dewar flask with liquid nitrogen and gently insert specimen holding basket. Wait until the bubbling of the liquid nitrogen ceases, and then top up with liquid nitrogen.
8. Prepare nitrogen slush for freezing the specimens using a purpose-built slusher.
9. Holding the mounted specimen by its carriers with fine forceps, plunge it rapidly under the surface of the nitrogen slush.

10. Quickly transfer the frozen specimen to the bench dewar flask and drop onto the holding basket under liquid nitrogen.
11. Push the basket down to ensure that the specimens are deep under the liquid nitrogen.
12. Check and set Balzers BAF 400T freeze-fracture device.
13. Press power switches of cold stage/knife cooling control unit, quartz crystal film thickness monitor and electron gun evaporation control unit to "on" position.
14. Switch on the liquid nitrogen valve/filling device and set valve to give a reading of 0.6–0.7 bar on the gauge.
15. Select appropriate specimen table and attach loading/retrieval rod to side-screw that operates the table's clamp. Check that the clamp functions smoothly when the side-screw is turned with the rod.
16. Place the table (with the loading rod still attached) in the polystyrene box so that the table rests on the bottom of the box.
17. Fill the box with liquid nitrogen and keep topping up until bubbling of liquid nitrogen ceases.
18. Precool the tips of fine forceps under liquid nitrogen in the bench dewar.
19. Transfer specimen sandwiches from the storage basket into the three slots using the precooled forceps.
20. Reattach loading/retrieval rod to table and twist to release table from stand.
21. Rest table, still attached to loading/retrieval rod, on the bottom of the liquid nitrogen-filled polystyrene box, place lid on box and transfer close to the side port of the freeze-fracture device.
22. When the temperature of the cold stage reaches −150°C, press pumping control to "off" position and open valve to bring chamber to atmospheric pressure with nitrogen gas.
23. When the side port to the counterflow loading system flaps open, immediately insert the specimen table through the port and clamp it onto the cold stage by twisting the loading/retrieval rod. Then twist back to release and withdraw the rod and, while gently holding the cover against the side port, shut the valve to the chamber and restart pumping.
24. For freeze-fracture without etching, dial 4.8 on the "table adj temp" control to raise the temperature of the cold stage to −110°C.
25. Observe table through the window of the chamber using the binocular microscope and optimize illumination. Use the left-hand manual control to raise the vertical plate of the microtome assembly to the preset height into contact with the post so as to flip the two halves of the device apart and fracture the specimens held within.
26. Move microtome assembly back (i.e., in reverse direction) to prefracture position (rear of chamber) and immediately turn up the filament control

to 80–100 mA to evaporate the platinum-carbon. Continue to deposit platinum-carbon until the thin film monitor reading reaches 4, then turn down filament control to 0.

27. Immediately turn to filament 2 (carbon backing) and set high voltage control to 2500 V. Turn up filament control progressively to 110–120 mA and continue to evaporate carbon until thin film monitor reading has reached 6.
28. On completion of carbon evaporation, turn down filament control and high voltage control to "0," filament selector to "0" and switch evaporator control unit to "off."
29. Raise the microtome assembly using the manual controls to set in rearmost and highest position in chamber.
30. Switch pump to "off" and open valve. When the port door flaps open, remove specimen table with the replicated specimens using the loading/retrieval rod.
31. Remove replicated specimens from the specimen table by gripping specimen carriers with fine forceps.
32. Dispense bleach into a glass well by Pasteur pipette and dilute with distilled water. Start by testing bleach at different concentrations on fixed but unreplicated specimens.
33. Remove replicated specimen from the carrier using fine forceps to grip the specimen, avoiding touching the replica itself. Place the replicated specimen upright on the surface of the glass dish, inspect with microscope, and allow any condensed moisture on the replica to dry.
34. Lower the specimen into the bleach solution gently, at an angle, keeping replica afloat or near surface if possible.
35. Leave replica or pieces of replica at room temperature for a period of 30 minutes to 24 hours.
36. Transfer replicas through a series of distilled water and collect replicas onto Formvar-coated grids.

6.6.4 Data analysis

FF-EM has contributed much of what we know about the structure of cell junctions, in particular, tight junction and gap junction (Goodenough & Paul, 2009; Tsukita, Furuse, & Itoh, 2001). Under FF-EM, tight junction is seen as a continuous network of "fibrils" or "strands" on the P-face and complementary empty grooves on the E-face of fractured membrane (Fig. 6.6.5). These "fibrils" or "strands" are visualized as linear parallel arrays of intramembrane particles of ~10 nm in diameter and separated by a distance of ~8 nm (Goodenough & Revel, 1970). Notably, the patterns of TJ strands are influenced by claudin composition and claudin interaction within the tight junction (Fig. 6.6.5) (Gong et al., 2015).

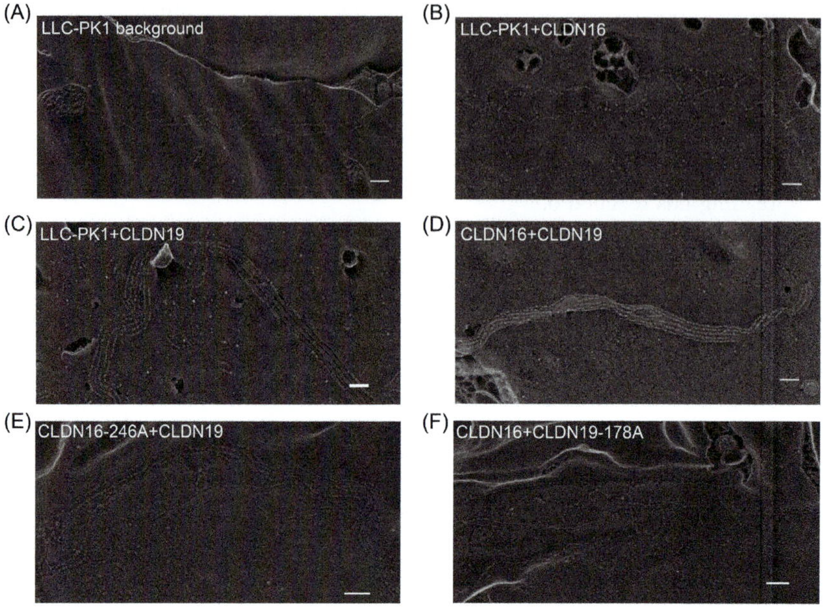

FIGURE 6.6.5 Freeze-fracture images of claudin-transfected LLC-PK1 cells. Freeze-fracture electron microscopy reveals TJ ultrastructure in LLC-PK1 cells with no transfection (A) or LLC-PK1 cells transfected with claudin-16 (B), claudin-19 (C), claudin-16 and -19 (D), claudin-16-246A mutant with claudin-19 (E), or claudin-16 with claudin-19-178A mutant (F). Notably, wildtype claudin-16 and -19 can interact, but mutant claudins (claudin-16-246A or claudin-19-178A) cannot. Coexpression of interaction-incompetent claudins has increased TJ strand number and complexity (E and F vs B–D). Bar: 100 nm. *Reproduced with permission from Gong, Y., Renigunta, V., Zhou, Y., Sunq, A., Wang, J., Yang, J., Renigunta, A., Baker, L. A., & Hou, J. (2015). Biochemical and biophysical analyses of tight junction permeability made of claudin-16 and claudin-19 dimerization.* Molecular Biology of the Cell *26(24):4333–46.*

6.6.5 Troubleshooting

To troubleshoot technical aspects of freeze-fracture experimentation, see Balzers' manual and other references (Rash, 1979; Severs & Shotton, 1995). The pattern of tight junction after freeze fracture may differ between fixed and unfixed specimens (Suzuki & Nagano, 1991). In the fixed sample, the TJ strands have a stronger affinity for the P-face of the lipid bilayer and appear as smoothly contoured ridges. In the unfixed sample, the TJ strands preferably associate with the E-face of the lipid bilayer and appear as rows of hemispherical particles. In the theory, fixation may alter the ultrastructure of tight junction or bring about artifacts. Rapidly frozen, fresh material is considered to be a *bona fide* biological state. On the other hand, fresh specimens cannot be treated with cryoprotectants such as glycerol before freezing. Replicas often show poor preservation of ultrastructure, with features indicating ice crystal damage. Membrane fracture faces may appear bumpy and distorted rather than smoothly contoured. Freeze fracture is easily obtained for tight junctions in tissue samples. For cell cultures, however, freeze fracture can only be used to study premature tight junctions in cells grown on cover glass. In order for cells to fully polarize and establish mature tight junctions, permeable supports such as Transwell have to be used in the culture. Transwell, as it is made of soft polycarbonate material, cannot allow freeze-fracturing the specimen mounted on top of it.

6.6.6 Concluding remarks

Freeze-fracture electron microscopy is the most valuable technique to understand tight junction ultrastructure. When combined with molecular tools, which allow proteins to be added into or removed from the tight junction, freeze-fracture electron microscopy can elegantly correlate the structure of tight junction with its molecular composition and interaction. Freeze-fracture electron microscopy is also useful to reveal the structural alteration in tight junctions from specimens treated with pharmacologic agents, infected with pathogens or inflicted with diseases. Even as a 50-year old technique, freeze-fracture electron microscopy can still find its place in cell biology in line with current trend that emphasizes direct visualization of the structure to function relationship of subcellular organelles.

References

Bozzola, J. J., & Russell, L. D. (1999). *Electron microscopy: Principles and techniques for biologists*. Jones & Bartlett Learning.

Gong, Y., Renigunta, V., Zhou, Y., Sunq, A., Wang, J., Yang, J., ... Hou, J. (2015). Biochemical and biophysical analyses of tight junction permeability made of claudin-16 and claudin-19 dimerization. *Molecular Biology of the Cell*.

Goodenough, D. A., & Paul, D. L. (2009). Gap junctions. *Cold Spring Harbor Perspectives in Biology*, *1*, a002576.

Goodenough, D. A., & Revel, J. P. (1970). A fine structural analysis of intercellular junctions in the mouse liver. *The Journal of Cell Biology*, *45*, 272–290.

Heuser, J. E., & Salpeter, S. R. (1979). Organization of acetylcholine receptors in quick-frozen, deep-etched, and rotary-replicated Torpedo postsynaptic membrane. *The Journal of Cell Biology*, *82*, 150–173.

Rash, J. E. (1979). *Freeze fracture: Methods, artifacts, and interpretations*. Raven Pr.

Severs, N. J. (2007). Freeze-fracture electron microscopy. *Nature Protocols*, *2*, 547–576.

Severs, N. J., & Shotton, D. M. (1995). *Rapid freezing, freeze fracture, and deep etching* (Vol. 2). Wiley-Liss.

Suzuki, F., & Nagano, T. (1991). Three-dimensional model of tight junction fibrils based on freeze-fracture images. *Cell and Tissue Research*, *264*, 381–384.

Tsukita, S., Furuse, M., & Itoh, M. (2001). Multifunctional strands in tight junctions. *Nature Reviews Molecular Cell Biology*, *2*, 285–293.

Chapter 6.7

Freeze-fracture replica immunolabeling technique

6.7.1 Background knowledge

The major goal of freeze-fracture replica immunolabeling (FRIL) is to perform simultaneous subcellular localization and biochemical identification of membrane proteins in cells or tissues. Initially, it was assumed that freeze-fracture replicas could not be immunolabeled because no biological material remained after the replicas were cleaned with bleach for viewing by electron microscopy. A serendipitous discovery was later made to show that sodium dodecyl sulfate (SDS), an ionic detergent, can remove the majority of biological components but spare the membrane proteins that are in direct contact with the replica (Fujimoto, 1995). Fixation of a specimen should be carried out with a protocol known to preserve the antibody binding site at the electron microscopic level (see Chapter 6.5: Transmission electron microscopy for immunolabeling application). The specimen is then rapidly frozen, fractured, replicated, and washed with SDS (see Chapter 6.6: Freeze-fracture electron microscopy). Immunolabeling of a replica is similar to what has been described for ultrathin section immunoelectron microscopy (see Chapter 6.5: Transmission electron microscopy for immunolabeling application).

6.7.2 Materials and reagents

6.7.2.1 Equipment

JEOL JEM-1400 120 kV TEM

Balzers BAF 400T freeze-fracture device

Formvar-coated grids from Polyscience Inc.

Polyvinyl alcohol adhesive (PVA) from Sigma Aldrich.

6.7.2.2 Cell model

LLC-PK1 epithelial cells; available from ATCC.

6.7.2.3 Cell culture medium

Complete medium

500 mL	Dulbecco's Modified Eagle's medium (DMEM) (high glucose/ + sodium pyruvate/ + glutamine); available from Gibco
50 mL	Fetal bovine serum (FBS); available from Gibco

5 mL	Penicillin/streptomycin; available from Gibco

6.7.2.4 Buffers

FRIL fixation buffer (1% paraformaldehyde in 0.1 M phosphate buffer)

16% Paraformaldehyde	6.25 mL
0.2 M phosphate buffer	50 mL
dH_2O	43.75 mL

Adjust pH to 7.4. Filter, sterilize, and store at 4°C.

0.2 M phosphate buffer

Na_2HPO_4	21.8 g
NaH_2PO_4	6.2 g

Bring to 1 L with dH_2O, stir to dissolve, and adjust pH to 7.4. Filter, sterilize, and store at 4°C.

20% Glycerol in 0.1 M phosphate buffer

Glycerol	20 mL
0.2 M phosphate buffer	50 mL
dH_2O	30 mL

Filter, sterilize, and store at 4°C.

25% Glycerol in 0.1 M phosphate buffer

Glycerol	25 mL
0.2 M phosphate buffer	50 mL
dH_2O	25 mL

Filter sterilize and store at 4°C.

30% Glycerol in 0.1 M phosphate buffer

Glycerol	30 mL
0.2 M phosphate buffer	50 mL
dH_2O	20 mL

Filter, sterilize, and store at 4°C.

Tris	**SDS buffer**
SDS	25 g
Sucrose	10 g
Tris	1.2 g

Bring to 1 L with dH_2O, stir to dissolve, and adjust pH to 8.0. Filter, sterilize, and store at 4°C.

BLOCK solution

NaCl	8 g
KCl	0.2 g
Na_2HPO_4	1.44 g
KH_2PO_4	0.24 g
Fetal calf serum	50 mL

Bring to 1 L with dH_2O, stir to dissolve, and adjust pH to 7.4.
Filter, sterilize, and store at 4°C.

6.7.3 Experimental procedure

1. Perform freeze fracture on cell monolayers or tissue blocks as described in Section 6.6.3.
2. At the end of freeze fracture experiment, clean the replicated specimens with Tris-SDS buffer instead of bleach solution at room temperature for 2 hours.
3. Transfer replicas through a series of 1 × PBS.
4. Prepare a clean staining surface by placing a piece of Parafilm at the bottom of a petri dish.
5. Place one drop of BLOCK buffer for each replica onto the Parafilm. Float the replicas on top of the buffer with the side containing digested tissue facing down.
6. Cover the petri dish with a lid and incubate for 30 minutes at room temperature.
7. Dilute primary antibodies (anti-TJ proteins) at 1:50–1:100 into the BLOCK solution. Place one drop of primary antibody solution for each replica onto the Parafilm. Float the replicas on top of the solution with the side containing digested tissue facing down.
8. Cover the petri dish with a lid and incubate 1 hour at room temperature.
9. Pick the replicas out of the antibody solution droplet with blunt-end forceps and dip into a series of glass beakers containing BLOCK buffer. Agitate the replicas several times up and down through each beaker.
10. Dilute secondary antibodies (antimouse, rat, or rabbit conjugated to colloidal gold) at 1:100–1:200 in BLOCK solution. Place one drop of secondary antibody solution for each replica onto the Parafilm. Float the replicas on top of the solution with the side containing digested tissue facing down.
11. Cover the petri dish with a lid and incubate 1 hour at room temperature.
12. Pick the replicas out of the antibody solution droplet with blunt-end forceps and dip into a series of glass beakers containing BLOCK buffer. Agitate the replicas several times up and down through each beaker.
13. Transfer the replicas through a series of distilled water and collect the replicas onto Formvar-coated grids.

6.7.4 Data analysis

FRIL is particularly suited to the studies of integral membrane proteins in cell junctions, for example, tight junction and gap junction, because these proteins are often clustered into membrane microdomains. For example, in

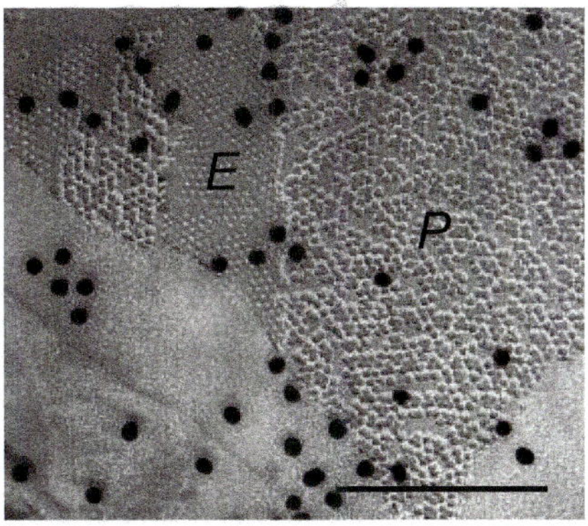

FIGURE 6.7.1 FRIL of gap junction. Freeze-fracture electron micrographs of the rat ependyma immunolabeled with colloidal gold-conjugated anti-connexin-43 antibody reveal the localization of connexin-43 proteins in the gap junction patches. E: E-face; P: P-face. Scale bars: 200 nm. *Reproduced with permission from Rash, J. E., & Yasumura, T. (1999). Direct immunogold labeling of connexins and aquaporin-4 in freeze-fracture replicas of liver, brain, and spinal cord: Factors limiting quantitative analysis.* Cell and Tissue Research, 296, *307–321.*

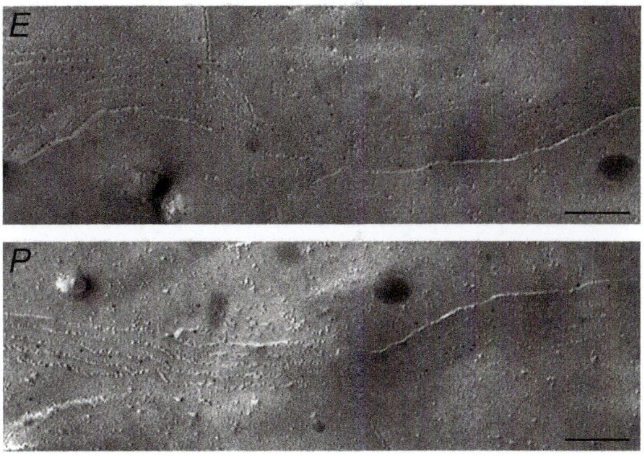

FIGURE 6.7.2 FRIL of tight junction. Freeze-fracture electron micrographs of LLC-PK1 cells immunolabeled with colloidal gold-conjugated anti-claudin-16 antibody reveal the localization of claudin-16 proteins in the tight junction strands. E: E-face; P: P-face. Scale bars: 200 nm. *Reproduced with permission from Hou, J., Renigunta, A., Konrad, M., Gomes, A. S., Schneeberger, E. E., Paul, D. L., Waldegger, S., & Goodenough, D. A. (2008). Claudin-16 and claudin-19 interact and form a cation-selective tight junction complex.* The Journal of Clinical Investigation, 118, *619–628.*

freeze-fracture replicas, gap junctions are characterized by patches of densely packed connexin protein particles of 9–10 nm in diameter (Fig. 6.7.1) (Rash & Yasumura, 1999). Tight junctions are seen as linear parallel "fibrils" or "strands" of claudin protein particles of ~10 nm in diameter (Fig. 6.7.2)

(Hou et al., 2008). The E-face of gap junction and tight junction appears as spaced pits, which are complementary to the protein particles on the P-face. Both the P- and E-faces of cell junctions can be immunolabeled with colloidal gold-conjugated antibodies (Figs. 6.7.1 and 6.7.2).

6.7.5 Troubleshooting

Formaldehyde is used as the primary fixative for FRIL because it produces a less cross-linked matrix than does glutaraldehyde, so creates less steric hindrance to antibody access. Formaldehyde alone, however, may give relatively poor preservation of ultrastructure. Ultrastructural preservation is markedly enhanced by the inclusion of glutaraldehyde at a concentration of less than 0.1%, which is unlikely to affect the antibody binding affinity. Replicas are "sticky," that is, they tend to absorb antibodies and antibody-coated colloidal gold particles nonspecifically. The nonspecific binding can be reduced or eliminated by blocking the sticking sites with albumin or fetal calf serum. Further reduction in nonspecific binding can be achieved by the addition of a nonionic detergent such as Tween-20 at 0.1%. Because ultrastructural information has been encoded into the replica, the use of detergent is permitted, in sharp contrast to ultrathin section immunoelectron microscopy. FRIL is difficult to apply to complex tissues. The loss of histological orientation in the replica prevents any meaningful correlation of labeled molecules with individual mapped cells in a complex tissue. Finally, the E-face labeling is unique to cell junction proteins (Fujimoto, 1995). Any label associated with E-face pit represents a form of "cryptic" labeling, because the labeled protein is not replicated in the E-face of a membrane bilayer. The only possible explanation is that *trans* interactions among cell junction proteins allow the binding partners underneath the E-face pits to be immunolabeled.

6.7.6 Concluding remarks

FRIL is regarded as the most conclusive way to demonstrate the protein localization in biological membranes. It combines molecular labeling with ultrastructural determination for subcellular organelles, in particular cell junctions. FRIL has contributed immensely to our understanding of the molecular organization of cell junctions, for example, tight junction and gap junction. When aided by genetic manipulation, the utility of FRIL can be expanded to include exploring protein interaction, trafficking and membrane stability in cell junctions.

References

Fujimoto, K. (1995). Freeze-fracture replica electron microscopy combined with SDS digestion for cytochemical labeling of integral membrane proteins. Application to the immunogold labeling of intercellular junctional complexes. *Journal of Cell Science*, *108*(Pt 11), 3443–3449.

Hou, J., Renigunta, A., Konrad, M., Gomes, A. S., Schneeberger, E. E., Paul, D. L., … Goodenough, D. A. (2008). Claudin-16 and claudin-19 interact and form a cation-selective tight junction complex. *The Journal of Clinical Investigation*, *118*, 619–628.

Rash, J. E., & Yasumura, T. (1999). Direct immunogold labeling of connexins and aquaporin-4 in freeze-fracture replicas of liver, brain, and spinal cord: Factors limiting quantitative analysis. *Cell and Tissue Research*, *296*, 307–321.

Chapter 7

Cell models of tight junction biology

Chapter 7.1

Cell culture

7.1.1 Primary culture and cell transformation

The process of initiating a culture from a human or animal tissue is known as primary culture. Primary culture may take the form of explant organ culture or single cell culture dissociated from organs by enzyme digestion. Primary cultures may be passaged for a finite number of cell doublings before senescence occurs. After a number of subcultures, a primary culture will either die out, referred to as a finite cell line, which is usually diploid, or transform into a continuous cell line. Transformed cells can be obtained from primary cultures by infecting them with oncogenic viruses or treating them with carcinogenic chemicals (Freshney & Freshney, 1996). It is often difficult to transform a primary culture from adult human tissues. In contrast, human neoplasms have generated many cell lines. It appears that the possession of a cancerous phenotype allows cells to overcome senescence. Transformed cells have the advantage of limitless growth, but they often retain very little of the original in vivo characteristics.

7.1.2 Subculture and propagation

Once a primary culture is subcultured or passaged, it is known as a cell line. The passage number is the number of times that the culture has been subcultured. It is often confused with the generation number, which denotes the number of doublings that the cell population has undergone. When the split ratio is 1:2, then the passage number equals the generation number. If the split ratio is greater than 1:2, then a power of 2, for example, 1:2, 1:4, 1:8 or

A Laboratory Guide to the Tight Junction. DOI: https://doi.org/10.1016/B978-0-12-818647-3.00007-6

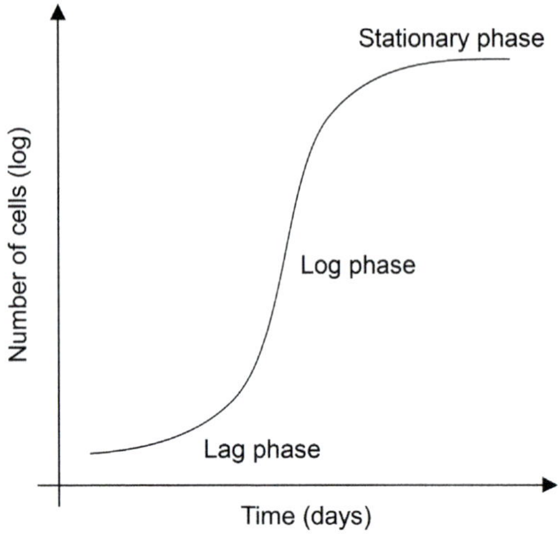

FIGURE 7.1.1 Cell growth curve. Lag phase: cells recover from subculture, attach to the substrate and start to spread. Log phase: cells grow exponentially and double at a characteristic rate defining the cell division time. Plateau phase: cell culture is confluent and cell growth slows or even ceases.

1:16, is often used to give the equivalent of one, two, three, or four generations, respectively. Following subculture, cells propagate through a characteristic growth pattern of the lag phase, the exponential or log phase, and the stationary or plateau phase (Fig. 7.1.1). The lag phase is the time following subculture during which there is little cell growth. It is a period of adaptation allowing cells to synthesize new proteins, attach to the substrate, and resume the cycle of cell division. The log phase is the period of exponential increase in cell number. The length of the log phase depends on the seeding density, the growth rate of the cells, and the availability of nutrients. The plateau phase is when the culture becomes confluent. At this stage, the cells stop proliferation and become immotile. The cessation of cell growth at the plateau phase was originally described as contact inhibition (Abercrombie & Heaysman, 1954). It has later been found that contact inhibition is not the only reason behind reduced cell growth. Depletion of nutrients, inhibition of cell spreading, and decrease in cell motility all play important roles (Dulbecco & Elkington, 1973; Stoker & Rubin, 1967; Stoker, 1973).

7.1.3 Anchorage independence

Many of the properties associated with cell transformation in vitro are the result of changes in cell adhesion. Transformed cells may lack specific groups of proteins important for cell—cell adhesion or cell-matrix adhesion, for example, cadherins, CAMs or integrins, which leads to detached cell growth and loss of contact inhibition (Hynes, 1992; Mege et al., 1988; Shapiro & Weis, 2009).

There is an obvious analogy between cell detachment in culture and tumor metastasis in vivo. The loss of contact inhibition may be detected morphologically by the formation of multiple cell layers or foci of rounded cells surrounded by normal cells. The loss of contact inhibition may also delimit the density control of cell growth. For example, cultures of human glioma can reach a higher saturation density than that of normal glial cells (Benda, Someda, Messer, & Sweet, 1971).

7.1.4 Cloning and selection

The traditional microbiological approach to the problem of cellular heterogeneity is to isolate pure cell strains by cloning. Although this technique is relatively easy for transformed cell lines, its success in primary cultures is limited due to cellular senescence. A few cell types have been successfully cloned from primary cultures. Kupffer cells were cloned from the primary culture of Chinese hamster liver (Clark & Pateman, 1978); Sertoli cells from the primary culture of rat testis (Zwain, Morris, & Cheng, 1991); juxtaglomerular cells from the primary culture of mouse kidney (Muirhead, Rightsel, Pitcock, & Inagami, 1990). Genetically modified cells can be selected by transfecting a drug resistance gene into the cells and then cloning the cells that are resistant to the drug, such as neomycin, hygromycin, or puromycin. Negative selection is also possible by using the Herpes simplex virus (HSV) TK gene, which activates Ganciclovir into a cytotoxic product. Transfected cells will be sensitized to the drug.

7.1.5 Gene transfer

In order to study the function of individual genes, the gene of interest can be cloned into a bacterial plasmid and then transfected into host cells by a variety of techniques, such as electroporation, calcium phosphate precipitation, and liposome encapsulation. Transfection may be transient or stable. Transient transfections are for short-term expression of the gene of interest, whereas stable transfections allow the gene of interest to integrate into the genome of host cells. The DNA used for stable transfection contains a selectable marker, such as the drug resistance gene, that confers resistance to neomycin, hygromycin, or puromycin. Viruses have a higher efficiency of gene transfer than DNA molecules. Among the commonly used viruses, retrovirus is able to permanently insert a gene of interest into the genome of host cells (Pear, Nolan, Scott, & Baltimore, 1993).

7.1.6 Cryopreservation

Cell lines in continuous culture are prone to variation due to clonal selection, genetic instability and cellular senescence. A primary culture or early-passage subculture should be frozen as what is called a token freeze. When a cloned cell strain is selected with the desired characteristics, then a seed stock should be frozen. The seed stock can be expanded to the distribution stock, which is then made available for general use. Storage in liquid nitrogen is currently the standard method of preserving cultured cells. The cell suspension, in the presence of a preservative such as glycerol or dimethyl sulfoxide, must be slowly frozen at ~1°C per minutes to −80°C, and rapidly transferred to liquid nitrogen (Ashwood-Smith & Farrant, 1980).

7.1.7 Contamination

Bacteria, yeasts, fungi, and mycoplasmas are all contaminants in cell culture. These microbial contaminations can be detected by light microscopic techniques (Segeritz & Vallier, 2017). Under a light microscope, bacteria appear as dark spheres, or spiral or rod-like structures (Fig. 7.1.2A). They may exist as single cells, or pairs, chains, or clusters of cells. Yeasts appear as round or ovoid particles that may bud off smaller particles (Fig. 7.1.2B). Fungi or molds produce thin filamentous mycelia and sometimes denser clumps of spores (Fig. 7.1.2C). Mycoplasmal infections can be tested by PCR, ELISA, or fluorescence microscopy. Fluorescent staining of mycoplasmal DNA by Hoechst 33258 is the simplest and the most reliable method used to reveal mycoplasmal infections in cell culture (Chen, 1977). The mycoplasmal DNA appears as a fine particulate or filamentous staining over the cytoplasm of cultured cells (Fig. 7.1.3). Microbial decontamination is difficult to achieve, so the general rule is that contaminated cultures are discarded immediately. Typically, there is an increase in contamination rate when the practice of aseptic techniques deteriorates, the incubators are less well maintained, or the culture media are not properly sterilized. The use of antibiotics can inhibit the growth of microbes but not completely kill them. The microbes

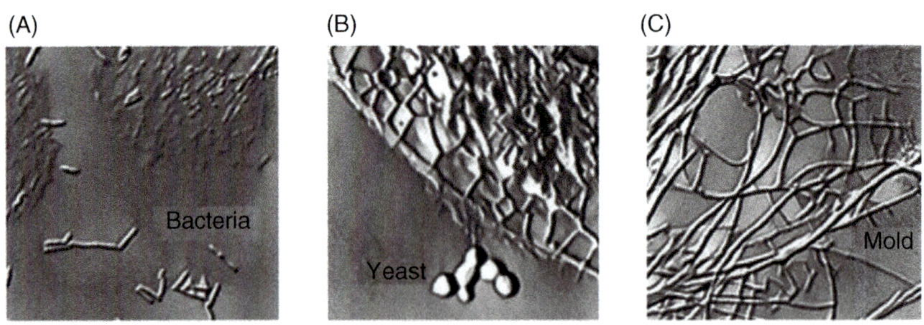

FIGURE 7.1.2 Microbial contaminants in cell culture. (A) The morphology of bacterial contamination under the microscope may vary from rod-like, spherical, to flagellated shapes. (B) Yeasts form multicellular ovoid particles. (C) Mold growth is marked by the production of multicellular, highly connected, thin filaments. *Reproduced with permission from Segeritz, C.-P., & Vallier, L. (2017). Cell culture: Growing cells as model systems in vitro. In* Basic science methods for clinical researchers *(pp. 151–172). Elsevier.*

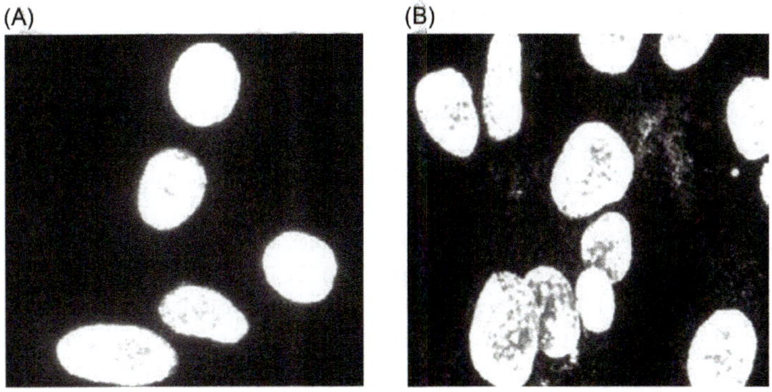

FIGURE 7.1.3 Mycoplasma. (A) Human normal fibroblasts stained with Hoechst 33258. (B) Human fibroblasts infected with mycoplasma and stained with Hoechst 33258. The nuclei fluoresce brightly from cellular DNA. Extranuclear fluorescence due to mycoplasma takes the bright particulate or filamentous form. *Reproduced with permission from Darin, N., Kadhom, N., Briere, J. J., Chretien, D., Bebear, C. M., Rotig, A., Munnich, A., & Rustin, P. (2003). Mitochondrial activities in human cultured skin fibroblasts contaminated by Mycoplasma hyorhinis.* BMC Biochemistry, 4, *15.*

may persist in the culture, undetected for most of the time, but periodically surfacing when culture conditions change. Therefore, it is essential that cultures are maintained in antibiotic-free conditions for at least some time in order to reveal the origin of potential contamination.

References

Abercrombie, M., & Heaysman, J. E. (1954). Observations on the social behaviour of cells in tissue culture. II. Monolayering of fibroblasts. *Experimental Cell Research*, *6*, 293–306.

Ashwood-Smith, M., & Farrant, J. (1980). *Low temperature preservation in medicine and biology*. University Park Press.

Benda, P., Someda, K., Messer, J., & Sweet, W. H. (1971). Morphological and immunochemical studies of rat glial tumors and clonal strains propagated in culture. *Journal of Neurosurgery*, *34*, 310–323.

Chen, T. R. (1977). In situ detection of mycoplasma contamination in cell cultures by fluorescent Hoechst 33258 stain. *Experimental Cell Research*, *104*, 255–262.

Clark, J. M., & Pateman, J. A. (1978). Long-term culture of Chinese hamster Kupffer cell lines isolated by a primary cloning step. *Experimental Cell Research*, *112*, 207–217.

Darin, N., Kadhom, N., Briere, J. J., Chretien, D., Bebear, C. M., Rotig, A., … Rustin, P. (2003). Mitochondrial activities in human cultured skin fibroblasts contaminated by *Mycoplasma hyorhinis*. *BMC Biochemistry*, *4*, 15.

Dulbecco, R., & Elkington, J. (1973). Conditions limiting multiplication of fibroblastic and epithelial cells in dense cultures. *Nature*, *246*, 197–199.

Freshney, R. I., & Freshney, M. G. (1996). *Culture of immortalized cells* (Vol. 4). Wiley-Liss.

Hynes, R. O. (1992). Integrins: versatility, modulation, and signaling in cell adhesion. *Cell*, *69*, 11–25.

Mege, R. M., Matsuzaki, F., Gallin, W. J., Goldberg, J. I., Cunningham, B. A., & Edelman, G. M. (1988). Construction of epithelioid sheets by transfection of mouse sarcoma cells with cDNAs for chicken cell adhesion molecules. *Proceedings of the National Academy of Sciences of the United States of America*, *85*, 7274–7278.

Muirhead, E. E., Rightsel, W. A., Pitcock, J. A., & Inagami, T. (1990). Isolation and culture of juxtaglomerular and renomedullary interstitial cells. *Methods in Enzymology*, *191*, 152–167.

Pear, W. S., Nolan, G. P., Scott, M. L., & Baltimore, D. (1993). Production of high-titer helper-free retroviruses by transient transfection. *Proceedings of the National Academy of Sciences*, *90*, 8392–8396.

Segeritz, C.-P., & Vallier, L. (2017). *Cell culture: growing cells as model systems in vitro. Basic science methods for clinical researchers* (pp. 151–172). Elsevier.

Shapiro, L., & Weis, W. I. (2009). Structure and biochemistry of cadherins and catenins. *Cold Spring Harbor Perspectives in Biology*, *1*, a003053.

Stoker, M. G. (1973). Role of diffusion boundary layer in contact inhibition of growth. *Nature*, *246*, 200–203.

Stoker, M. G., & Rubin, H. (1967). Density dependent inhibition of cell growth in culture. *Nature*, *215*, 171–172.

Zwain, I. H., Morris, P. L., & Cheng, C. Y. (1991). Identification of an inhibitory factor from a Sertoli clonal cell line (TM4) that modulates adult rat Leydig cell steroidogenesis. *Molecular and Cellular Endocrinology*, *80*, 115–126.

Chapter 7.2

Culture of epithelial cells

7.2.1 Background knowledge

7.2.1.1 Epithelial phenotypes

Epithelial cells display an apicobasal polarity (Simons & Fuller, 1985). In these cells, the plasma membrane partitions into two distinct domains—the apical membrane and the basolateral membrane, each consisting in a unique set of proteins. Tight junction is found at the intersection between the apical and the lateral membranes, which prevent the lateral diffusion of membrane proteins from one domain to another. Tight junction also joins each cell in an epithelium to form a diffusion barrier, which limits the diffusion of molecules between the apical and the basolateral spaces (Hou, 2018). Immediately basal to the tight junction is adherens junction, which mediates cell adhesion via cadherin-based protein interactions (Meng & Takeichi, 2009). The more basal cell junctions include desmosome and gap junction, which connect the lateral membranes of adjacent cells and create intercellular ion channels to communicate between the adjacent cells, respectively (Delva, Tucker, & Kowalczyk, 2009; Goodenough & Paul, 2009). A key factor for optimal expression of the epithelial phenotypes in vitro is the polarity of nutrient uptake (Simons & Fuller, 1985). In vivo, nutrients reach the epithelial cells from the basolateral side, which faces the blood supply, however, when epithelial cells are cultured on glass or plastic substrate, they are forced to feed from the apical surface, which faces the culture medium. Hence, the apicobasal polarity is not possible to establish. To polarize properly, the epithelial cells must be grown on permeable supports, which expose the basolateral membrane to nutrients and growth factors (Cereijido, Robbins, Dolan, Rotunno, & Sabatini, 1978; Misfeldt, Hamamoto, & Pitelka, 1976).

7.2.1.2 Epithelial cell lines

Typically, a primary culture can only undergo a small number of passages before senescence takes place. There are, however, several epithelial cell lines that spontaneously become immortal (Table 7.2.1). Among them, the Madin-Darby canine kidney (MDCK) cell is the best-studied epithelial cell line (Rindler et al., 1979). Morphologically, the MDCK cell resembles a cuboidal epithelium with microvilli on the apical surface. Two different strains of the MDCK cell are now in use (Richardson, Scalera, & Simmons, 1981). Strain I cells derive from a low-passage culture and these cells form a tight epithelium with transepithelial resistance (TER) above 2,000 Ω cm^2.

TABLE 7.2.1 Examples of commonly used immortalized epithelial cell lines.

Cell line	Organ source	Tumor origin	Transformation	Reference
MDCK	Dog kidney	No	No	Rindler, Chuman, Shaffer, and Saier (1979)
LLC-PK1	Pig kidney	No	No	Hull, Cherry, and Weaver (1976)
OK	Opossum kidney	No	No	Koyama, Goodpasture, Miller, Teplitz, and Riggs (1978)
mIMCD-3	Mouse kidney	No	SV40	Rauchman, Nigam, Delpire, and Gullans (1993)
HK-2	Human kidney	No	HPV-16	Ryan et al. (1994)
Eph4	Mouse mammary gland	Yes	Yes	Reichmann, Ball, Groner, and Friis (1989)
Caco-2	Human colon	Yes	Yes	Hidalgo, Raub, and Borchardt (1989)
HT-29	Human colon	Yes	Yes	Le Bivic, Hirn, and Reggio (1988)

Strain II cells form a leaky epithelium with low TER of 100–200 Ω cm^2. MDCK strain II cells have been used primarily for studies of the transport physiology of tight junction because of their low TER values. Transcytosis, on the other hand, is often studied in MDCK strain I cells owing to their high electrical resistance.

7.2.2 Materials and reagents

7.2.2.1 Equipment

Corning Transwell (Table 3.2.1).

7.2.2.2 Cell model

MDCK type II epithelial cells; available from ATCC.

7.2.2.3 Cell culture medium

Complete medium

500 mL	Dulbecco's Modified Eagle's medium (DMEM) (high glucose/ + sodium pyruvate/ + glutamine); available from Gibco
50 mL	Fetal bovine serum (FBS); available from Gibco
5 mL	Penicillin/streptomycin; available from Gibco

1 × Phosphate-buffered saline (PBS); available from Gibco
0.05% Trypsin/0.02% EDTA; available from Gibco

7.2.3 Experimental procedure

7.2.3.1 Growing MDCK cells on plastic substrate

1. Warm all solutions to 37°C. When cells are ready for splitting, aspirate the medium from the 75-cm^2 flask containing a confluent layer of MDCK cells. Add 10 mL of 1 × PBS, rinse and aspirate the solution.
2. Add 3 mL of trypsin-EDTA solution and incubate at 37°C for 15–20 minutes. Observe the cells under a light microscope every 5 minutes until they are rounded up but not yet detached from the flask.
3. Add 10 mL of prewarmed complete medium to the flask and resuspend the cells by gently pipetting up and down for several times.
   (*Caution: do not hit or shake the flask because mechanical stress may damage the cells.*)
4. Centrifuge the cell suspension at 250 × *g* for 5 minutes.
5. Aspirate the supernatant and resuspend the cell pellet with 10 mL of prewarmed complete medium.
6. Plate 2 mL of the cell suspension in a new 75-cm^2 flask containing 15 mL of prewarmed complete medium.

7.2.3.2 Seeding MDCK cells on permeable Transwell filter

1. Trypsinize MDCK cells as described in Section 7.2.3.1.
2. After centrifugation, resuspend the cell pellet with 2 mL instead of 10 mL of prewarmed complete medium.
3. Count the total number of cells in the suspension with a hemocytometer.
4. Load the cells into the Transwell filter insert (12-mm diameter) at 1×10^6 per insert. Add culture medium to the interior and the exterior compartments of the Transwell.
5. Culture for 9 consecutive days and change medium every 3 days.
6. Record the transepithelial resistance of each Transwell with an epithelial ohmmeter (see Chapter 3.4: Epithelial ohmmeter).

7.2.4 Data analysis

The epithelial cells grown on the Transwell filter cannot be observed in the inverted microscope because the filter is not transparent. To assess the integrity of the epithelial cell layer on the filter, two approaches are often used, for example, electrophysiological recording and immunofluorescence labeling. The transepithelial resistance in MDCK strain II cells increases for approximately 3–4 days after seeding before it reaches a peak of 100 Ω cm^2. After peaking, the resistance value slightly decreases to enter a plateau phase of 90 Ω cm^2 (Fig. 7.2.1). The development of electrical resistance is due to the formation of tight junction as a diffusion barrier. The tight junction itself can be immunolabeled with anti-TJ protein antibodies and visualized by confocal microscopy (Chapter 5.4: Confocal microscopy for cells on Transwell).

7.2.5 Troubleshooting

It is recommended that MDCK cells not be used for more than 20 passages. New seed stock cells should be thawed from liquid nitrogen to establish additional distribution stocks of cells. The seeding density is important for cell growth on the Transwell filter. Cells are seeded onto the Transwell filter at a higher density than that achieved by confluent cells on plastic substrate. The cells on the Transwell filter form tight junctions within 24 hours and reach maximal tightness in 4 days. During this time, the density of cells increases to more than five times that achieved on plastic substrate. In the theory, the size of Transwell filter is irrelevant to how cells form tight junctions. In practice, however, epithelia tend to be tighter on smaller Transwells. Conceivably, larger Transwells may contain more cell lumps or gaps due to cell detachment. Larger Transwells are more difficult to handle

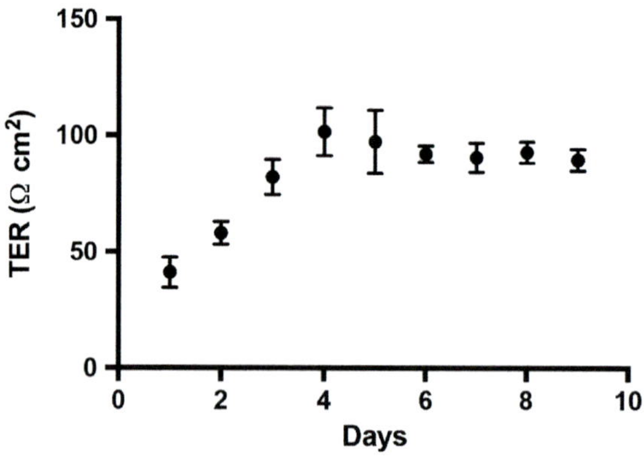

FIGURE 7.2.1 Time course of TER recording in MDCK cells. MDCK strain II cells were seeded onto Transwell at Day 0 and recorded every day from Day 1 to Day 9. The TER value peaked at Day 4 and then reached a plateau of 90 Ω cm^2. $N = 4$ Transwells.

and require more frequent change of medium. The risk of contamination is also higher when the area for cell growth increases.

7.2.6 Concluding remarks

For most studies on epithelial cell biology, cultured cells have been used. These models are superior to primary cultures obtained from tissues because they can be grown under a defined condition and can be easily manipulated, both pharmacologically and genetically. The cell population in a cloned cell line is homogeneous. The tight junction proteins can be purified and analyzed from such a homogeneous culture. The structure and function of tight junction can be investigated in cultured cells with a large collection of imaging and recording tools designed for in vitro research.

References

Cereijido, M., Robbins, E. S., Dolan, W. J., Rotunno, C. A., & Sabatini, D. D. (1978). Polarized monolayers formed by epithelial cells on a permeable and translucent support. *The Journal of Cell Biology*, *77*, 853–880.

Delva, E., Tucker, D. K., & Kowalczyk, A. P. (2009). The desmosome. *Cold Spring Harbor Perspectives in Biology*, *1*, a002543.

Goodenough, D. A., & Paul, D. L. (2009). Gap junctions. *Cold Spring Harbor Perspectives in Biology*, *1*, a002576.

Hidalgo, I. J., Raub, T. J., & Borchardt, R. T. (1989). Characterization of the human colon carcinoma cell line (Caco-2) as a model system for intestinal epithelial permeability. *Gastroenterology*, *96*, 736–749.

Hou, J. (2018). *The paracellular channel – biology, physiology and disease*. Academic Press.

Hull, R. N., Cherry, W. R., & Weaver, G. W. (1976). The origin and characteristics of a pig kidney cell strain, LLC-PK. *In vitro*, *12*, 670–677.

Koyama, H., Goodpasture, C., Miller, M. M., Teplitz, R. L., & Riggs, A. D. (1978). Establishment and characterization of a cell line from the American opossum (*Didelphys virginiana*). *In vitro*, *14*, 239–246.

Le Bivic, A., Hirn, M., & Reggio, H. (1988). HT-29 cells are an in vitro model for the generation of cell polarity in epithelia during embryonic differentiation. *Proceedings of the National Academy of Sciences of the United States of America*, *85*, 136–140.

Meng, W., & Takeichi, M. (2009). Adherens junction: molecular architecture and regulation. *Cold Spring Harbor Perspectives in Biology*, *1*, a002899.

Misfeldt, D. S., Hamamoto, S. T., & Pitelka, D. R. (1976). Transepithelial transport in cell culture. *Proceedings of the National Academy of Sciences of the United States of America*, *73*, 1212–1216.

Rauchman, M. I., Nigam, S. K., Delpire, E., & Gullans, S. R. (1993). An osmotically tolerant inner medullary collecting duct cell line from an SV40 transgenic mouse. *The American Journal of Physiology*, *265*, F416–F424.

Reichmann, E., Ball, R., Groner, B., & Friis, R. R. (1989). New mammary epithelial and fibroblastic cell clones in coculture form structures competent to differentiate functionally. *The Journal of Cell Biology*, *108*, 1127–1138.

Richardson, J. C., Scalera, V., & Simmons, N. L. (1981). Identification of two strains of MDCK cells which resemble separate nephron tubule segments. *Biochimica et Biophysica Acta, 673*, 26–36.

Rindler, M. J., Chuman, L. M., Shaffer, L., & Saier, M. H., Jr. (1979). Retention of differentiated properties in an established dog kidney epithelial cell line (MDCK). *The Journal of Cell Biology, 81*, 635–648.

Ryan, M. J., Johnson, G., Kirk, J., Fuerstenberg, S. M., Zager, R. A., & Torok-Storb, B. (1994). HK-2: an immortalized proximal tubule epithelial cell line from normal adult human kidney. *Kidney International, 45*, 48–57.

Simons, K., & Fuller, S. D. (1985). Cell surface polarity in epithelia. *Annual Review of Cell Biology, 1*, 243–288.

Chapter 7.3

Calcium switch assay

7.3.1 Background knowledge

Cereijido and coworkers first demonstrated that calcium chelation with EGTA opened the paracellular pathway in epithelial cells with an experimental approach that later became a standard assay known as the calcium switch assay (Cereijido, Robbins, Dolan, Rotunno, & Sabatini, 1978). Electron microscopy revealed that EGTA treatment disrupted the apical cell junctions including adherens junction and tight junction, and widened the intercellular cleft (Cereijido et al., 1978). It is clear from the crystal structure of cadherin that cell adhesion in the adherens junction depends upon the presence of Ca^{++} (Meng & Takeichi, 2009). While cell adhesion in the tight junction is not Ca^{++} dependent (Kubota et al., 1999), a crosstalk mechanism has long been speculated to operate between adherens junction and tight junction (Campbell, Maiers, & DeMali, 2017). Disruption of adherens junction by calcium chelation results in the breakdown of tight junction barrier within minutes to hours (Cereijido et al., 1978; Tobey, Argote, Hosseini, & Orlando, 2004). Calcium replenishment can restore the barrier function of tight junction, indicating that tight junction breakdown is a reversible process (Martinez-Palomo, Meza, Beaty, & Cereijido, 1980).

7.3.2 Materials and reagents

7.3.2.1 Equipment

Corning Transwell (Table 3.2.1).

7.3.2.2 Cell model

MDCK type II epithelial cells; available from ATCC.

7.3.2.3 Cell culture medium

Complete medium

500 mL	Dulbecco's Modified Eagle's medium (DMEM) (high glucose/ + sodium pyruvate/ + glutamine); available from Gibco
50 mL	Fetal bovine serum (FBS); available from Gibco
5 mL	Penicillin/streptomycin; available from Gibco

TABLE 7.3.1 Ca^{++}-switch buffers.

Content	Physio-buffer	Chelating-buffer
Calculated osmolality (mosm/kg)	300	300
NaCl	145	145
$CaCl_2$	2	0
$MgCl_2$	1	1
Glucose	10	10
EGTA	0	3
HEPES	10	10

Concentration is in mmol/L; pH is adjusted to 7.4 by NaOH or HCl.

7.3.2.4 Ca^{++}-switch buffers

The Ca^{++}-switch buffers are based upon the Ringer solution (Table 7.3.1). HEPES is used as a buffering agent. Glucose is included as an energy source. EGTA is used to chelate Ca^{++}.

7.3.3 Experimental procedure

1. Trypsinize MDCK cells as described in Section 7.2.3.1.
2. After centrifugation, resuspend the cell pellet with 2 mL of prewarmed complete medium.
3. Count the total number of cells in the suspension with a hemocytometer.
4. Load the cells into the Transwell filter insert (12-mm diameter) at 1×10^6 per insert. Add culture medium to the interior and the exterior compartments of the Transwell.
5. Culture for 9 consecutive days and change medium every 3 days.
6. Aspirate medium from both the interior and the exterior compartments of the Transwell.
7. Add prewarmed Physio-buffer to both the interior and the exterior compartments of the Transwell.
8. Record the transepithelial resistance of each Transwell with an epithelial ohmmeter (see Chapter 3.4: Epithelial ohmmeter).
9. Aspirate Physio-buffer from both the interior and the exterior compartments of the Transwell.
10. Add prewarmed Chelating-buffer to both the interior and the exterior compartments of the Transwell.

11. Record the transepithelial resistance of each Transwell with an epithelial ohmmeter every 10 minutes for 30 minutes. Incubate the Transwells at 37°C.
12. Aspirate Chelating-buffer from both the interior and the exterior compartments of the Transwell.
13. Add prewarmed Physio-buffer to both the interior and the exterior compartments of the Transwell.
14. Record the transepithelial resistance of each Transwell with an epithelial ohmmeter every hour for 4 hours. Incubate the Transwells at 37°C.

7.3.4 Data analysis

Chelation of Ca^{++} caused a sharp drop in transepithelial resistance (TER) within the first 10 minutes, followed by a phase of slower decrease to near $0\ \Omega\ cm^2$ (Fig. 7.3.1). Replacement with normal Ca^{++}-containing solution slowly restored the transepithelial resistance to approximately 60% of the baseline level by 4 hours (Fig. 7.3.1). The slow recovery of TER after Ca^{++} replenishment likely reflects the process of de novo biogenesis of tight junction. The localization of TJ proteins during the calcium switch assay can be analyzed by the light microscopic approaches described in Chapter 5, Light microscopy for tight junction. The integrity of tight junction during the calcium switch assay can be analyzed by the electron microscopic approaches described in Chapter 6, Electron microscopy for tight junction.

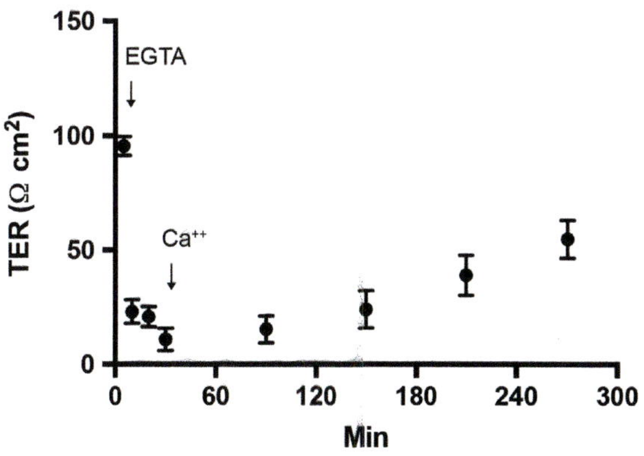

FIGURE 7.3.1 Calcium switch assay in MDCK cells. MDCK strain II cells were seeded onto Transwell and cultured for 9 consecutive days. The TER was first recorded in Physio-buffer at time 0. Immediately after recording, the Physio-buffer was replaced with the Chelating-buffer and the TER was recorded every 10 min for 30 min. At 30 min, the Chelating-buffer was replaced with the Physio-buffer and the TER was recorded every hour for 4 h. $N = 4$ Transwells.

7.3.5 Troubleshooting

The short-term effect of Ca^{++} chelation appears to be reduction in electrical resistance but no ultrastructural alteration in tight junction itself (Martinez-Palomo et al., 1980). Prolonged depletion of Ca^{++} (i.e., >15 minutes) may disassemble the tight junction and create gaps between the cell junction membranes (Cereijido et al., 1978). The recovery phase of tight junction structure and function is therefore dependent upon the extent of initial tight junction breakdown (Martinez-Palomo et al., 1980). The longer the tight junction is left open, the longer it will take to reseal the tight junction. Notably, the TER response to calcium switch differs among cells and tissues. A series of preliminary studies have to be undertaken to plot the baseline TER response for each cell or tissue model before genetic or pharmacologic testing is under way.

7.3.6 Concluding remarks

Calcium switch assay is the simplest approach that can be used to open and close the tight junction in a controllable manner. It offers the opportunity to study the factors that participate in the disassembly and reassembly processes of tight junction. This system can also give valuable morphological information on the ultrastructural alteration associated with the disintegration and reconstitution of tight junction. With all that being said, it is important to acknowledge that calcium switch assay is not specific for tight junction. It acts by triggering a chain reaction that starts at the adherens junction. As a result, any mechanistic explanation must include aspects from both types of cell junction.

References

Campbell, H. K., Maiers, J. L., & DeMali, K. A. (2017). Interplay between tight junctions & adherens junctions. *Experimental Cell Research*, *358*, 39–44.

Cereijido, M., Robbins, E. S., Dolan, W. J., Rotunno, C. A., & Sabatini, D. D. (1978). Polarized monolayers formed by epithelial cells on a permeable and translucent support. *The Journal of Cell Biology*, *77*, 853–880.

Kubota, K., Furuse, M., Sasaki, H., Sonoda, N., Fujita, K., Nagafuchi, A., & Tsukita, S. (1999). Ca(2 +)-independent cell-adhesion activity of claudins, a family of integral membrane proteins localized at tight junctions. *Current Biology: CB*, *9*, 1035–1038.

Martinez-Palomo, A., Meza, I., Beaty, G., & Cereijido, M. (1980). Experimental modulation of occluding junctions in a cultured transporting epithelium. *The Journal of Cell Biology*, *87*, 736–745.

Meng, W., & Takeichi, M. (2009). Adherens junction: molecular architecture and regulation. *Cold Spring Harbor Perspectives in Biology*, *1*, a002899.

Tobey, N. A., Argote, C. M., Hosseini, S. S., & Orlando, R. C. (2004). Calcium-switch technique and junctional permeability in native rabbit esophageal epithelium. *American Journal of Physiology Gastrointestinal and Liver Physiology*, *286*, G1042–G1049.

Chapter 7.4

Retrovirus-mediated transgene expression

7.4.1 Background knowledge

7.4.1.1 Retrovirus-mediated gene transfer

Retroviruses are among the most competent vehicles available for introducing genes into mammalian cells. Efficiencies of gene transduction by retroviruses can reach 100%, and once integrated, the viral genome becomes a permanent component in the host cell's genome. Retroviral infection occurs seamlessly as a sequence of normal cellular events, which poses no harm to the host cell. These features make retrovirus a superior vehicle for gene transfer as compared to other techniques, such as calcium phosphate transfection, electroporation, lipid fusion, and microinjection.

7.4.1.2 Recombinant retroviral vector

Most recombinant retroviral viruses are derived from the Moloney murine leukemia virus (Fig. 7.4.1). A large portion of the viral genome encoding the *gag*, *pol*, and *env* genes are replaced with one or more mammalian gene expression cassettes. The expression cassettes typically contain a suitable promoter, such as CMV or ubiquitin promoter; a selectable marker gene, such as *neo*, *GFP*, or *lacZ*; and a multiple cloning site allowing the gene of interest to be inserted (Morgenstern & Land, 1990). The recombinant viruses are replication defective because they do not express the structural and enzymatic proteins required to package the virions. This feature can prevent the host cells from developing the pathology induced by a progressing retroviral life cycle.

7.4.1.3 Packaging of retrovirus

Recombinant retroviral genome is constructed into a bacterial plasmid DNA, known as vector. When introduced into cells, the retroviral genome is transcribed, and the transcripts are packaged into infectious virions (Fig. 7.4.2). As the *gag*, *pol*, and *env* genes are missing from the recombinant vector, the packaging function of these proteins must be provided by a helper virus or by additional vectors that express the proteins. The cells that can reliably package replication-defective recombinant retroviruses are

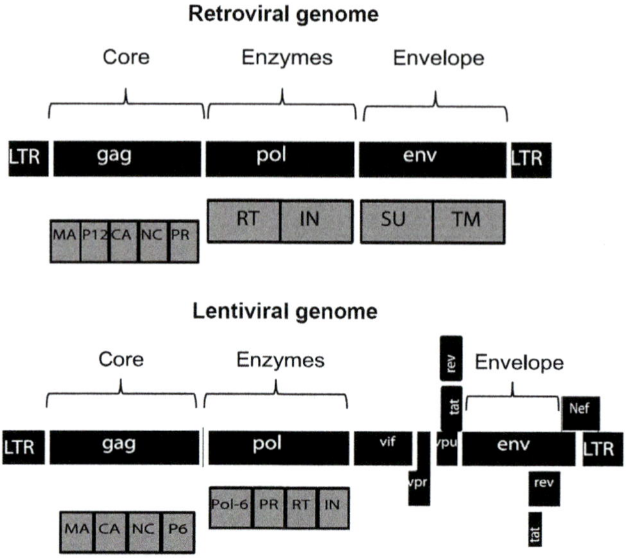

FIGURE 7.4.1 Retroviral genome. Architectures of retroviral genome and lentiviral genome are illustrated. Lentivirus is a genus of more complex retrovirus. Note that both types of viruses are the RNA virus, because it is the RNA molecules, instead of the DNA molecules, that are packaged into the virions. The core components of retroviral genome consist in *gag*, *pol* and *env* genes flanked by two long terminal repeat (LTR) sequences. The lentiviral genome includes an additional gene vital for its replication, known as the *rev* gene. *Reproduced with permission from Vargas, J. E., Chicaybam, L., Stein, R. T., Tanuri, A., Delgado-Canedo, A., & Bonamino, M. H. (2016). Retroviral vectors and transposons for stable gene therapy: Advances, current challenges and perspectives.* Journal of Translational Medicine, 14, *288.*

known as the producer cells (Miller, Miller, Garcia, & Lynch, 1993). The producer cells are established upon two types of cells—HEK293 and NIH 3T3.

7.4.1.4 Pseudotyping with VSV-G protein

The viral *env* gene encodes the envelope protein, which determines the range of infectivity (tropism) of retrovirus. Viral envelopes are classified according to the receptors used to enter the host cells. For example, ecotropic virus can recognize a receptor found on only mouse and rat cells. Amphotropic virus recognizes a receptor found on a broad range of mammalian cell types. Dualtropic virus recognizes two different receptors found on a broad range of mammalian cell types. Pantropic virus, when pseudo-typed with the envelope glycoprotein from the vesicular stomatitis virus (VSV-G), can infect all types of mammalian cells and some types of nonmammalian cells (Fig. 7.4.2) (Burns, Friedmann, Driever, Burrascano, & Yee, 1993). Unlike other viral envelope proteins, VSV-G mediates viral

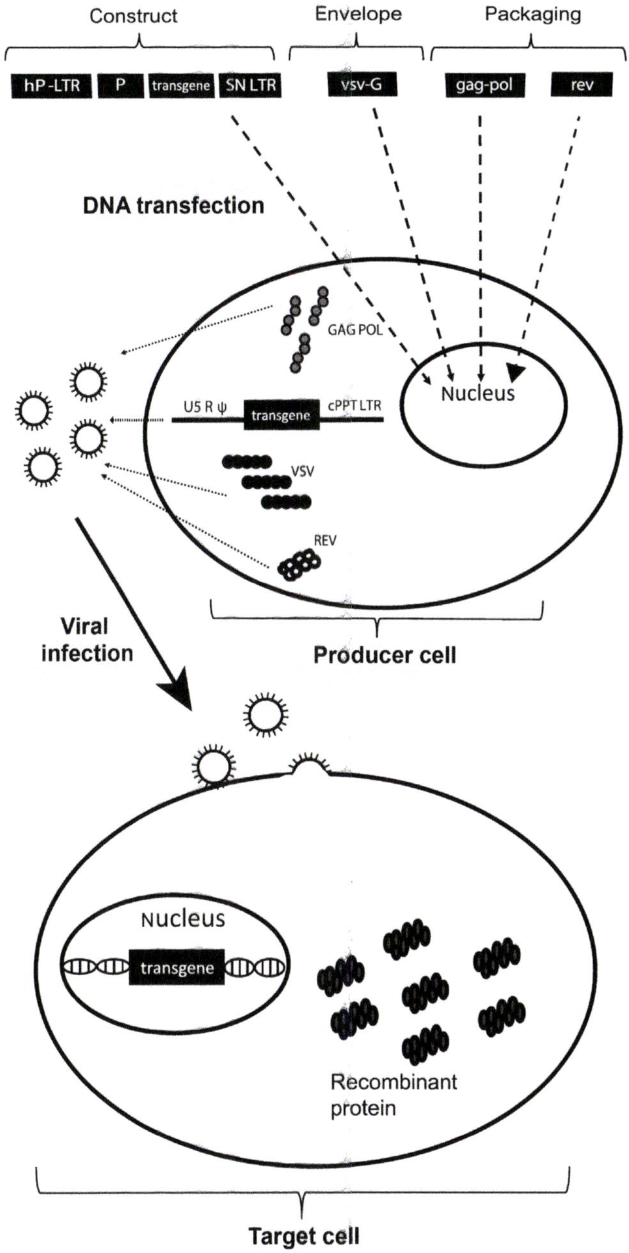

FIGURE 7.4.2 Packaging and infection of retrovirus. Recombinant retroviral vectors are cotransfected with the *gag-pol*, *vsvG*, and *rev* (in the case of lentivirus) genes into the producer cells. These vectors contain a transgene expression cassette in place of the structural and enzymatic genes in the retroviral genome, which ensures the packaged viruses are replication (*Continued*)

entry through lipid binding and plasma membrane fusion (Emi, Friedmann, & Yee, 1991). Because expression of the VSV-G protein is toxic to cells, pantropic virus can only be packaged by transiently transfecting the *VSV-G* gene into the producer cells.

7.4.2 Materials and reagents

7.4.2.1 Cell model for ectopic gene expression

HEK293 cells (*producer cell*); available from ATCC.
MDCK strain II cells (*host cell*); available from ATCC.

7.4.2.2 Plasmids

pQCXIN retroviral expression vector (pQCXIN-cldn16); available from Clontech.
pGag-pol expression vector; available from Addgene.
pVSV-G expression vector; available from Addgene.

7.4.2.3 Cell culture medium

Complete medium

500 mL	Dulbecco's Modified Eagle's medium (DMEM) (high glucose/ + sodium pyruvate/ + glutamine); available from Gibco
50 mL	Fetal bovine serum (FBS); available from Gibco
5 mL	Penicillin/streptomycin; available from Gibco

7.4.2.4 Buffers

$CaCl_2$ (2.5 M)	
27.7 g	$CaCl_2$

Bring to 100 mL with dH_2O, and stir to dissolve.
Filter, sterilize, and store at 4°C

◀ defective. A packaging signal (ψ) allows the transcripts to be packaged into virions with high efficiency. The envelope protein, VSV-G, determines the tropism of packaged retroviruses. Viral infection begins by specific binding to the plasma membrane of the target cells. Following endocytosis and uncoating of viral particles, the viral genome of RNA is reverse-transcribed into DNA and integrated into the genome of the target cells. The target cells now express the transgene into recombinant proteins. *Reproduced with permission from Vargas, J. E., Chicaybam, L., Stein, R. T., Tanuri, A., Delgado-Canedo, A., & Bonamino, M. H. (2016). Retroviral vectors and transposons for stable gene therapy: Advances, current challenges and perspectives.* Journal of Translational Medicine, 14, *288*.

2× HBS buffer	
140 mM	NaCl
1.5 mM	Na_2HPO_4
50 mM	HEPES

Bring to 100 mL with dH_2O, stir to dissolve, and adjust pH to 7.0. Filter, sterilize, and store at 4°C

TNE buffer	
130 mM	NaCl
50 mM	Tris
1 mM	EDTA

Bring to 100 mL with dH_2O, stir to dissolve, and adjust pH to 7.8. Filter, sterilize, and store at 4°C.

7.4.3 Experimental procedure

Day 1

1. Grow HEK293 cells to confluence in one T225 flask.
2. Trypsinize the T225 flask, centrifuge the cells at 250 × *g* for 5 minutes at room temperature, and resuspend cells in 5 mL DMEM complete medium.
3. Prepare four 150 mm culture dishes, and add 20 mL DMEM complete medium to each dish.
4. Add 1 mL resuspended cells to each dish.
5. Return the cells to incubator and culture for overnight (12–16 hours) at 37°C.

Day 2

1. Take the dishes from the incubator, and change fresh medium.
   (*Because HEK293 cells are fragile, tilt the flask when removing the medium and adding fresh medium so that the cell monolayer is not disturbed.*)
2. Mix the following DNAs (made w/Endo-free Qiagen midi-Kits) in a 15 mL centrifuge tube.

20 μg	pQCXIN-cldn16
20 μg	pGag-pol
20 μg	pVsvG
x μL	nuclease-free H_2O
1.9 mL	total volume

3. Add 100 μL 2.5 M $CaCl_2$ to DNA mixture. Vortex to mix.
4. In a separate 15 mL centrifuge tube, add 1 mL of 2 × HBS.

5. Add DNA mixture (Step 2) to 2 × HBS (Step 3) dropwise with a Pasteur pipette. Flick the HBS tube while adding DNA mixture. When finishing adding, vortex gently (using low speed) for 20 seconds.
   (This is sufficient to transfect one 150 mm dish.)
6. Let the DNA-HBS mixture stand for 20 minutes at room temperature.
7. Add 2 mL DNA-HBS mixture dropwise to a 150 mm flask.
   (When adding the mixture, hold the pipette stable while swirling the flask slowly so that the drops can be distributed evenly in different areas of the flask. After adding the mixture, continue swirling the flask for 20 seconds so that the DNA mixture can fully dissolve into the medium.)
8. Return the flasks into the incubator for 12–16 hours.

Day 3

1. Take the dishes of transfected HEK293 cells out of the incubator, and change fresh medium. Return the flasks into the incubator for 48 hours.

Day 4

1. Trypsinize a dish or flask of target cells—MDCK type II cells.
2. Load 1×10^4 trypsinized cells into a well of the 6-well plate so that on day 5, the cells will reach ~30% confluence.

Day 5

1. Sterilize 50 mL ultra-speed centrifuge tubes by UV or autoclaving.
2. Harvest viral supernatant and centrifuge at 500 × *g* for 5 minutes at 4°C.
3. Transfer supernatant to ultra-speed centrifuge tubes and centrifuge at 50,000 × *g* for 2 hours at 4°C.
4. After centrifuge, a white-to-yellow pellet can be seen at the bottom of the tube.
5. Aspirate the supernatant and add 200 μL of ice-cold TNE buffer to cover the pellet. Resuspend the pellet by pipetting up and down for 50 times.
6. Take the target cells from incubator (~30% confluence) and change with fresh ice-cold medium.
7. Add dissolved virus to the target cells at 1 μL of virus for one well of cells.
   (The virus can be serially diluted from 1:1 to 1:10^6 to determine the titer).
8. Return cells to incubator for 24 hours.

Day 6

1. Take the infected cells from incubator and aspirate the medium.
2. Wash the cells once with 1 × PBS.
3. Add fresh medium.
4. Return to incubator for 24 hours.

Day 7

1. Trypsinize the cells from the 6-well plate.
   (The trypsin step is essential because cell division is required for the integration of retrovirus into host cell's genome.)
2. Load the cells into T25 flask. Add fresh medium.
3. Return to incubator for 24 hours.

Day 8

1. Aspirate the medium and add fresh medium.
2. Add drugs such as neomycin to select the infected cells.
3. Determine the titer of retrovirus.

7.4.4 Data analysis

The viral titer corresponds to the number of colonies present at the highest dilution that contains colonies, multiplied by the dilution factor. For example, the presence of four colonies after 10^6 dilution from 1 μL viral stock represents a viral titer of 4×10^6 colony forming units (cfu) per μL.

$$4 \times 10^6\ \mathrm{cfu}/\mu L = 4 \times 10^9\ \mathrm{cfu}/\mu L$$

High-level and homogeneous expression of claudin proteins in target cells can be achieved at a viral titer of 1×10^9 cfu/mL (Fig. 7.4.3). Because the transgene carried by the virus is integrated into host cell's genome, the expression of transgene is stable over time and cell division. This feature has proven to be particularly important for epithelial cells grown on permeable Transwell filters (see Chapter 7.2: Culture of epithelial cells). Because it may take from 5 to 10 days for epithelial cells to polarize and develop tight junctions on the Transwell filters, transient gene expression approaches, such as DNA transfection and microinjection, are not suited for this application. Compared to other approaches to establish stable cell lines, such as clonal selection, retrovirus offers a unique advantage in that it bypasses the step of subculture. Subculture is not only time consuming but may also introduce phenotypic variation.

7.4.5 Troubleshooting

7.4.5.1 Low viral titer

While retrovirus is stable at 4°C, it is surprisingly unstable at 37°C. To maintain the titer of virus, it is important to keep the viral stocks on ice or frozen before infection. Freezing at −80°C can preserve the virus for a long period. However, each freeze-thaw cycle will result in a 20%−50% drop in the viral titer. Several factors may influence the production of virus, for example, the viability of producer cell, the time point for harvest, and the size of viral

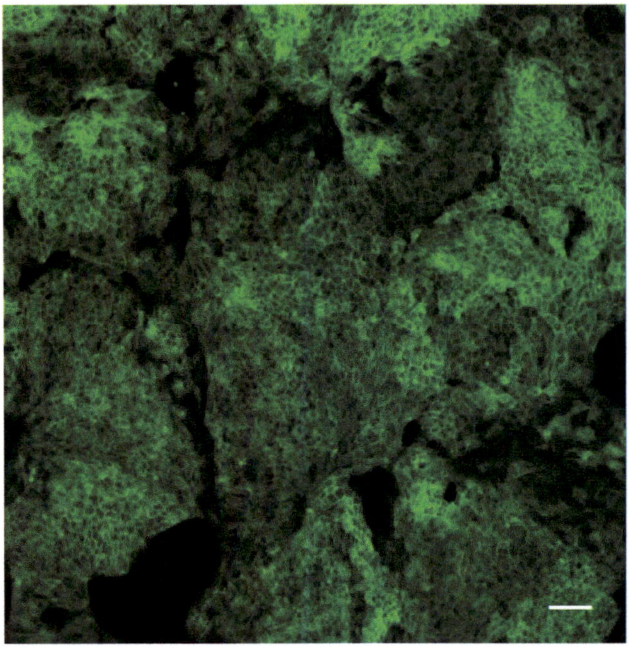

FIGURE 7.4.3 Retrovirus-mediated claudin expression in MDCK strain II cells. Immunofluorescence microscopic image (at low magnification) shows that over 95% infected MDCK II cells express claudin-16 proteins in a homogenous manner without further clonal or antibiotic selection. Bar: 100 μm. *Reproduced with permission from Hou, J., Paul, D. L., & Goodenough, D. A. (2005). Paracellin-1 and the modulation of ion selectivity of tight junctions.* Journal of Cell Science, 118, *5109–5118.*

genome. In the theory, there is a limit to the size of viral genome that can be packaged into virions. If the transgene is too large, that is, >8 kb, then the packaging efficiency for virus drops significantly. Generally speaking, the highest production of virus is from the time when the producer cells reach confluence to 5 days after. The viral supernatant can be collected every 24 hours without significant loss of titer by replacing with fresh cell culture medium. Although serum is not required for the production of virus, it does play a vital role in cell viability. The type of serum has to be independently tested for each kind of producer cell to sustain the optimal level of metabolism.

7.4.5.2 Low transduction efficiency

Besides a good titer, the most important factor to ensure high transduction efficiency of retrovirus is to let the target cells divide because retrovirus can only gain access to the nucleus during mitosis (Miller, Adam, & Miller, 1990). The viral transduction efficiency can also be enhanced by the use of polycations such as Polybrene and DEAD-dextran, which promote the attachment of virus to the plasma membrane of target cells. Finally, increasing the infection time is a simple and effective way to improve the transduction efficiency.

7.4.6 Concluding remarks

Retroviral gene transfer is a useful technique for introducing stable expression of tight junction proteins to epithelial or endothelial cells. Because tight junctions are formed only by mature and polarized cells, the ectopic expression of TJ proteins must match the developmental course of tight junction during cell growth. Transient or stable DNA transfection is less likely to allow homogeneous gene expression in a large cell population. Retrovirus, owing to its ability to integrate into host cell's genome, can fulfill the need of stability and homogeneity in the expression of transgenes. Retrovirus, however, is not without limitation. Large genes, which encode membrane receptors, kinases, and enzymes, are difficult to transduce into cells via recombinant retroviruses.

References

Burns, J. C., Friedmann, T., Driever, W., Burrascano, M., & Yee, J. K. (1993). Vesicular stomatitis virus G glycoprotein pseudotyped retroviral vectors: Concentration to very high titer and efficient gene transfer into mammalian and nonmammalian cells. *Proceedings of the National Academy of Sciences of the United States of America, 90*, 8033–8037.

Emi, N., Friedmann, T., & Yee, J. K. (1991). Pseudotype formation of murine leukemia virus with the G protein of vesicular stomatitis virus. *Journal of Virology, 65*, 1202–1207.

Miller, A. D., Miller, D. G., Garcia, J. V., & Lynch, C. M. (1993). Use of retroviral vectors for gene transfer and expression. *Methods in Enzymology, 217*, 581–599.

Miller, D. G., Adam, M. A., & Miller, A. D. (1990). Gene transfer by retrovirus vectors occurs only in cells that are actively replicating at the time of infection. *Molecular and Cellular Biology, 10*, 4239–4242.

Morgenstern, J. P., & Land, H. (1990). Advanced mammalian gene transfer: High titre retroviral vectors with multiple drug selection markers and a complementary helper-free packaging cell line. *Nucleic Acids Research, 18*, 3587–3596.

Chapter 7.5

Retrovirus-mediated RNA interference

7.5.1 Background knowledge

7.5.1.1 Concept of RNA interference

The term RNA interference (RNAi) describes a sequence-specific gene silencing event in metazoans (Fire et al., 1998). The RNAi pathway is initiated by the enzyme Dicer, which cleaves long double-stranded RNA molecules into short double-stranded fragments of 21-nucleotide RNA molecules, known as the small interfering RNA (siRNA). Each siRNA is unwound into two single-stranded RNA molecules, the passenger strand and the guide strand. The passenger strand is degraded and the guide strand is incorporated into the RNA-induced silencing complex (RISC). The RISC silences the gene expression by cleaving the mRNA transcript when the guide strand of siRNA pairs with the complementary sequence in the mRNA transcript of the gene (Fig. 7.5.1) (Hammond, Caudy, & Hannon, 2001).

7.5.1.2 RNA interference as a tool to study loss of gene function

In mammalian cells, long (>100 bp) double-stranded RNA molecules may trigger the interferon response (Stark, Kerr, Williams, Silverman, & Schreiber, 1998). The 21-nucleotide intermediate molecule in the RNAi pathway—siRNA, on the other hand, can silence gene expression at no expense of secondary deleterious effects (Elbashir et al., 2001). One important concern in a RNAi experiment is the off-target silencing of genes with similar nucleotide identity, such as a gene family member or a gene that adventitiously has a similar stretch of sequence. A good test for the specificity of RNAi-induced gene silencing is through an add-back experiment. The add-back experiment exploits alternative codon usage to introduce an mRNA molecule that encodes the same protein but resists the siRNA. Mutagenic studies have shown that 3-point differences within the 21-nucleotide target region are sufficient to confer siRNA resistance to an mRNA molecule (Elbashir et al., 2001; Lassus, Rodriguez, & Lazebnik, 2002).

7.5.1.3 Sequence selection for RNA interference

Individual siRNA molecules may lack the silencing efficacy because they are unable to incorporate into the RISC, bind to the target mRNA, or trigger the mRNA cleavage (Dykxhoorn, Novina, & Sharp, 2003). Selecting the ideal siRNA sequence is therefore vital to the success of a gene silencing

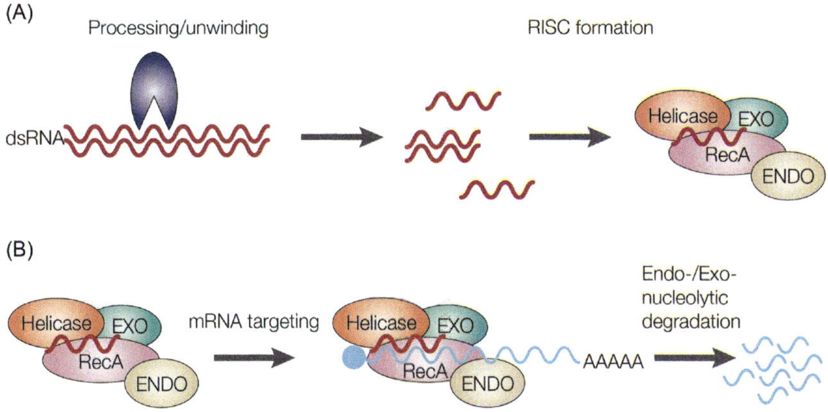

FIGURE 7.5.1 Molecular mechanism of RNA interference. (A) The input double-stranded RNA is processed by the enzyme Dicer into 21–23-nucleotide short sequences known as the small interfering RNA (siRNA). (B) Each siRNA is unwound into two single-stranded RNA molecules, the passenger RNA and the guide RNA. The guide RNA is incorporated into a nuclease complex, known as the RNA-induced silencing complex (RISC), which acts to destroy the mRNAs that are recognized by the guide RNA through base-pairing interactions. Endo, endonucleolytic nuclease; Exo, exonucleolytic nuclease; RecA, homology-searching activity related to *Escherichia coli* recA.

experiment. There are several empirical rules that can be written into a computer program to facilitate the selection of siRNA sequences. For example, GC content must be lower than 40%; AA dinucleotides are found at the 5′-end; no inverted repeat is allowed; and some bases are preferred at positions 3, 10, 13, and 19 in the passenger strand (Elbashir et al., 2001; Reynolds et al., 2004; Schwarz et al., 2003). The commercial programs available for siRNA selection are listed below.

https://dharmacon.horizondiscovery.com/design-center/
https://www.invivogen.com/sirnawizard/
https://www.genscript.com/design_center.html

7.5.2 Materials and reagents

7.5.2.1 Cell model for ectopic gene expression

HEK293 cells (*producer cell*); available from ATCC.
MDCK strain II cells (*host cell*); available from ATCC.

7.5.2.2 Plasmids

pSIREN retroviral siRNA expression vector; available from Clontech.
pGag-pol expression vector; available from Addgene.
pVSV-G expression vector; available from Addgene.

7.5.2.3 Cell culture medium

Complete medium

500 mL	Dulbecco's Modified Eagle's medium (DMEM) (high glucose/ + sodium pyruvate/ + glutamine); available from Gibco
50 mL	Fetal bovine serum (FBS); available from Gibco
5 mL	Penicillin/streptomycin; available from Gibco

7.5.2.4 Buffers

$CaCl_2$ (2.5 M)	
27.7 g	$CaCl_2$

Bring to 100 mL with dH_2O, and stir to dissolve.
Filter, sterilize, and store at 4°C

2 × HBS buffer	
140 mM	NaCl
1.5 mM	Na_2HPO_4
50 mM	HEPES

Bring to 100 mL with dH_2O, stir to dissolve, and adjust pH to 7.0.
Filter, sterilize, and store at 4°C.

TNE buffer	
130 mM	NaCl
50 mM	Tris
1 mM	EDTA

Bring to 100 mL with dH_2O, stir to dissolve, and adjust pH to 7.8.
Filter, sterilize, and store at 4°C.

7.5.3 Experimental procedure

The protocol to generate retroviruses to deliver siRNA molecules as transgene is similar to that described in Chapter 7.4, Retrovirus-mediated transgene expression. Briefly, the following DNA plasmids are transfected into the producer cells—HEK293 cells.

20 μg	pSIREN-Cldn2-siRNA
20 μg	pGag-pol
20 μg	pVsvG

```
Position  19   3   10   13
          T    A   G    C

RNAi Oligo positions and Oligos

360

RNAi Oligo score ------------> 15

No of GC        9

GC content      0.47

No of AT in the last 5 nucleotides          3

Sense oligos with restriction sites  5--->3

GATCCGCATGAGATGCACTGTCTTTTCAAGAGAAAGACAGTGCATCTCATGCTTTTTTGCTAGCG

Antisense oligos with restriction sites   5--->3

AATTCGCTAGCAAAAAAGCATGAGATGCACTGTCTTTCTCTTGAAAAGACAGTGCATCTCATGCG

     : Passenger strand;        : Guide strand.
```

FIGURE 7.5.2 Rational selection of siRNA sequence. A computer algorithm automatically selects the siRNA sequence with high probability of gene silencing.

After incubation, the supernatant is collected to harvest the virions. The target cells—MDCK II cells—are infected by viruses at the titer of 1×10^9 cfu/mL.

7.5.4 Data analysis

A computer program in the Appendix is used to predict the efficacy of siRNA sequence. The program scans the mRNA sequence of a target gene and selects the siRNA sequence with the highest silencing probability (expressed as an integer score from −25 to +25). An example of siRNA selection is shown here for a target sequence in the dog claudin-2 gene with a probability score of 15 (Fig. 7.5.2). Once an siRNA sequence is chosen, it can be cloned into DNA vector and packaged into retrovirus to infect the target cells, for example, MDCK II cells. The efficacy of gene silencing is assessed by Western blot and immunofluorescence microscopy in the target cells (Fig. 7.5.3). To ensure the authenticity of gene silencing, at least two independent siRNA sequences must be tested for each target gene.

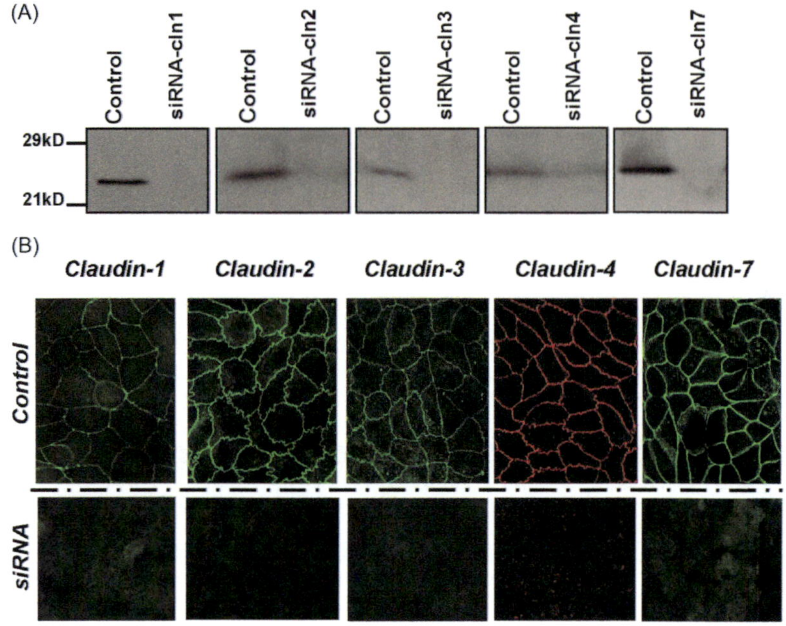

FIGURE 7.5.3 Claudin gene silencing in MDCK II cells. (A) Western immunoblots of claudin protein in MDCK II cells expressing siRNA against claudin-1, −2, −3, −4, or −7, respectively. (B) Confocal microscopy showing the tight junction staining of claudin protein in MDCK II cells expressing siRNA against claudin-1, −2, −3, −4, or −7, respectively. Note that the immunostaining of claudin protein in siRNA-expressing cells disappears, giving way to the background level. *Reproduced with permission from Hou, J., Gomes, A. S., Paul, D. L., & Goodenough, D. A. (2006). Study of claudin function by RNA interference.* The Journal of Biological Chemistry, 281, *36117–36123.*

7.5.5 Troubleshooting

7.5.5.1 Expression level of siRNA

The capacity of the RNAi machinery must be carefully watched to prevent saturation. Although the gene silencing effect can be titrated by varying the virus titer, exceedingly high expression levels of siRNA might activate the interferon pathway (Bridge, Pebernard, Ducraux, Nicoulaz, & Iggo, 2003). Microarray analyses of interferon genes are encouraged for every siRNA experiment. When designing a RNAi experiment, a negative control made of scrambled siRNA must be included in addition to an untreated control. Untreated samples allow determining the baseline levels of cell viability, phenotype, and target gene expression. Scrambled siRNAs are designed to have no known target in the cells being studied. They are important for distinguishing sequence-specific silencing from nonspecific effects, for example, the interferon response in the RNAi experiment.

7.5.5.2 Expression level of target gene

The expression of a target gene can be regulated on two levels: mRNA transcription and protein translation. RNAi is effective on the mRNA level against most cellular genes. However, there are exceptions. Some genes, such as those encoding intermediate filament protein vimentin and human coagulation trigger tissue factor, are particularly difficult to silence with siRNA (Harborth, Elbashir, Bechert, Tuschl, & Weber, 2001; Holen, Amarzguioui, Wiiger, Babaie, & Prydz, 2002). The half-life of an mRNA transcript, its abundance and its secondary structure are all important variables to consider in a RNAi experiment. When the mRNA of a target gene is cleaved by siRNA, translation of the target protein is dramatically reduced, which has been referred to as protein knockdown. The total cellular abundance level of the target protein is influenced not only by its neosynthesis but also by its degradation. Tight junction proteins, in particular, have a very long half-life. As a result, knockdown of tight junction proteins appears slower than unbound cytosolic proteins.

7.5.6 Concluding remarks

RNAi, when combined with retroviral transduction approach, can generate knockdown cells for virtually every gene in the genome. The principal advantages of the retroviral RNAi system, namely the unlimited host range, the high transduction efficiency, the speed and ease with which a population of knockdown cells can be generated, and the sustained expression of siRNA during epithelial polarization and tight junction formation, make retrovirus-mediated RNAi an applicable and convenient tool for the study of loss-of-function effects of tight junction genes.

```
/// For Siren RNAi vector, determine the sore of RNAi inserts
#include <iostream>

using std::cout;
using std::cin;
using std::ios;
using std::cerr;
using std::endl;

#include <iomanip>

using std::setw;
using std::setiosflags;
using std::resetiosflags;
using std::setprecision;

#include <fstream>

using std::ifstream;
using std::ofstream;

#include <cstdlib>

#include <cstring>

#include <string>
using std::string;

#include <iterator>
#include <algorithm>

#include<sstream>

#include<cctype>

#include<cmath>

void complement (string &);

int main()
{
// file handler
   const int NameSize=50;
   int LineLength=60;
   char InFileName[NameSize], RNAiName[NameSize];
```

```
    cout << "Enter the name of sequence file.\n";
    cin >> InFileName;

    cout << "Enter the length of each line in the output file (eg 60).\n";
    cin >> LineLength;

    strcpy(RNAiName, "RNAiScoreV3");
    strcat(RNAiName, InFileName);

    ifstream InFile ( InFileName, ios::in );
    ofstream RNAi ( RNAiName, ios::out );

    ofstream RNAiTemp ("RNAiScoreV3Temp.txt", ios::out);
    ifstream InFileTemp ("RNAiScoreV3Temp.txt", ios::in);

// read in sequence
    const int SIZE=500;
    char StrandOneLine[SIZE];
    string sStrandDNAPlus;
    string StrandDNAPlus, StrandDNAMinus;
    int m;
    string::iterator itInput;


    while ( InFile.getline (StrandOneLine, SIZE))
          if (isalnum(StrandOneLine[0]))                //genebank format
                sStrandDNAPlus.append (StrandOneLine);   // no # et al

    for (m=0; m<sStrandDNAPlus.length(); m++)
          if (isalpha(sStrandDNAPlus.at(m)))
              StrandDNAPlus.append(sStrandDNAPlus, m, 1); // no ' ' /
number between nucleotide


    for (itInput = StrandDNAPlus.begin(); itInput != StrandDNAPlus.end();
itInput++)
          *itInput =static_cast<char>(toupper(*itInput)); // to upper
letter


// complement
    StrandDNAMinus=StrandDNAPlus;
    complement (StrandDNAMinus);

// input for word
    char WordOneLine[SIZE];
    string Word;
    int WordLength;

    cout<<"Type the leading nucleic acid sequence. Normally AA"<<endl;
    cin>>WordOneLine;

    Word.assign(WordOneLine);
    string::iterator itWord;
```

```
    for (itWord = Word.begin(); itWord != Word.end(); itWord++)
          *itWord =static_cast<char>(toupper(*itWord)); // to upper letter

    WordLength=Word.length();


// scan for word
    long WordPosition;

// find word on top strand
    WordPosition=StrandDNAPlus.find(Word);
    while (WordPosition<string::npos)
    {
          RNAiTemp<<(WordPosition+1)<<endl;
          WordPosition=StrandDNAPlus.find(Word,WordPosition+1);
    };


// output every LineLength characters
    long DNAlength=StrandDNAPlus.length();
    long i=LineLength;
    int outCounter;

    RNAi<< "> The length of the sequence:  "
                 << DNAlength << endl << endl;

    while (i<DNAlength)
    {
          for (outCounter=0; outCounter<LineLength; outCounter++)
                RNAi<<StrandDNAPlus[outCounter+i-LineLength];
          RNAi<<endl;
          for (outCounter=0; outCounter<ceil(LineLength/10.0);
outCounter++)
                RNAi<<"|         ";
          RNAi<<endl;
          for (outCounter=0; outCounter<LineLength; outCounter++)
                RNAi<<StrandDNAMinus[outCounter+i-LineLength];
          RNAi<<endl<<endl;
          for (outCounter=0; outCounter<ceil(LineLength/10.0);
outCounter++)
                RNAi<<setw(10)<<setiosflags(ios::left)<<(outCounter*10+i-
LineLength+1);// start at 1 not 0
          RNAi<<endl;
          for (outCounter=0; outCounter<LineLength; outCounter++)
                RNAi<<'-';
          RNAi<<endl<<endl;
          i+=LineLength;
    };

// output for the tail of sequence
    for (outCounter=i-LineLength; outCounter<DNAlength; outCounter++)
          RNAi<<StrandDNAPlus[outCounter];
    RNAi<<endl;
```

```
   for (outCounter=0; outCounter<ceil((DNAlength-i+LineLength)/10.0);
outCounter++)
         RNAi<<"|          ";
   RNAi<<endl;
   for (outCounter=i-LineLength; outCounter<DNAlength; outCounter++)
         RNAi<<StrandDNAMinus[outCounter];
   RNAi<<endl<<endl;
   for (outCounter=0; outCounter<ceil((DNAlength-i+LineLength)/10.0);
outCounter++)
               RNAi<<setw(10)<<setiosflags(ios::left)<<(outCounter*10+i-
LineLength+1);
   RNAi<<endl<<endl;



// construct oligo
   long OligoPosition;
   string SenseOligo;
   string AntisenseOligo;
   string Oligo;
   int OligoLength;
   int OligoScore;
   int GC, AT;


   cout<<"Type the length of RNAi oligo. Normally 19"<<endl;
   cin>> OligoLength;


   while ((InFileTemp>>OligoPosition)&&(DNAlength-(OligoPosition-
1+WordLength)>OligoLength))
   {
         Oligo.append(StrandDNAPlus, OligoPosition-1+WordLength,
OligoLength);  // start after AA


         OligoScore=0;

// criterion 1- GC% 36-52

      GC=0;                                  // in string, string.at(i) is
used for receiving value
      for (i=0; i<OligoLength; i++)
      {
            switch (Oligo[i]) {
            case 'G':
            case 'C':
                  GC++;
                  break;

            default:
                  break;};
      };
```

```
     if ((static_cast<double>(GC)/OligoLength>0.36) &&
           (static_cast<double>(GC)/OligoLength<0.52))
           OligoScore+=5;

// criterion 2- at least 3 A or T at last 5 positions

    AT=0;

     for (i=OligoLength-1; i>OligoLength-6; i--)
     {
           switch (Oligo[i]) {
           case 'A':
           case 'T':
                 AT++;
                 break;

           default:
                 break;};
     };

     if (AT>=3) OligoScore+=5;

// criterion 3- A at position 19 && A at 3 && T at postion 10 &&  not G, C
at 19 && not G at 13

     switch (Oligo[19-1]) {
           case 'A':
          OligoScore+=5;
              break;

           case 'T':
                 break;

           case 'G':
                 OligoScore-=4;
                 break;

           case 'C':
                 OligoScore-=4;
                 break;

           default:
                 break;
     };

    if (Oligo[3-1]=='A') OligoScore+=5;
     if (Oligo[10-1]=='T') OligoScore+=5;
     if (Oligo[13-1]=='G') OligoScore-=4;

     RNAi<<"Position  19   3   10    13     "<<endl<<"          "<<
           Oligo[19-1]<<"    "<<Oligo[3-1]<<"    "<<Oligo[10-1]<<"
"<<Oligo[13-1]<<endl<<endl;
```

```
// construct oligo
        SenseOligo.assign("CG");
         SenseOligo += Oligo;
         SenseOligo.append("TTCAAGAGA");

// reverse - strand
         complement (Oligo);
         std::reverse(Oligo.begin(), Oligo.end());//algorith function

         SenseOligo += Oligo;
         SenseOligo.append("TTTTTTGCTAGCG");

         AntisenseOligo=SenseOligo;

// reverse - strand
         complement (AntisenseOligo);
         std::reverse(AntisenseOligo.begin(),
AntisenseOligo.end());//algorith function

         SenseOligo.insert(0, "GATC");
         AntisenseOligo.insert(0, "AATT");


         RNAi<<endl;
         RNAi<<"RNAi Oligo positions and Oligos"<<endl<<endl;

         RNAi<< OligoPosition+WordLength<<endl<<endl;

         RNAi<<"RNAi Oligo score ------------> "<< OligoScore<<endl<<endl;
         RNAi<<"No of GC         "<<GC<<endl<<endl;
         RNAi<<"GC content
"<<setiosflags(ios::fixed|ios::showpoint)<<setprecision(2)
                <<static_cast<double>(GC)/OligoLength<<endl<<endl;

         RNAi<<"No of AT in the last 5 nucleotides
"<<AT<<endl<<endl;

         RNAi<<"Sense oligos with restriction sites  5--->3"<<endl<<endl;
         RNAi<<SenseOligo<<endl<<endl;

         RNAi<<"Antisense oligos with restriction sites   5---
>3"<<endl<<endl;
         RNAi<<AntisenseOligo<<endl<<endl;

         RNAi<<endl;
         RNAi<<"----------------------------"<<endl<<endl;

         Oligo.erase();
         SenseOligo.erase();
         AntisenseOligo.erase();

    };
```

```
    InFile.close();
    RNAi.close();
    RNAiTemp.close();
    InFileTemp.close();


    return 0;

}


void complement (string &substrand)
{
      string::iterator position;
      for (position = substrand.begin(); position != substrand.end();
position++)
            switch (*position) {
            case 'A':
            case 'a':
                  *position='T';
                  break;

            case 'C':
            case 'c':
                  *position='G';
                  break;

            case 'G':
            case 'g':
                  *position='C';
                  break;

            case 'T':
            case 't':
                  *position='A';
                  break;

            default:
                  *position='N';
                  break;
      };
}
```

References

Bridge, A. J., Pebernard, S., Ducraux, A., Nicoulaz, A. L., & Iggo, R. (2003). Induction of an interferon response by RNAi vectors in mammalian cells. *Nature Genetics*, *34*, 263–264.

Dykxhoorn, D. M., Novina, C. D., & Sharp, P. A. (2003). Killing the messenger: Short RNAs that silence gene expression. *Nature Reviews Molecular Cell Biology*, *4*, 457–467.

Elbashir, S. M., Harborth, J., Lendeckel, W., Yalcin, A., Weber, K., & Tuschl, T. (2001). Duplexes of 21-nucleotide RNAs mediate RNA interference in cultured mammalian cells. *Nature*, *411*, 494–498.

Fire, A., Xu, S., Montgomery, M. K., Kostas, S. A., Driver, S. E., & Mello, C. C. (1998). Potent and specific genetic interference by double-stranded RNA in *Caenorhabditis elegans*. *Nature*, *391*, 806–811.

Hammond, S. M., Caudy, A. A., & Hannon, G. J. (2001). Post-transcriptional gene silencing by double-stranded RNA. *Nature Reviews Genetics*, *2*, 110–119.

Harborth, J., Elbashir, S. M., Bechert, K., Tuschl, T., & Weber, K. (2001). Identification of essential genes in cultured mammalian cells using small interfering RNAs. *Journal of Cell Science*, *114*, 4557–4565.

Holen, T., Amarzguioui, M., Wiiger, M. T., Babaie, E., & Prydz, H. (2002). Positional effects of short interfering RNAs targeting the human coagulation trigger Tissue Factor. *Nucleic Acids Research*, *30*, 1757–1766.

Lassus, P., Rodriguez, J., & Lazebnik, Y. (2002). Confirming specificity of RNAi in mammalian cells. *Science's STKE: Signal Transduction Knowledge Environment*, *2002*, pl13.

Reynolds, A., Leake, D., Boese, Q., Scaringe, S., Marshall, W. S., & Khvorova, A. (2004). Rational siRNA design for RNA interference. *Nature Biotechnology*, *22*, 326–330.

Schwarz, D. S., Hutvagner, G., Du, T., Xu, Z., Aronin, N., & Zamore, P. D. (2003). Asymmetry in the assembly of the RNAi enzyme complex. *Cell*, *115*, 199–208.

Stark, G. R., Kerr, I. M., Williams, B. R., Silverman, R. H., & Schreiber, R. D. (1998). How cells respond to interferons. *Annual Review of Biochemistry*, *67*, 227–264.

Chapter 8

Mouse models of tight junction physiology

Chapter 8.1

Mouse genetics and transgenics

8.1.1 Laboratory mouse

The laboratory mouse, *Mus musculus*, is the principal mammalian species used in biomedical research as a model organism for human physiology. Mice are easy to breed. They deliver relatively large progenies with a short generation time. They tolerate inbreeding rather well compared to other mammalian species. During the past 20 years, hundreds of mutations, most of them with deleterious alleles, have been introduced to the mouse genome, allowing the identification of causal genes for many human diseases. Another important advantage to be credited to the mouse is that it seems to be one of the rare, maybe the only species, from which it is possible to grow pluripotent embryonic stem (ES) cells in vitro (Evans, 2008). The ES cells can be genetically engineered in a number of ways and still retain the capacity to participate in the formation of the germ cells once injected into a developing embryo. Finally, the complete sequence of the mouse genome is available, which allows comparisons with the human genome and annotations concerning the function of genes.

8.1.2 Mouse strain

8.1.2.1 Inbred mouse strain

Inbred strains are defined as those derived from 20 or more consecutive sister to brother matings. A strain's degree of inbreeding is designated by the letter F followed by the number of generations of filial breeding (e.g., F20). Heterozygosity can be completely eliminated by F60 (Herbert III, 2012).

A Laboratory Guide to the Tight Junction. DOI: https://doi.org/10.1016/B978-0-12-818647-3.00008-8

Most commonly used inbred strains, for example, BALB/c, C57BL/6, DBA/2, and 129/Sv, have been inbred for more than 200 generations. Continual inbreeding produces mice that are genetically uniform, being homozygous at virtually every locus. In addition to genetic and phenotypic uniformity, commonly used inbred strains are well characterized with regard to their physiology, how they respond to experimental perturbations, and what to expect as they age. The use of inbred strains can significantly reduce the number of mice needed for each experiment.

8.1.2.2 Congenic mouse strain

Congenic strains are inbred strains carrying a mutant gene or polymorphic allele from a different inbred strain. They are derived by successively mating mice carrying a mutant gene from a donor strain to mice of a recipient strain (Herbert III, 2012). A strain is considered congenic after 10 generations of backcrossing to a recipient strain. A fully congenic strain is expected to be identical to the recipient strain at every locus except for the transferred locus and a linked segment of chromosome.

8.1.2.3 Hybrid mouse strain

The deliberate crossing of mice of two inbred strains generates hybrid mice. F1 hybrids (e.g., B6D2F1, derived from crossing of C57BL/6 strain to DBA/2 strain) are similar to inbred strains in that they are genetically and phenotypically uniform. In contrast to most inbred strains, F1 hybrids display a hybrid vigor, that is, increased disease resistance, better survival under stress, greater natural longevity, and larger litter size. They are useful as hosts for tissue transplants, embryo donors, and research models for radiation, behavior, and bioassay. Many strains carrying targeted mutations are established on a mixed C57BL/6 and 129/Sv background. Because of the genetic heterozygosity, there are no matching wild-type controls. F2 hybrids (e.g., B6D2F1 × B6D2F1) are frequently used physiological controls for these strains. Like the targeted mutant mice, the genetic background of F2 hybrid mice varies among littermates. As a result, F2 hybrid mice provide only an approximate match to the genetic background of the F1 hybrid parents.

8.1.2.4 Outbred mouse strain

Commonly used outbred strains include CD-1, ICR, and Swiss Webster. Most outbred mice exhibit hybrid vigor similar to or exceeding that of F1 hybrids. Compared to inbred mice, they have longer life span, higher disease resistance, higher reproductive performance, and lower neonatal mortality. Outbred mice are particularly suited as stud males or foster mothers in transgenic experiments.

8.1.3 Mouse genome

The diploid chromosomal complement of inbred strains is $2N = 40$, where $N = 19$ autosomes plus X or Y sex chromosome. Unlike human chromosomes, the autosomes and the X chromosome from mouse cells are telocentric, that is, the centromere is at one end of a single-armed chromosome. The mouse Y chromosome is acrocentric, that is, the chromosome has a short p arm as well as a long q arm, similar to the structure found in human chromosomes. The genomes of two species can be compared by means of an Oxford Grid (Fig. 8.1.1) (Andersson et al., 1996). Humans and mice share virtually the same set of genes. Almost every gene found in one species has a closely related counterpart in the other. Of the approximately 4000 genes that have been studied, less than 10 are found in one species but not in the other. On average, the protein-coding regions of human and mouse genomes are 85% identical. These regions are evolutionarily conserved because they are required for function. In contrast, the noncoding regions are much less similar (only 50% or less identical).

8.1.4 Random mutagenesis in laboratory mouse

Ethyl nitrosourea (ENU) is an alkylating agent that can induce point mutations in DNA, namely base pair substitutions such as A/T → T/A transversions and A/T → G/C transitions (Justice, Noveroske, Weber, Zheng, & Bradley, 1999). Optimal doses of ENU can induce mutations with the frequency of one mutation per gene in every gamete (Hitotsumachi, Carpenter, & Russell, 1985). Due to the nature of ENU-induced mutations, the mutant phenotypes represent a unique resource for genetic research, because they (1) reflect the consequences of single-gene changes independent of positional effects; (2) provide a fine-structure dissection of protein function; (3) display a range of mutation types from partial or complete loss of function to hypermorphic or dominant-negative effect; and (4) reveal gene functions in an unbiased manner. Using mice to model human diseases, phenotype-driven ENU screens have facilitated the discovery of genes important for many inherited diseases, such as cataracts, deafness, hypertension, diabetes, and neurodegeneration (Hrabe de Angelis et al., 2000; Soewarto et al., 2000).

8.1.5 Transgenesis in laboratory mouse

The classic way of generating transgenic mice is by direct injection of DNA into one of the two haploid pronuclei present in newly fertilized oocytes (1-cell embryos) (Gordon, Scangos, Plotkin, Barbosa, & Ruddle, 1980). Depending on circumstances, between 5% and 25% of injected embryos incorporate some of the foreign DNA into their chromosomes. These are the

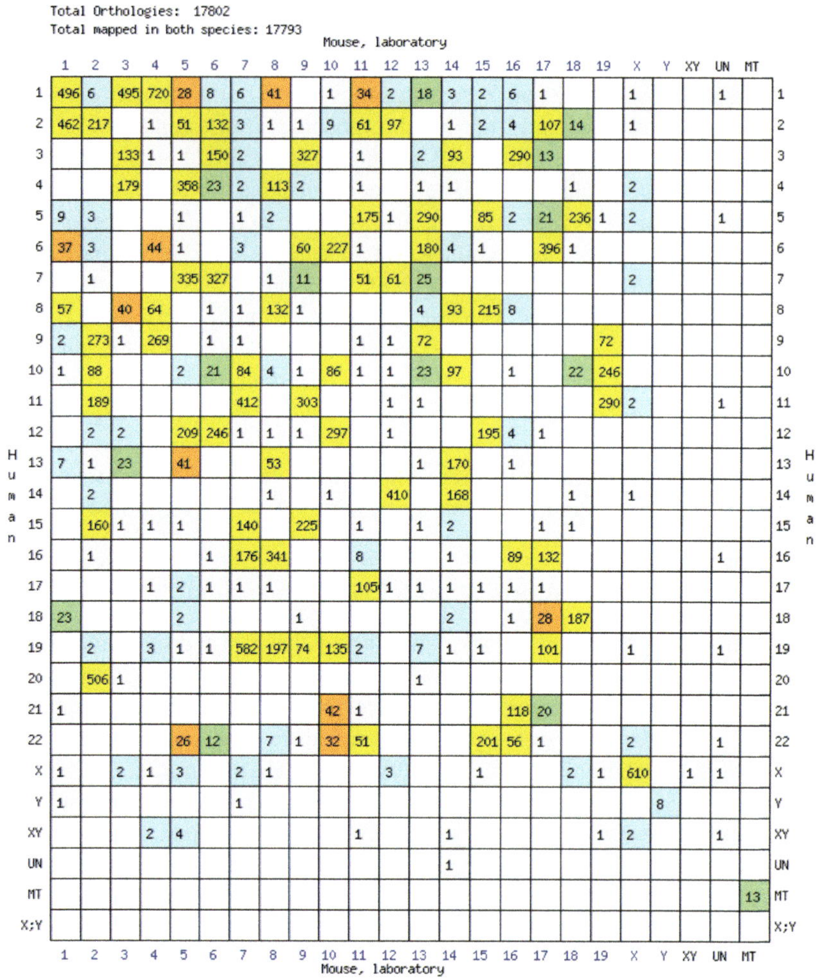

FIGURE 8.1.1 Human–mouse genome grid. Each row or column represents a chromosome from one species. Each homologous locus is plotted in the relevant cell, each cell being referenced by its chromosome number in each of the two species. A cell with a large number of homologous loci denotes a conserved segment of DNA between the two species.

transgenic founders. When DNA is introduced in this way, it integrates at random chromosomal sites, irrespective of whether or not it possesses regions of homology with any of the chromosomes. In most cases, integration occurs at a single site on one chromosome in a given embryo, but integration at two or more sites has also been observed. About 75% of transgenic founder mice are mosaic, which is to say that they contain both normal cells and cells that carry the transgene (Wilkie, Brinster, & Palmiter, 1986). In principle, transgenic mosaicism may arise either by insertion of the transgene into a region of the chromosome that has replicated at least once, or by the loss of the transgene from one of the daughter chromosomes following insertion. Once the transgene is integrated into the chromosome, it can be faithfully transmitted through the germ cells and consequently, each founder mouse, provided that it is fertile, is capable of giving rise to a unique transgenic line (Gordon & Ruddle, 1981). Because the transgene is incorporated into only one allele of a given locus in a diploid embryo, transgenic mice generated by pronuclear injection are considered to be hemizygous and designated with the symbol of "Tg/ + " or "Tg/0."

8.1.6 Gene targeting in laboratory mouse

8.1.6.1 Manipulation of mouse embryonic stem cells

ES cells were first derived from the inner cell mass of mouse embryo at the blastocyst stage (Evans & Kaufman, 1981). There is a general agreement that ES cells are most easily obtained from the inbred 129/Sv mouse strain, less easily from the inbred C56BL/6 mouse strain, and with great difficulty from other mouse strains. The *Oct3/4* gene promoter is active in the ES cells but becomes inactive when they differentiate. Leukemia inhibitory factor (LIF), when added to the culture medium, can inhibit ES cell differentiation (Smith et al., 1988). Isolation of ES-like cells has been reported from a variety of species, such as rabbit, monkey, and cow, but there is no evidence of germline transmission from ES cells other than those derived from a few mouse strains. Cloned DNA can be introduced into ES cells by electroporation. Genetically modified ES cells are injected into blastocysts, which are then implanted into the uterine horns of recipient dams. The chimeric pups (F0) are made up of host cells and transgenic cells derived from the injected ES cells. The ES cells may or may not colonize the germline. When germline colonization does occur, the chimeras produce three different types of gamete, assuming one allele of an autosomal gene is targeted for ablation: (1) those derived from the host blastocyst, (2) those with an ES cell genome and carrying the modified gene, and (3) those with an ES cell genome but not carrying the genetic modification. When the chimeras are mated to normal mice, the pups (F1) to which the type (2) gametes contribute are heterozygous for the ablated gene, and designated with the symbol of " + / − ." Homozygotes (F2), which are designated with the symbol of " − / − ," can be obtained by mating the heterozygous F1 mice together.

8.1.6.2 Homologous recombination

Homologous recombination is a type of genetic recombination in which nucleotide sequences are exchanged between two similar or identical molecules of DNA. It is most widely used by cells to repair harmful

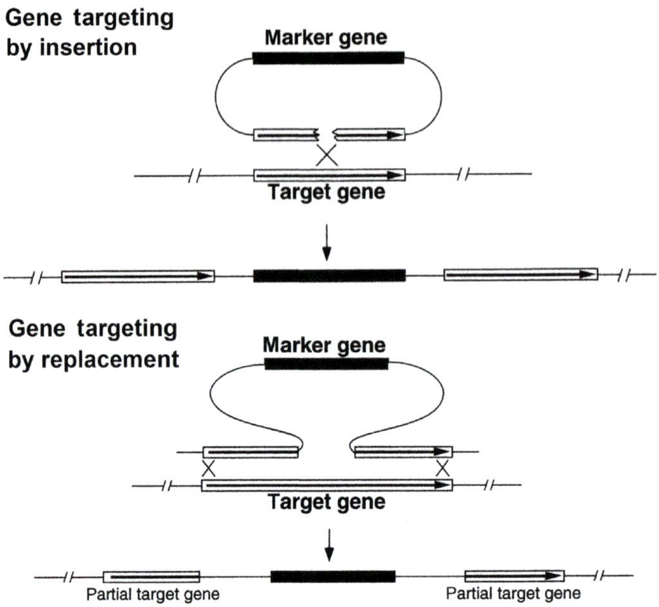

FIGURE 8.1.2 Gene targeting strategy. In the insertion mode, the target gene is duplicated and the marker gene is inserted between them. In the replacement mode, the marker gene replaces part of the target gene, causing interruption or deletion in the transcription of the target gene. *Reproduced with permission from Rong, Y. S., & Golic, K. G. (2000). Gene targeting by homologous recombination in Drosophila.* Science (New York, NY), 288, *2013–2018.*

breaks that occur on both strands of DNA, known as double-strand breaks (DSB). DNA integration by homologous recombination provides a way of introducing mutations to the mouse genome at preselected loci, which is referred to as gene targeting. There are two modes of DNA integration by homologous recombination (Thomas & Capecchi, 1987). In the insertion mode, foreign DNA is added to the chromosome with no loss of the preexisting chromosomal DNA. In the replacement mode, foreign DNA replaces part of the chromosomal DNA (Fig. 8.1.2). Gene knockout (KO) is the most commonly used strategy of homologous recombination by replacement. This method involves creating a DNA construct containing a drug resistance gene in place of the target gene. The construct also contains a minimum of 2 kb of homologous sequence flanking the target gene. After the construct is electroporated into ES cells, the desired recombination event can be selected by resistance to the drug. The length of the region of homology with the target sequence has a strong influence on the frequency of homologous recombination. Elongating the homologous sequence from 2 to 14 kb can increase the targeting frequency by more than 100-fold (Deng & Capecchi, 1992).

References

Andersson, L., Archibald, A., Ashburner, M., Audun, S., Barendse, W., Bitgood, J., ... Burt, D. (1996). Comparative genome organization of vertebrates. The first international workshop on comparative genome organization. *Mammalian Genome: Official Journal of the International Mammalian Genome Society*, *7*, 717–734.

Deng, C., & Capecchi, M. R. (1992). Reexamination of gene targeting frequency as a function of the extent of homology between the targeting vector and the target locus. *Molecular and Cellular Biology*, *12*, 3365–3371.

Evans, M. (2008). Embryonic stem cells: The mouse source—vehicle for mammalian genetics and beyond (Nobel lecture). *Chembiochem: A European Journal of Chemical Biology*, *9*, 1690–1696.

Evans, M. J., & Kaufman, M. H. (1981). Establishment in culture of pluripotential cells from mouse embryos. *Nature*, *292*, 154–156.

Gordon, J. W., & Ruddle, F. H. (1981). Integration and stable germ line transmission of genes injected into mouse pronuclei. *Science (New York, NY)*, *214*, 1244–1246.

Gordon, J. W., Scangos, G. A., Plotkin, D. J., Barbosa, J. A., & Ruddle, F. H. (1980). Genetic transformation of mouse embryos by microinjection of purified DNA. *Proceedings of the National Academy of Sciences of the United States of America*, *77*, 7380–7384.

Herbert, C., III (2012). *Origins of inbred mice*. Elsevier.

Hitotsumachi, S., Carpenter, D. A., & Russell, W. L. (1985). Dose-repetition increases the mutagenic effectiveness of *N*-ethyl-*N*-nitrosourea in mouse spermatogonia. *Proceedings of the National Academy of Sciences of the United States of America*, *82*, 6619–6621.

Hrabe de Angelis, M. H., Flaswinkel, H., Fuchs, H., Rathkolb, B., Soewarto, D., Marschall, S., ... Jung, M. (2000). Genome-wide, large-scale production of mutant mice by ENU mutagenesis. *Nature Genetics*, *25*, 444–447.

Justice, M. J., Noveroske, J. K., Weber, J. S., Zheng, B., & Bradley, A. (1999). Mouse ENU mutagenesis. *Human Molecular Genetics*, *8*, 1955–1963.

Smith, A. G., Heath, J. K., Donaldson, D. D., Wong, G. G., Moreau, J., Stahl, M., & Rogers, D. (1988). Inhibition of pluripotential embryonic stem cell differentiation by purified polypeptides. *Nature*, *336*, 688–690.

Soewarto, D., Fella, C., Teubner, A., Rathkolb, B., Pargent, W., Heffner, S., ... Hrabe de Angelis, M. (2000). The large-scale Munich ENU-mouse-mutagenesis screen. *Mammalian Genome: Official Journal of the International Mammalian Genome Society*, *11*, 507–510.

Thomas, K. R., & Capecchi, M. R. (1987). Site-directed mutagenesis by gene targeting in mouse embryo-derived stem cells. *Cell*, *51*, 503–512.

Wilkie, T. M., Brinster, R. L., & Palmiter, R. D. (1986). Germline and somatic mosaicism in transgenic mice. *Developmental Biology*, *118*, 9–18.

Chapter 8.2

Transgenic overexpression by DNA injection

8.2.1 Background knowledge

8.2.1.1 Transcription regulation

8.2.1.1.1 General transcription machinery

Mammalian gene regulation involves a complicated interplay between transcription activators, repressors, RNA polymerases, transcription factors (TFs), and chromatin. The general transcription machinery consists of RNA polymerase II (Pol II) and TF-II subunits (Orphanides, Lagrange, & Reinberg, 1996). Pol II and TFs mediate basal transcription on a core promoter. Immediately upstream of the core promoter is the regulatory promoter, and farther away either upstream or downstream are enhancer sequences. Regulatory promoters and enhancers bind proteins known as activators, which activate gene transcription by recruiting Pol II and TFs to the core promoter. To activate a gene, the chromatin encompassing the gene and its promoters must be remodeled to permit transcription. High-order chromatin structures comprising networks of attached nucleosomes must be decondensed, specific nucleosomes over gene-specific enhancers and promoters must be made accessible to activators, and, finally, nucleosomes within the gene itself must be dissembled to permit binding of pol II and TFs.

8.2.1.1.2 Core promoter architecture

A typical core promoter contains the following DNA sequence elements (Fig. 8.2.1).

1. The TATA motif. This sequence element, with the consensus TATAAA, is located 25–30 bp upstream of the transcription start site. The TATA motif is capable of independently directing a low level of transcription by Pol II on naked DNA templates in vitro or transfected DNA templates in vivo (Conaway & Conaway, 1993).

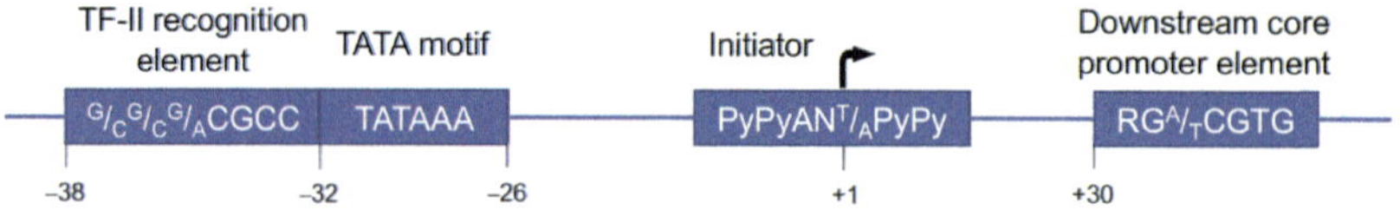

FIGURE 8.2.1 Sequence elements in a typical core promoter of eukaryotic gene.

2. The initiator element. The second type of core promoter element is the initiator sequence (Smale et al., 1998). The initiator sequence directly overlaps the transcription start site and carries out the same function as the TATA motif. The functional activity of the initiator depends on a loose consensus sequence of PyPyAN$^{T}/_{A}$PyPy.
3. The downstream core promoter element. The downstream core promoter element is a 7-nucleotide consensus sequence RG$^{A}/_{T}$CGTG located approximately 30 bp downstream of the transcription start site (Burke & Kadonaga, 1997). The downstream core promoter element is often found in TATA-less promoters and acts in conjunction with the initiator element to prime gene transcription.
4. The TF-II recognition element. The TF-II recognition element is a sequence with the consensus $^{G}/_{C}{}^{G}/_{C}{}^{G}/_{A}$CGCC located just upstream of the TATA box, which mediates specific DNA binding by the TF-II subunits (Lagrange, Kapanidis, Tang, Reinberg, & Ebright, 1998).

8.2.1.2 Transgene design

8.2.1.2.1 Promoter and regulatory elements

If a transgene needs to be expressed in all cells, heterologous promoters, such as the viral promoter CMV or those derived from the housekeeping genes (e.g., ubiquitin, histone, and β-actin), are often used. If tissue-specific expression is required, the promoters that confer tissue specificity must be chosen. Classic promoter studies can be applied to map the regulatory elements that determine the spatiotemporal expression pattern of a particular gene in tissues. The promoter including its regulatory elements is tested in vivo by fusing it to the reporter genes, such as β-galactosidase, luciferase, or GFP. If the expression of a transgene needs to match that of an endogenous gene, the full-length promoter of the endogenous gene, including 5′-end, 3′-end, and internal regulatory regions, must be used. Nevertheless, the exact locations of these elements are not always known and might be many kb away from the transcription start site. If the promoter becomes too large (i.e., >100 kb) for plasmid-based cloning, a number of artificial chromosomes, such as bacterial artificial chromosome (BAC) and yeast artificial chromosome (YAC), can be used to encompass the large DNA sequence.

8.2.1.2.2 Intron–exon boundaries, Kozak sequence, and polyadenylation

Although cDNA-based transgenes can express in vivo, the expression levels are often low. Preserving the intron–exon boundaries may increase the transgene expression to some extent. If the sizes of the introns are too large, the inclusion of only one generic intron has been shown to increase the transgene expression significantly (Choi, Huang, Gorman, & Jaenisch, 1991). A Kozak sequence with the consensus $^{A}/_{G}$NNATGG is needed to initiate the translation of the transgene (Kozak, 1989). Polyadenylation is a process of adding a poly(A) tail to a transcribed mRNA molecule. The poly(A) tail is important for the nuclear export, translation, and stability of mRNA (Richter, 1999). Most human polyadenylation sites contain the AAUAAA sequence (Beaudoing, Freier, Wyatt, Claverie, & Gautheret, 2000), but variants of the sequence exist with weaker activities (Liu et al., 2007).

8.2.1.3 Pronuclear injection of mouse embryo

Mouse mating usually occurs during the night, and 1-cell embryos can be recovered from the oviduct in the next morning. The 1-cell embryos contain two pronuclei, one derived from the egg and the other from the sperm. With the help of micromanipulators, transgene DNA (~1000 molecules in 1-pL volume) can be injected into one of the pronuclei through a fine glass needle (Fig. 8.2.2). The injected embryos are then introduced into a recipient dam via the oviduct to allow developing to term. This process is remarkably efficient. Up to 60% of embryos survive the injection, and up to 30% of the embryos transferred to the oviduct survive after birth. Up to 20% of pups born have incorporated the transgene into the genome (Gordon, Scangos, Plotkin, Barbosa, & Ruddle, 1980).

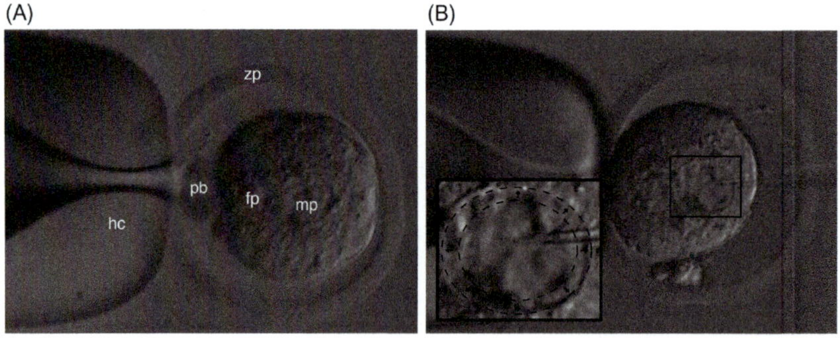

FIGURE 8.2.2 Pronuclear injection of DNA. (A) The holding capillary stabilizes one embryo in the center. The position of the embryo is in the same plane of focus so that the opening of the holder, the polar body, and the female and male pronuclei align horizontally. (B) When the male pronucleus is penetrated with the injection capillary and the DNA (1 pL) is injected, the male pronucleus swells (inset). The female pronucleus can also be injected, but it is more difficult to penetrate as it is smaller than the male pronucleus. *fp*, Female pronucleus; *hc*, holding capillary; *mp*, male pronucleus; *pb*, polar body; *zp*, zona pellucida. *Reproduced with permission from Ittner, L. M., & Gotz, J. (2007). Pronuclear injection for the production of transgenic mice.* Nature Protocols, 2, *1206–1215.*

8.2.2 Materials and reagents

8.2.2.1 Equipment

NIKON Ti-U light microscope with Hoffman modulation contrast.
Narishige micromanipulators.
Harvard Apparatus PLI-100 compressed air driven Pico-injector.

8.2.2.2 Buffers

Microinjection buffer	
Tris	10 mM
EDTA	0.1 mM

Bring to 100 mL with dH_2O, stir to dissolve and adjust pH to 7.4.
Filter, sterilize, and store at 4°C.

8.2.3 Experimental procedure

8.2.3.1 Transgene release

The transgene is released from vector sequences by restriction endonuclease digestion. For transgenes of regular size (i.e., 1–10 kb), 15 μg of plasmid DNA is sufficient to digest and purify for microinjection. Removal of prokaryotic plasmid sequences should be carried out as completely as possible. Transgene is separated from the plasmid backbone sequence by electrophoresis. Transgene fragments are excised from agarose gels and purified by using Qiagen Gel Extraction kit. Transgene DNA is eluted with microinjection buffer at the concentration of 5–10 ng/μL. The DNA solution is rapidly frozen at −20°C until the time of injection.

8.2.3.2 Pronuclear injection of DNA and production of transgenic mice

The detailed protocol for pronuclear DNA injection and production of transgenic mice has been described elsewhere (Hogan, Costantini, & Lacy, 1986; Ittner & Gotz, 2007). In brief, an inverted light microscope equipped with Hoffman modulation contrast is needed to visualize the 3D appearance of pronuclei in live mouse embryos. The injection needle is orientated by a micromanipulator to the same focal plane as the pronuclei. A gas-driven pico-injector delivers the transgene DNA into the male pronucleus. The injected embryos are then implanted into the oviduct of foster mother and allowed to develop to term.

8.2.4 Data analysis

In an example of overexpressing the claudin-14 gene in the thick ascending limb (TAL) epithelium of the kidney, the mouse claudin-14 gene was cloned downstream of a proven TAL-specific gene promoter, Tamm–Horsfall protein (THP) (Fig. 8.2.3) (Gong & Hou, 2014). A pair of genotyping primers was designed to flank the junction between the THP promoter and the open reading frame (ORF) of the claudin-14 gene, which allowed amplification of the transgenic allele (Fig. 8.2.3). Because the transgene does not contain 5′- or 3′-untranslated region (UTR) of the claudin-14 gene, two pairs of primers were designed to differentiate the transgenic from endogenous claudin-14 expression. The primer pair "qPCRorf" allowed for amplification of the total claudin-14 transcripts, whereas the primer pair "qPCR3utr" selectively amplified the endogenous claudin-14 transcripts (Fig. 8.2.3). The total claudin-14 mRNA level (normalized to β-actin mRNA) was 4.73-fold higher in transgenic mouse kidneys than wild-type mouse kidneys (Fig. 8.2.4A). The endogenous claudin-14 mRNA levels were, instead, decreased by 49% in transgenic mouse kidneys (Fig. 8.2.4B). The transgenic expression of claudin-14 on the mRNA level was, therefore, estimated to be 9.27-fold relative to the endogenous claudin-14 expression. The total claudin-14 proteins were 18.35-fold higher in transgenic mouse kidneys than wild-type mouse kidneys (Fig. 8.2.4C). Because the transgene contains no 3′-UTR of the claudin-14 gene where microRNAs can bind (Gong et al., 2012), the transgenic claudin-14 transcript is a more efficient template for translation

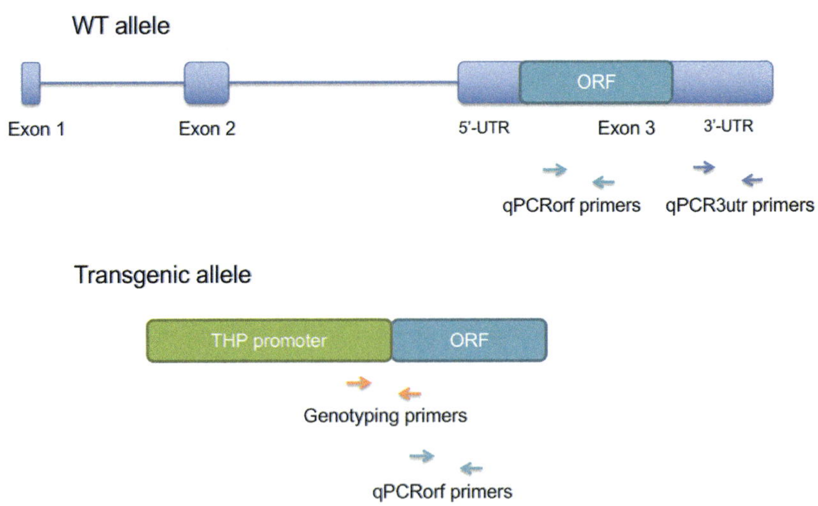

FIGURE 8.2.3 Diagram of claudin-14 gene alleles in wild-type and transgenic loci. The locations of genotyping primers and qPCR primers are shown below each allele. ORF: open reading frame. UTR: untranslated region.

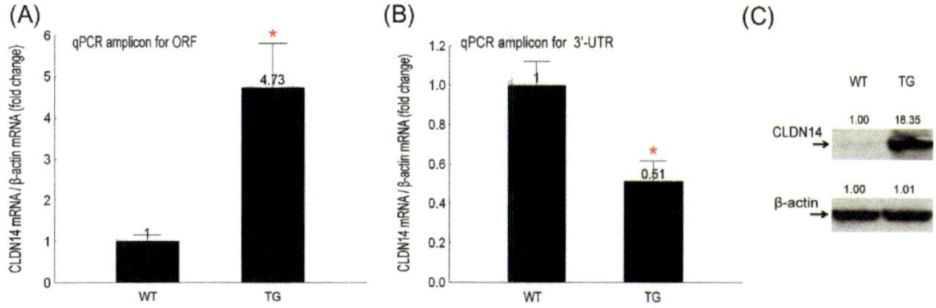

FIGURE 8.2.4 Ectopic versus endogenous claudin-14 gene expression. (A–B) The total claudin-14 transcripts (A) and the endogenous claudin-14 transcripts (B) are compared between transgenic and wild-type mouse kidneys. *, $P < .05$, $n = 6$. (C) The total claudin-14 proteins are compared between transgenic and wild-type mouse kidneys. *Reproduced with permission from Gong, Y., & Hou, J. (2014). Claudin-14 underlies Ca++-sensing receptor-mediated Ca++ metabolism via NFAT-microRNA-based mechanisms.* Journal of the American Society of Nephrology: JASN, 25, *745–760.*

compared to the endogenous claudin-14 transcript. This example also highlights the intricate feedback regulation of the expression of endogenous gene by ectopic transgene. Transgenic expression of claudin-14 causes renal loss of Ca^{++}, which in turn reduces serum Ca^{++} and downregulates endogenous claudin-14 expression via Ca^{++}-sensing receptor (CaSR) (Gong & Hou, 2014).

8.2.5 Troubleshooting

8.2.5.1 Unwanted transgene expression

Ectopic transgene expression due to chromosomal position effect or copy number variation can be very troublesome when it is unwanted. Chromosomal position effect refers to the observation that a transgene when integrated into different chromosomal sites exhibits different levels of expression (Lacy, Roberts, Evans, Burtenshaw, & Costantini, 1983). It may also affect the tissue distribution of expression and the time of the onset of expression during development. Before chromosomal integration, multiple copies of the transgene can be concatenated into a transgene array by a process known as end-joining (Burdon & Wall, 1992). The copy number difference in the transgene array may cause significant variation in the expression of transgene.

8.2.5.2 Transgene silencing

Heterochromatin refers to the chromatin regions that are condensed during interphase and transcriptionally inactive, whereas euchromatin refers to the chromatin regions that are decondensed and transcriptionally active.

Constitutive heterochromatin is found in all cells; facultative heterochromatin is found only in some types of cells. Silencing of transgene expression in mice has been shown to correlate with the proximity of integration site to heterochromatin (Festenstein et al., 1996). Methylation of a gene is another way to reduce its transcriptional activity. The binding of transcription factors to the gene promoters can be significantly weakened or even eliminated when the promoter sequences are methylated at CpG sites (Tate & Bird, 1993). There are numerous examples that large transgene arrays attract methylation. The copy number of a transgene correlates positively to methylation and negatively to expression (Garrick, Fiering, Martin, & Whitelaw, 1998).

8.2.5.3 Transgenic mosaicism

A mouse is mosaic when it is made up of at least two cell types that derive from a single zygote but differ in their genomes. Up to 65% of transgenic founder mice are mosaic (Whitelaw, Springbett, Webster, & Clark, 1993). Such a high level of mosaicism is consistent with the interpretation that DNA integration occurs in the chromosome that has already divided. In the theory, transgenes are faithfully transmitted through the germline. Because of the mosaicism in founder mice, the transgenic lines descended from the same founder are different in both genotype and phenotype. They must be defined as independent pedigrees.

8.2.6 Concluding remarks

Transgenic expression of tight junction genes under the direction of tissue or cell type specific promoters is an efficient way to investigate the function of tight junction proteins in vivo in live mice. This approach allows increasing the dosage of tight junction gene expression in a target organ or introducing mutant forms of tight junction proteins to specific epithelium or endothelium. More sophisticated transgenic techniques are available to deliver transgenes in a controllable manner. Together, transgenesis by pronuclear DNA injection is a simple and effective method to study the gain-of-function or dominant-negative effects of tight junction genes.

References

Beaudoing, E., Freier, S., Wyatt, J. R., Claverie, J. M., & Gautheret, D. (2000). Patterns of variant polyadenylation signal usage in human genes. *Genome Research*, *10*, 1001–1010.

Burdon, T. G., & Wall, R. J. (1992). Fate of microinjected genes in preimplantation mouse embryos. *Molecular Reproduction and Development*, *33*, 436–442.

Burke, T. W., & Kadonaga, J. T. (1997). The downstream core promoter element, DPE, is conserved from Drosophila to humans and is recognized by TAFII60 of Drosophila. *Genes & Development*, *11*, 3020–3031.

Choi, T., Huang, M., Gorman, C., & Jaenisch, R. (1991). A generic intron increases gene expression in transgenic mice. *Molecular and Cellular Biology*, *11*, 3070–3074.

Conaway, R. C., & Conaway, J. W. (1993). General initiation factors for RNA polymerase II. *Annual Review of Biochemistry*, *62*, 161–190.

Festenstein, R., Tolaini, M., Corbella, P., Mamalaki, C., Parrington, J., Fox, M., ... Kioussis, D. (1996). Locus control region function and heterochromatin-induced position effect variegation. *Science (New York, NY)*, *271*, 1123–1125.

Garrick, D., Fiering, S., Martin, D. I., & Whitelaw, E. (1998). Repeat-induced gene silencing in mammals. *Nature Genetics*, *18*, 56–59.

Gong, Y., & Hou, J. (2014). Claudin-14 underlies Ca++-sensing receptor-mediated Ca++ metabolism via NFAT-microRNA-based mechanisms. *Journal of the American Society of Nephrology: JASN*, *25*, 745–760.

Gong, Y., Renigunta, V., Himmerkus, N., Zhang, J., Renigunta, A., Bleich, M., & Hou, J. (2012). Claudin-14 regulates renal Ca(+)(+) transport in response to CaSR signalling via a novel microRNA pathway. *The EMBO Journal*, *31*, 1999–2012.

Gordon, J. W., Scangos, G. A., Plotkin, D. J., Barbosa, J. A., & Ruddle, F. H. (1980). Genetic transformation of mouse embryos by microinjection of purified DNA. *Proceedings of the National Academy of Sciences of the United States of America*, *77*, 7380–7384.

Hogan, B., Costantini, F., & Lacy, E. (1986). *Manipulating the mouse embryo: A laboratory manual* (Vol. 34). NY: Cold spring harbor laboratory Cold Spring Harbor.

Ittner, L. M., & Gotz, J. (2007). Pronuclear injection for the production of transgenic mice. *Nature Protocols*, *2*, 1206–1215.

Kozak, M. (1989). The scanning model for translation: An update. *The Journal of Cell Biology*, *108*, 229–241.

Lacy, E., Roberts, S., Evans, E. P., Burtenshaw, M. D., & Costantini, F. D. (1983). A foreign beta-globin gene in transgenic mice: Integration at abnormal chromosomal positions and expression in inappropriate tissues. *Cell*, *34*, 343–358.

Lagrange, T., Kapanidis, A. N., Tang, H., Reinberg, D., & Ebright, R. H. (1998). New core promoter element in RNA polymerase II-dependent transcription: Sequence-specific DNA binding by transcription factor IIB. *Genes & Development*, *12*, 34–44.

Liu, D., Brockman, J. M., Dass, B., Hutchins, L. N., Singh, P., McCarrey, J. R., ... Graber, J. H. (2007). Systematic variation in mRNA 3'-processing signals during mouse spermatogenesis. *Nucleic Acids Research*, *35*, 234–246.

Orphanides, G., Lagrange, T., & Reinberg, D. (1996). The general transcription factors of RNA polymerase II. *Genes & Development*, *10*, 2657–2683.

Richter, J. D. (1999). Cytoplasmic polyadenylation in development and beyond. *Microbiology and Molecular Biology Reviews: MMBR*, *63*, 446–456.

Smale, S. T., Jain, A., Kaufmann, J., Emami, K. H., Lo, K., & Garraway, I. P. (1998). The initiator element: A paradigm for core promoter heterogeneity within metazoan protein-coding genes. *Cold Spring Harbor Symposia on Quantitative Biology*, *63*, 21–31.

Tate, P. H., & Bird, A. P. (1993). Effects of DNA methylation on DNA-binding proteins and gene expression. *Current Opinion in Genetics & Development*, *3*, 226–231.

Whitelaw, C. B., Springbett, A. J., Webster, J., & Clark, J. (1993). The majority of G0 transgenic mice are derived from mosaic embryos. *Transgenic Research*, *2*, 29–32.

Chapter 8.3

Lentivirus-mediated gene knockdown

8.3.1 Background knowledge

8.3.1.1 Lentivirus-mediated transgenesis

Retroviruses such as Moloney murine leukemia virus have been widely used to generate transgenic mice. When injected into mouse embryos, retroviruses can randomly integrate into the genome to disrupt functional genes, a technique being known as insertional mutation or gene trapping (Soriano, Gridley, & Jaenisch, 1987). The retroviral genes, however, are silenced by epigenetic mechanisms in the host cells (Jahner & Jaenisch, 1985). Lentivirus is a genus of retroviruses that cause chronic inflammatory diseases in mammals. The best-known lentivirus is human immunodeficiency virus (HIV). Unlike other retroviruses that are silenced in the mouse genome, lentivirus delivered transgenes are stably integrated into the genome, actively transcribed by the mouse cells and faithfully transmitted through the germline (Lois, Hong, Pease, Brown, & Baltimore, 2002). The lentiviral genome is more complex than the retroviral genome (Fig. 7.4.1). The packaging process is rather similar between the two types of viruses (Fig. 7.4.2).

8.3.1.2 RNA interference in live mice

Transgenic RNA interference (RNAi) approach has several advantages over traditional gene knockout technology for the following reasons. (1) It is simpler because there is no need for complicated ES cell culture and targeting steps and no need to genotype large numbers of mutant cell lines. (2) It is faster because there is no need to screen mosaic mice and subsequent extensive crossing to obtain homozygotes. (3) It is more versatile because RNAi can provide a range of phenotypes depending on the level of interference and yield information about gene dosage effects that cannot be obtained by gene knockout experiments (Stein, Svoboda, & Schultz, 2003). The selection and validation of small interfering RNA (siRNA) molecules should be performed in cultured mouse cells as described in Chapter 7.5. The most effective siRNA molecules are then packaged into lentivirus for the generation of transgenic mice.

8.3.2 Materials and reagents

8.3.2.1 Equipment

NIKON Ti-U light microscope with Hoffman modulation contrast.
Narishige micromanipulators.
Harvard Apparatus PLI-100 compressed air driven Pico-injector.

8.3.2.2 Cell model for ectopic gene expression

HEK293 cells (*producer cell*); available from ATCC.
Mouse L cells (*host cell*); available from ATCC.

8.3.2.3 Plasmids

pLL3.7 lentiviral siRNA expression vector; available from Addgene.
pREV expression vector; available from Addgene.
pRRE expression vector; available from Addgene.
pVSV-G expression vector; available from Addgene.

8.3.2.4 Cell culture medium

Complete medium	
500 mL	Dulbecco's Modified Eagle's medium (DMEM) (high glucose/ + sodium pyruvate/ + glutamine); available from Gibco
50 mL	Fetal bovine serum (FBS); available from Gibco
5 mL	Penicillin/streptomycin; available from Gibco

8.3.2.5 Buffers

$CaCl_2$ (2.5 M)	
27.7 g	$CaCl_2$

Bring to 100 mL with dH_2O, and stir to dissolve.
Filter, sterilize, and store at 4°C.

2× HBS buffer	
140 mM	NaCl
1.5 mM	Na_2HPO_4
50 mM	HEPES

Bring to 100 mL with dH_2O, stir to dissolve, and adjust pH to 7.0.
Filter, sterilize, and store at 4°C.

TNE buffer	
130 mM	NaCl
50 mM	Tris
1 mM	EDTA

Bring to 100 mL with dH_2O, stir to dissolve, and adjust pH to 7.8.
Filter, sterilize, and store at 4°C.

8.3.3 Experimental procedure

8.3.3.1 Lentivirus production

Day 1

1. Grow HEK293 cells to confluence in one T225 flask.
2. Trypsinize the T225 flask, centrifuge the cells at 250x*g* for 5 minutes at room temperature, and resuspend cells in 5 mL DMEM complete medium.
3. Prepare four 150 mm culture dishes, and add 20 mL DMEM complete medium to each dish.
4. Add 1 mL resuspended cells to each dish.
5. Return the cells to incubator and culture overnight (12–16 hours) at 37°C.

Day 2

1. Take the dishes from the incubator and change the fresh medium.
   (*Because HEK293 cells are fragile, tilt the flask when removing the medium and adding fresh medium so that the cell monolayer is not disturbed.*)
2. Mix the following DNAs (made w/ Endo-free Qiagen midi-Kits) in a 15 mL centrifuge tube.

20 μg	pLL3.7-Cldn16-siRNA
20 μg	pREV
20 μg	pRRE
20 μg	pVSV-G
x μL	Nuclease-free H_2O
1.9 mL	Total volume

3. Add 100 μL 2.5 M $CaCl_2$ to DNA mixture. Vortex to mix.
4. In a separate 15 mL centrifuge tube, add 1 mL of 2 × HBS.
5. Add DNA mixture (Step 2) to 2 × HBS (Step 3) dropwise with a Pasteur pipette. Flick the HBS tube while adding a DNA mixture. When finishing adding, vortex gently (using low speed) for 20 seconds.
   (*This is sufficient to transfect one 150 mm dish.*)
6. Let the DNA-HBS mixture stand for 20 minutes at room temperature.
7. Add 2 mL DNA-HBS mixture dropwise to a 150 mm flask.
   (*When adding the mixture, hold the pipette stable while swirling the flask slowly so that the drops can be distributed evenly in different areas of the flask. After adding the mixture, continue swirling the flask for 20 seconds so that the DNA mixture can fully dissolve into the medium.*)
8. Return the flasks into the incubator for 12–16 hours.

Day 3

1. Take the dishes of transfected HEK293 cells out of the incubator and change fresh medium. Return the flasks into the incubator for 48 hours.

Day 4

1. Trypsinize a dish or flask of target cells—mouse L cells.
2. Load 1×10^4 trypsinized cells into a well of the 6-well plate so that on day 5, the cells will reach ~30% confluence.

Day 5

1. Sterilize 50-mL ultra-speed centrifuge tubes by UV or autoclaving.
2. Harvest viral supernatant and centrifuge at 500x*g* for 5 minutes at 4°C.
3. Transfer supernatant to ultra-speed centrifuge tubes and centrifuge at 50,000 × *g* for 2 hours at 4°C.
4. After centrifuge, a white-to-yellow pellet can be seen at the bottom of the tube.
5. Aspirate the supernatant and add 200 μL of ice-cold TNE buffer to cover the pellet. Resuspend the pellet by pipetting up and down for 50 times.
6. Take the target cells from the incubator (~30% confluence) and change with fresh ice-cold medium.
7. Add dissolved virus to the target cells at 1 μL of virus for one well of cells.
   (*The virus can be serially diluted from 1:1 to 1:10^6 to determine the titer.*)
8. Return cells to incubator for 24 hours.

Day 6

1. Take the infected cells from the incubator and aspirate the medium.
2. Wash the cells once with 1 × PBS.
3. Add fresh medium.
4. Return to the incubator for 24 hours.

Day 7

1. Trypsinize the cells from the 6-well plate.
   (*The trypsin step is dispensable because lentivirus can infect both dividing and nondividing cells.*)
2. Load the cells into T25 flask. Add fresh medium.
3. Return to the incubator for 24 hours.

Day 8

1. Aspirate the medium and add fresh medium.
2. Add drugs such as neomycin to select the infected cells.
3. Determine the titer of retrovirus.

8.3.3.2 Perivitelline injection of lentivirus and production of transgenic mice

The detailed protocol for perivitelline injection of lentivirus and production of transgenic mice has been described elsewhere (Barde, Verp, Offner, & Trono, 2011). In brief, the packaged virions are injected beneath the zona pellucida of 1-cell embryos to allow fusion with the plasma membrane and delivery of the genetic material (Fig. 8.3.1). The injected embryos are then implanted into the oviduct of the foster mother and allowed to develop to term.

8.3.4 Data analysis

The viral titer corresponds to the number of colonies present at the highest dilution that contains colonies, multiplied by the dilution factor. For example, the presence of 1 colony after 10^4 dilutions from 1 μL viral stock represents a viral titer of 1×10^4 colony forming units (cfu) per μL.

$$1 \times 10^4 \text{ cfu/mL} = 1 \times 10^7 \text{ cfu/mL}$$

Note that the titer of lentivirus is approximately 100-fold lower than the titer of Moloney murine leukemia retrovirus. Successful transgenesis can be achieved at a viral titer of 1×10^7 cfu/mL. Compared to pronuclear DNA injection, lentivirus-mediated transgenesis is relatively easier to perform and can lead to higher percentages of positive animals. Up to 70% of founder mice express the transgene (Hou et al., 2007). The founder mice exhibit multicopy integration in the genome. The copy number varies from 1 to 10 (Hou et al., 2007). The onset of transgenic expression is as early as the blastocyst stage (Fig. 8.3.2). All major organs express the transgene (Fig. 8.3.3).

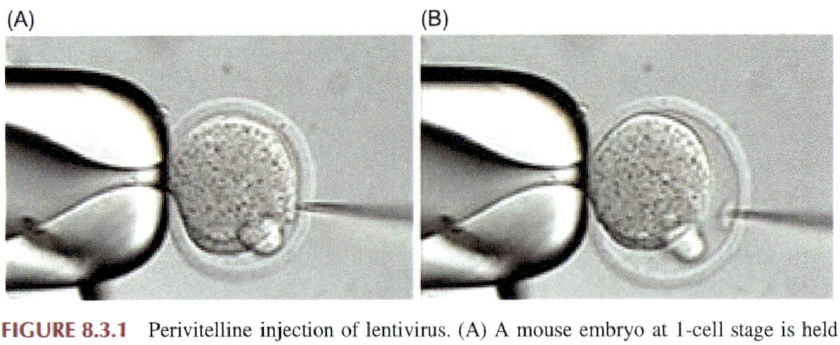

FIGURE 8.3.1 Perivitelline injection of lentivirus. (A) A mouse embryo at 1-cell stage is held on the left by the holding capillary and on the right the injection capillary contains the lentivirus. (B) The injection capillary gently penetrates the zona pellucida and delivers 10 pL of lentivirus at a titer of 1×10^7 cfu/mL. Notice the visible swelling in the perivitelline space. *Reproduced with permission from Barde, I., Verp, S., Offner, S., & Trono, D. (2011). Lentiviral vector mediated transgenesis.* Current Protocols in Mouse Biology, 1, *169–184.*

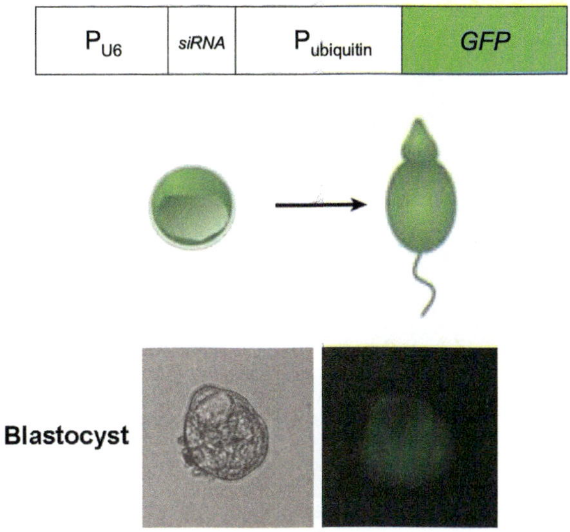

FIGURE 8.3.2 Lentivirus-mediated transgenesis. Diagram of lentiviral vector consisting of the siRNA expression cassette (driven by the U6 promoter) and the GFP expression cassette (driven by the ubiquitin promoter) is shown above. Expression of transgene (GFP) in a blastocyst-stage embryo is shown below. *Reproduced with permission from Hou, J., Shan, Q., Wang, T., Gomes, A. S., Yan, Q., Paul, D. L., Bleich, M., & Goodenough, D. A. (2007). Transgenic RNAi depletion of claudin-16 and the renal handling of magnesium.* The Journal of Biological Chemistry, 282, *17114–17122.*

Transgenic expression is stable over the entire life span of the animal (Hou et al., 2007). Knockdown of the target gene—claudin-16 is effective ($>$100-fold reduction) on the levels of mRNA and protein (Hou et al., 2007). Immunolabeling of claudin-16 proteins is completely eliminated in the kidney epithelia of transgenic mice, indicating a lack of cellular mosaicism in founder mice (Fig. 8.3.4).

8.3.5 Troubleshooting

8.3.5.1 Silencing of recombinant lentivirus

The expression of a transgene is observed in all cells of a given type and nearly all cell types, which indicates that lentiviral gene silencing is not a major problem (Lois et al., 2002). However, there are exceptions. The expression is silenced in pancreatic fibroblasts and endothelial cells (Fig. 8.3.5), hepatic endothelial and Kupffer cells (Fig. 8.3.6) and germ cells from ovary and testis (Fig. 8.3.7) (Hou et al., 2007). More selective silencing can be observed in the kidney. For example, a single cell within a distal convoluted tubule shows gene silencing, whereas the rest of the cells of

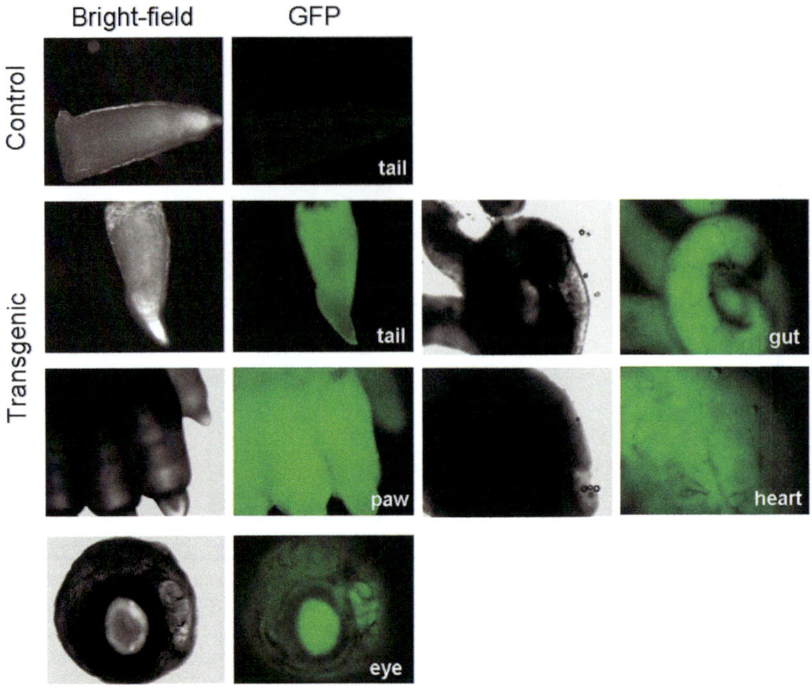

FIGURE 8.3.3 Expression of transgene in major organs. Gallery of the expression of transgene (GFP) in major organs of a transgenic mouse is shown in bottom panels. The control mouse is shown in top panels. *Reproduced with permission from Hou, J., Shan, Q., Wang, T., Gomes, A. S., Yan, Q., Paul, D. L., Bleich, M., & Goodenough, D. A. (2007). Transgenic RNAi depletion of claudin-16 and the renal handling of magnesium.* The Journal of Biological Chemistry, 282, *17114–17122.*

the same tubule are active in the expression of the transgene (Fig. 8.3.8) (Hou et al., 2007).

8.3.5.2 Toxicity of siRNA expression

The toxicity of RNAi is on many levels. When siRNA expression is high, cells might activate the interferon response as a part of innate immunity. If an siRNA molecule has off-target effects, then it might compromise the functions of the off-target genes, thereby inflicting toxicity on organ systems beyond the target tissue. Liver injury is a serious side effect of RNAi in live animals. When siRNA molecules are expressed at high levels in the hepatocytes, they compete with endogenous microRNAs for binding to nuclear karyopherin exportin-5, which inadvertently downregulates liver-derived microRNAs (Grimm et al., 2006). The competition model also explains why efforts to establish transgenic mice expressing multicopy of siRNA sequences at high levels have remained unsuccessful (Cao, Hunter, Strnatka, McQueen, & Erickson, 2005).

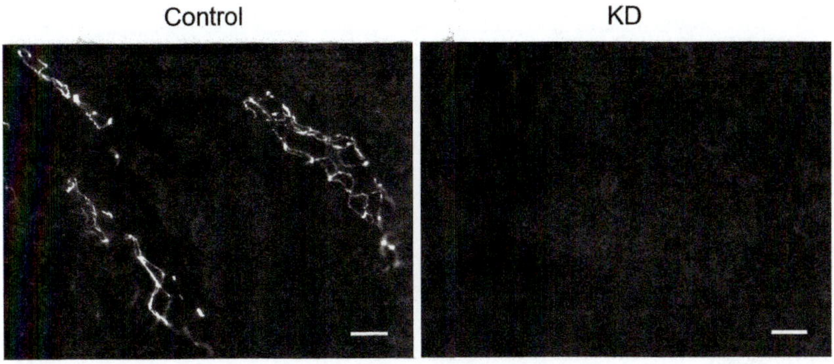

FIGURE 8.3.4 Immunolabeling of claudin-16 proteins in transgenic mouse kidney. Sagittal cryostat section (10 μm) from the control mouse kidney shows claudin-16 localization in the tight junctions of tubular epithelia in the medullary ray (i.e., the thick ascending limb). In knockdown (KD) mouse kidney, claudin-16 labeling disappears, giving way to a background level. Bar: 10 μm. *Reproduced with permission from Hou, J., Shan, Q., Wang, T., Gomes, A. S., Yan, Q., Paul, D. L., Bleich, M., & Goodenough, D. A. (2007). Transgenic RNAi depletion of claudin-16 and the renal handling of magnesium. The* Journal of Biological Chemistry, 282, *17114–17122.*

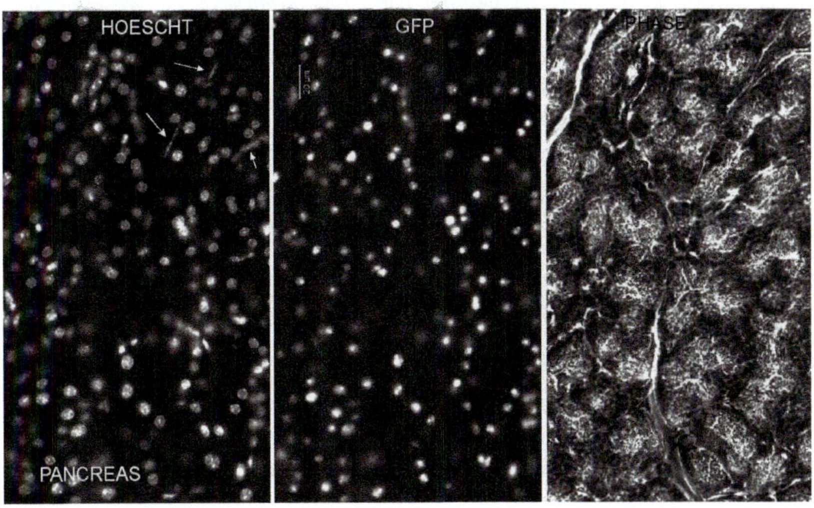

FIGURE 8.3.5 Silencing of lentivirus in the pancreas. Arrows indicate fibroblasts or endothelial cells in the connective tissue of the pancreas are not seen with GFP expression. Bar: 20 μm. *Reproduced with permission from Hou, J., Shan, Q., Wang, T., Gomes, A.S., Yan, Q., Paul, D. L., Bleich, M., & Goodenough, D. A. (2007). Transgenic RNAi depletion of claudin-16 and the renal handling of magnesium.* The Journal of Biological Chemistry, 282, *17114–17122.*

8.3.5.3 Knockdown versus knockout

Because RNAi often does not reduce protein levels to zero, the residual amount of targeted proteins may be enough to perform the gene's function without

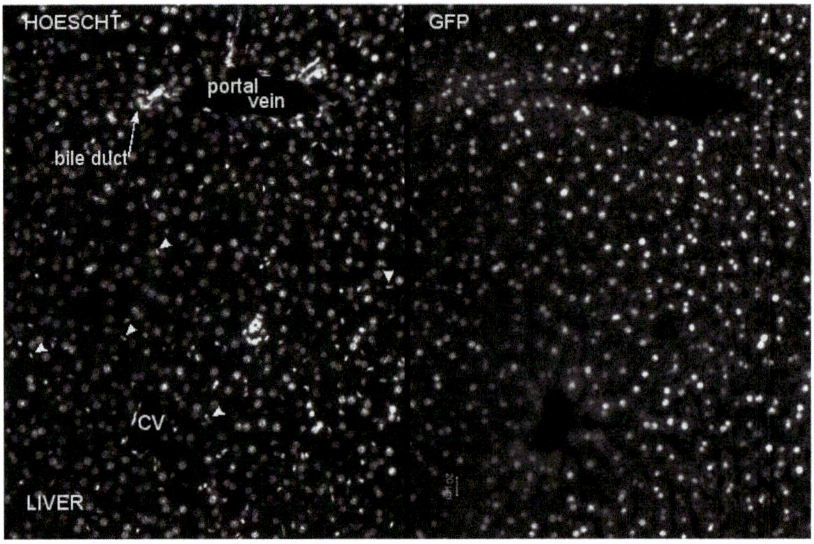

FIGURE 8.3.6 Silencing of lentivirus in the liver. Arrowheads indicate endothelial or Kupffer cells in the liver are not expressing the GFP. Bar: 20 μm. *Reproduced with permission from Hou, J., Shan, Q., Wang, T., Gomes, A. S., Yan, Q., Paul, D. L., Bleich, M., & Goodenough, D. A. (2007). Transgenic RNAi depletion of claudin-16 and the renal handling of magnesium.* The Journal of Biological Chemistry, 282, *17114–17122.*

resulting in a phenotypic change. It is also important to distinguish whether RNAi fails to work because the chosen gene is RNAi resistant or because of some technical issues such as ineffective siRNA molecule. On the other hand, knockdown phenotypes may be more physiologic than knockout phenotypes. When a target gene is completely removed, cells have no choice but to activate compensatory mechanisms. A knockout phenotype includes, in essence, not only the loss-of-function effects of a target gene but also the compensatory effects due to activation of related genes. From a pathologic point of view, knockdown phenotypes might better resemble diseases. Many types of Mendelian diseases are caused by missense mutations, which often reduce the target protein levels rather than completely remove the proteins as nonsense mutations.

8.3.6 Concluding remarks

Conventional DNA injection-based methods are successful in generating transgenic mice, but the transgenic rate is low and the expression of the transgene is often mosaic. Lentivirus-based transgenic approaches overcome these limitations, in part because of its ability to incorporate transgenes into the genome with high efficiency. When combined with RNAi technology, recombinant lentivirus allows the generation of knockdown mice for tight junction genes. The knockdown strategy resembles the knockout strategy in that both aim to reveal the phenotype caused by the loss of a gene. Because knockdown mice retain a residue function

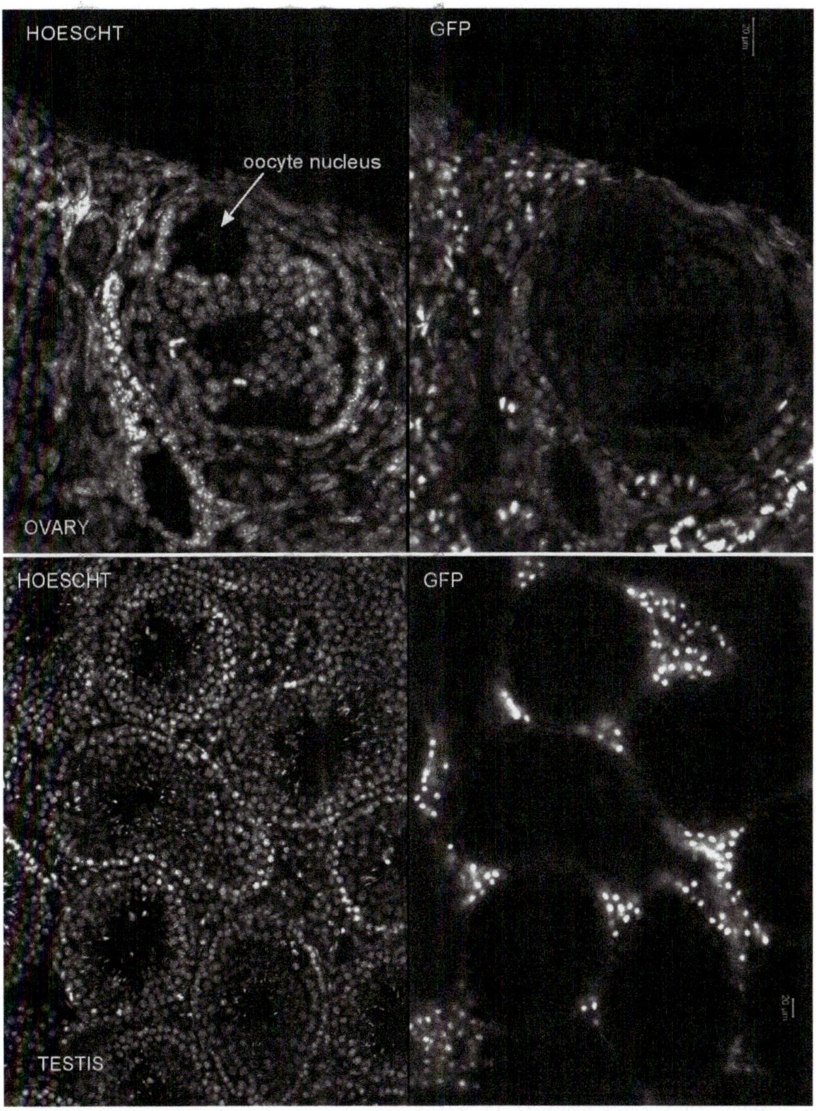

FIGURE 8.3.7 Silencing of lentivirus in the germ cells. Note that oocytes in the ovary or spermatogonia and spermatozoa in the testis are with no GFP expression. Bar: 20 μm. *Reproduced with permission from Hou, J., Shan, Q., Wang, T., Gomes, A. S., Yan, Q., Paul, D. L., Bleich, M., & Goodenough, D. A. (2007). Transgenic RNAi depletion of claudin-16 and the renal handling of magnesium.* The Journal of Biological Chemistry, 282, *17114–17122.*

of the target gene, the knockdown phenotype might better describe what the gene does in the absence of compensatory mechanisms. Gene compensation is common in tight junction biology. Therefore, lentivirus-mediated RNAi technology provides an important tool for tight junction research.

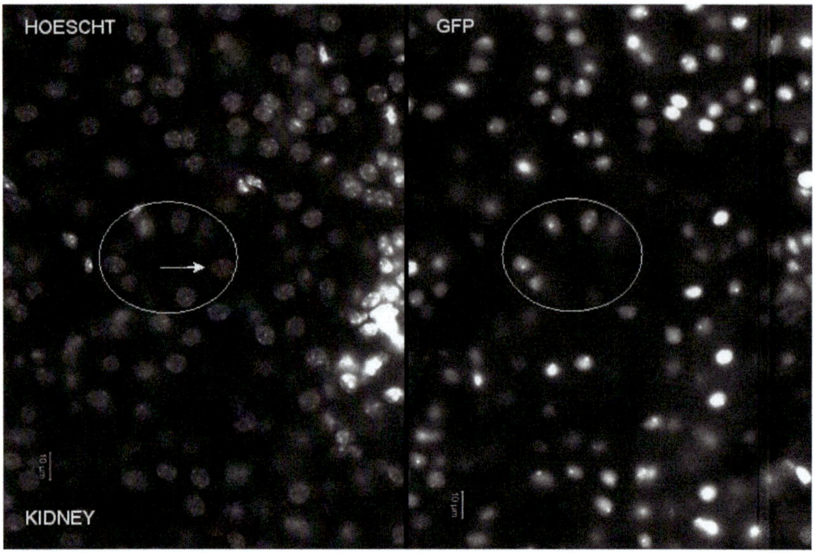

FIGURE 8.3.8 Silencing of lentivirus in the distal tubular cell of the kidney. Arrow indicates a cell in the distal convoluted tubule (*circled*) is without GFP expression. Bar: 10 μm. *Reproduced with permission from Hou, J., Shan, Q., Wang, T., Gomes, A. S., Yan, Q., Paul, D. L., Bleich, M., & Goodenough, D. A. (2007). Transgenic RNAi depletion of claudin-16 and the renal handling of magnesium.* The Journal of Biological Chemistry, 282, *17114–17122.*

References

Barde, I., Verp, S., Offner, S., & Trono, D. (2011). Lentiviral vector mediated transgenesis. *Current Protocols in Mouse Biology*, *1*, 169–184.

Cao, W., Hunter, R., Strnatka, D., McQueen, C. A., & Erickson, R. P. (2005). DNA constructs designed to produce short hairpin, interfering RNAs in transgenic mice sometimes show early lethality and an interferon response. *Journal of Applied Genetics*, *46*, 217–225.

Grimm, D., Streetz, K. L., Jopling, C. L., Storm, T. A., Pandey, K., Davis, C. R., ... Kay, M. A. (2006). Fatality in mice due to oversaturation of cellular microRNA/short hairpin RNA pathways. *Nature*, *441*, 537–541.

Hou, J., Shan, Q., Wang, T., Gomes, A. S., Yan, Q., Paul, D. L., ... Goodenough, D. A. (2007). Transgenic RNAi depletion of claudin-16 and the renal handling of magnesium. *The Journal of Biological Chemistry*, *282*, 17114–17122.

Jahner, D., & Jaenisch, R. (1985). Retrovirus-induced de novo methylation of flanking host sequences correlates with gene inactivity. *Nature*, *315*, 594–597.

Lois, C., Hong, E. J., Pease, S., Brown, E. J., & Baltimore, D. (2002). Germline transmission and tissue-specific expression of transgenes delivered by lentiviral vectors. *Science (New York, NY)*, *295*, 868–872.

Soriano, P., Gridley, T., & Jaenisch, R. (1987). Retroviruses and insertional mutagenesis in mice: Proviral integration at the Mov 34 locus leads to early embryonic death. *Genes & Development*, *1*, 366–375.

Stein, P., Svoboda, P., & Schultz, R. M. (2003). Transgenic RNAi in mouse oocytes: A simple and fast approach to study gene function. *Developmental Biology*, *256*, 187–193.

Chapter 8.4

Conditional gene knockout by homologous recombination

8.4.1 Background knowledge

8.4.1.1 Site-specific recombination system

Two of the most widely used site-specific recombination systems are (1) Cre-loxP from bacteriophage P1 (Austin, Ziese, & Sternberg, 1981) and (2) FLP-FRT from *Saccharomyces cerevisiae* (Andrews, Proteau, Beatty, & Sadowski, 1985). The mechanism of action of these recombination systems is the same. A recombinase recognizes a 34-bp site, known as loxP in the case of Cre and FRT in the case of FLP (Fig. 8.4.1). Each site consists of a core and two flanking sequences. The flanking sequences are identical but inverted, which are bound by the recombinase. The core sequence is asymmetrical. It confers orientation to the site by determining which sides of the two participating duplexes are brought together. Excision, insertion, and chromosomal rearrangements have been performed with the recombination systems in bacteria, plants and eukaryotes (Torres & Kühn, 1997). Rearrangement of DNA sequences through the use of site-specific recombination has proven particularly useful for the development of conditional gene knockout mouse models.

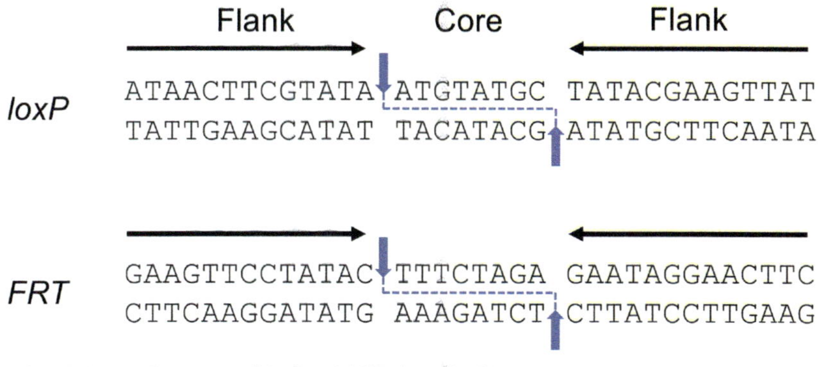

FIGURE 8.4.1 Sequences of loxP and FRT sites. The black arrows draw attention to the identical inverted flanking regions; the blue arrows show the points where each of the duplexes exchanges with a second identical duplex.

8.4.1.2 Design of targeting vector

The homologous recombination between the targeting vector and the genomic sequence in ES cells requires extensive stretches of homology to proceed with fidelity and efficiency (Deng & Capecchi, 1992). Because strain-specific polymorphisms may interfere with homologous recombination, the 5′- and 3′-homologous sequences in the targeting vector must be isogenic to the genomic sequences in the ES cells. The homologous DNA is often distributed asymmetrically in the 5′- and 3′-flanking regions of homology (also known as "homologous arms") on the targeting vector so that one of the homologous arms is considerably shorter to make it more convenient to perform diagnostic screens (Fig. 8.4.2). The lengths of the homologous arms are critical. The minimal length for the short arm of is 2 kb. The long arm ranges from 5 to 10 kb. The resistance gene to a drug (i.e., neomycin) is placed between the homologous arms to select for correct integration, which is known as positive selection. The DTA toxin gene is placed outside the homologous arms to select against random integration, which is known as negative selection (McCarrick, Parnes, Seong, Solter, & Knowles, 1993). Conventional knockout vectors are designed for the drug resistance gene to replace the endogenous gene or its key exons. In conditional knockout mice, the endogenous gene must function properly until recombined by the enzyme Cre, therefore, any modification including the drug resistance gene has to be removed from the genome before an experiment commences.

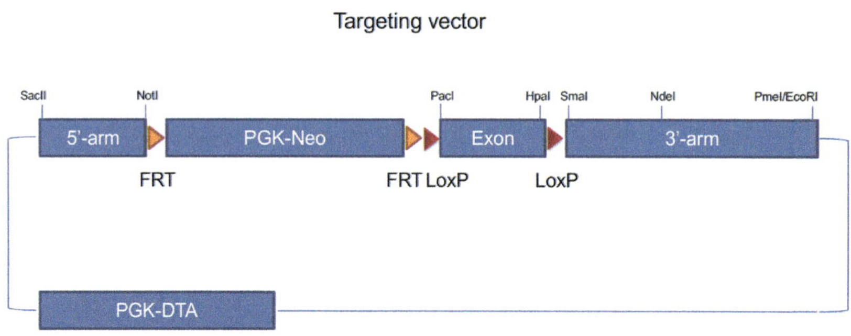

FIGURE 8.4.2 Targeting vector for conditional gene knockout. A targeting vector typically contains two homologous arms that flank the endogenous gene to be replaced with the resistance gene to a drug, that is, neomycin. Neomycin resistance gene is expressed by the PGK promoter in the PGK-Neo expression cassette. The PGK-Neo cassette is flanked by two FRT sites, which facilitates its cleavage after the targeting vector is correctly integrated. The endogenous gene or its key exon is flanked by two loxP sites to ensure a knockout effect in the presence of Cre. The DTA toxin is expressed by the PGK promoter in the PGK-DTA expression cassette to prevent random insertion of the targeting vector to the genome. A restriction endonuclease site must be placed in the 5′-end of the 5′-arm (SacII) or the 3′-end of the 3′-arm (PmeI) to allow linearizing the targeting vector.

A common strategy to remove the drug resistance gene is by flanking it with two FRT sites and introducing the FLP enzyme to the targeted ES cells or the ES cell-derived transgenic mice. The loxP sites must be inserted in a manner that recombination will result in the excision of the endogenous gene or its key exon to render the gene inactive. It is also important to note that the distance between the loxP sites and the drug resistance gene may influence the efficiency of generating clones that contain all elements of the targeting vector.

8.4.2 Materials and reagents

8.4.2.1 Equipment

Harvard Apparatus BTX HT200 Electroporator.
Harvard Apparatus "flat pack" 1.8 mm gap cuvette for BTX electroporator.
NIKON Ti-U light microscope with Hoffman modulation contrast.
Narishige micromanipulators.

8.4.2.2 Cell line

EDJ#22 ES cell line (Strain 129/Sv); available from ATCC.
Mouse embryonic fibroblast (MEF) feeder cells (γ-irradiated, neomycin resistant); available from Applied StemCell.

8.4.2.3 Cell culture medium

ES cell medium

DMEM	400 mL	Gibco
Defined ESC Grade FBS	85 mL	Hyclone
100 × L-Glutamine	5 mL	Corning
100 × NEAA	5 mL	Corning
100 × Pen/Strep	5 mL	Gibco
HEPES	6.25 mL	Corning
β-ME	5 μL	Sigma
LIF	50 μL	Millipore
Volume	500 mL	

MEF medium

DMEM	435 mL	Gibco
FBS	50 mL	Hyclone
100 × L-Glutamine	5 mL	Corning
100 × NEAA	5 mL	Corning
100 × Pen/Strep	5 mL	Gibco
HEPES	6.25 mL	Corning
β-ME	5 μL	Sigma
Volume	500 mL	

2 × ES cell freezing medium	
70 mL	ES cell medium
10 mL	Defined ESC Grade FBS
20 mL	DMSO

8.4.2.4 Buffer

Electroporation buffer	
HEPES	20 mM
NaCl	137 mM
KCl	5 mM
Na_2HP0_4	0.7 mM
Dextrose	6 mM

Bring to 100 mL with dH_2O, stir to dissolve, and adjust pH to 7.0. Filter, sterilize, and store at 4°C. On the day of transfection, aliquot 8.5 mL of electroporation buffer and add 1.5 mL of defined ESC grade FBS; set the buffer on ice until ready to use.

8.4.3 Experimental procedure

8.4.3.1 Plating embryonic stem cells

1. Plate (1) 10-cm dish with irradiated MEF feeder cells at 1.5×10^6 in 10 mL MEF medium.
2. Incubate a minimum of 2 hours, but preferably overnight at 37°C.
3. Remove the MEF medium from the 10-cm dish and replace it with 10 mL of ES cell medium.
4. Remove a vial ES cell from liquid nitrogen and quickly thaw by holding the vial with hand.
5. Directly add cells into the 10-cm dish pre-seeded with MEF feeder cells and incubate at 37°C.
6. Check cell growth under the scope and replenish ES media daily.
   a. Let cell grow 2 − 3 days to reach 75% confluence.
   b. Never over-grow beyond 75% confluence because over-growth will cause ES cell differentiation.

8.4.3.2 DNA electroporation

1. The day before the transfection, plate (10) 10-cm dishes with irradiated MEF cells at 1.5×10^6 in 10 mL of MEF medium.
2. Check if ES cell density has reached 70%−75% confluence. Also check ES cell morphology for any sign of ES cell differentiation.
   - If they are too sparse, continue culture for one more day.
   - If they are too dense, make sure they are still undifferentiated.

3. Replenish the ES cell culture dish with ES cell medium 2 hours before electroporation. After 2 hours of incubation, aspirate ES cell medium and rinse ES cells with 10 mL of 1 × PBS.
4. Aspirate 1 × PBS, add 3 mL of trypsin-EDTA, roll the dish back and forth to let trypsin-EDTA cover the entire dish and incubate for 10 minutes at 37°C.
   a. Check the disaggregation progress of cells under a light microscope every 5 minutes.
5. When cells start to detach, pipette the cells with a 5-mL pipette up and down gently to disrupt the cell clumps.
6. Check under the microscope to determine if the clumped cells are now single cells. If they are not single cells, repeat the disaggregation process.
7. When the single-cell suspension is achieved, add resuspended ES cells to a 15-mL centrifuge tube containing 10 mL of ES cell medium.
8. Centrifuge the cells at 250x*g* for 5 minutes.
9. Aspirate the medium and resuspend the pellet in 5 mL of ES cell medium by pipetting up and down very gently.
10. Count the ES cells with a hemocytometer.
11. Transfer an appropriate volume of resuspended ES cells into a sterile centrifuge tube to obtain 3×10^6 cells.
12. Repeat this step for each electroporation preparation.
13. Centrifuge the cells at 250 x *g* for 5 minutes.
14. Carefully remove the ES cell medium with an Eppendorf pipette.
    a. Do not aspirate the medium with a vacuum to prevent accidental removal of pelleted cells.
    b. Ensure that the pipette's set volume is not exceeded to prevent aspirating cells into the pipette.
15. Resuspend the cell pellet in 1 mL of ice-cold Electroporation Buffer with a sterile Pasteur pipette attached to a rubber head.
16. Add 20 μg of a linearized targeting vector to the cell suspension and gently mix with the Pasteur pipette.
    a. Linearize the targeting vector by restriction endonuclease digestion according to Chapter 8.2.3.1.
    b. Dissolve the linearized DNA with TE buffer at 1 μg/μL.
17. Transfer the ES cell + DNA mixture into a sterile ice-cold "flat-back" cuvette with a Pasteur pipette and place it firmly between the two electrodes of electroporator.
    a. Check the solution in the cuvette to ensure there are no air bubbles. Air bubbles may cause an electric arc.
18. Zap the cells at the appropriate settings for ES cells.
    a. Voltage: 185 V.
    b. Resistance (ohms): 350 Ω.
    c. Capacitance (uFarads): 500 μF.

19. Carefully transfer the zapped cell suspension from the cuvette to a 15-mL centrifuge tube that contains 5 mL of ES cell medium.
    a. Use a sterile, cotton-plugged Pasteur pipette to reach the bottom of the cuvette.
20. Aspirate the MEF medium from the 10 dishes pre-seeded with MEF and replenish with 10 mL of ES cell medium and add 0.5 mL of zapped ES cells to each dish.
21. Return dishes to the incubator and let ES cells grow for 1 day.
22. Aspirate the ES cell medium and replenish with fresh ES cell medium containing 600 μg/mL of neomycin (G418). Return dishes to incubator and change medium every day. Let cells grow 2–5 days until it is ready to pick single colonies.

    (*Note that neomycin selection should be applied until individual ES cell clones are picked, genotyped and frozen to ensure that no wild-type ES cells contaminate the clones.*)

8.4.3.3 Embryonic stem cell screening

1. The day before the colony picking, plate (10) 24-well plates with irradiated MEF cells at 7.5×10^4 in each well.
2. Under a light microscope, check the size of ES cell colonies and the ES cell morphology for any sign of differentiation.
    a. Good clones have tight borders and are closely packed.
    b. The presence of large cells with well-defined membrane structures in a colony indicates the differentiation of ES cells.
3. Replenish the 24-well plates with ES cell medium (+neomycin) 2 hours before picking the colonies.
4. Place the plate under the microscope and focus on the ES cell clone to be picked. Use a 200 μL barrier tip to gently detach the ES cell clone from the surrounding MEFs. With a pipet set to 50 μL, depress the plunger button and pluck the clone using a forward scooping suction motion. Place each clone into one of the 24 wells containing ES cell medium (+neomycin). Number the wells from 1 to 240. Each well corresponds to an independent ES cell clone.
5. Return plates to the incubator and change medium every day. Let cells grow 2–5 days until the individual wells of the 24-well plates contain 50–100 small healthy undifferentiated colonies.
6. To freeze individual clones, rinse each well with 0.5 mL of 1x PBS, and then add 0.2 mL of Trypsin/EDTA to each well. Place the plate in the incubator for 10 minutes. Add 0.5 mL of ES cell medium to each well and dissociate the cells with a 1000 μL pipet by pipetting up and down four to five times.
7. Place 500 μL of the dissociated cells into a Nunc vial (1.8 mL) that is numbered the same as the well. Add 0.5 mL of 2× ES freezing medium

to each vial. Place the numbered cryotubes in a freezer box and put in −80°C freezer.

8. After freezing all clones, refeed each well in the 24-well plates with 1.0 mL of ES cell medium. Allow the cells to grow to confluence and harvest the cells for genotyping.

8.4.3.4 Embryonic stem cell injection to blastocyst

1. After genotyping, thaw and plate the correctly targeted ES cell clone as described in Chapter 8.4.3.1.
2. On the day of blastocyst injection, trypsinize the ES cells and resuspend the ES cells at 1.0×10^6 in 2 mL of ES cell medium.
3. The detailed protocol for blastocyst injection and production of transgenic mice has been described elsewhere (Hogan, Costantini, & Lacy, 1986). In brief, use the holding capillary to orient the blastocyst so that the inner cell mass is in the 6 o'clock position with respect to the injection capillary. Move the injection capillary so it is in the same focal plane with the zona pellucida and the outer trophoblast layers of the blastocyst. Insert the injection capillary inside the blastocoel cavity and gently expel the ES cells, being careful not to touch the inner cell mass (Fig. 8.4.3). The injected blastocysts are then implanted into the uterus of the foster mother and allowed to develop to term.

8.4.4 Data analysis

8.4.4.1 Screening of targeted embryonic stem cell clones

Fig. 8.4.4A shows the wild-type claudin-4 locus, the targeting vector, and the targeted locus. On the targeting vector, the 5′-arm homologous region is 3 kb long; the 3′-arm homologous region is 5 kb long. The exon 1 (only exon) of claudin-4 gene is flanked by two loxP sites. The PGK-neo expression cassette is flanked by two FRT sites. Once correctly targeted, the PGK-neo cassette will be removed by the FLP enzyme. The exon 1 will be removed by the Cre enzyme, generating conditional knockouts. Integration of the targeting vector to claudin-4 locus is investigated by a PCR strategy. The primer pair (F1/R1) allows verifying the 5′-arm recombination. The primer F1 is located outside the 5′-arm homologous region whereas the primer R1 is located in the neo gene, which is not present in the mouse genome (Fig. 8.4.4B). Out of 200 neomycin-resistant clones, six clones (#16, #26, #47, #70, #94, and #170) showed correct amplicon size (Fig. 8.4.4B). The primer pair (F2/R2) is used to detect the integration of exon 1 and two loxP sites. The primer F2 is located in the neo gene whereas the primer R2 is located downstream of second loxP site, which is very often deleted after recombination. A unique restriction endonuclease site—Fse1 has been

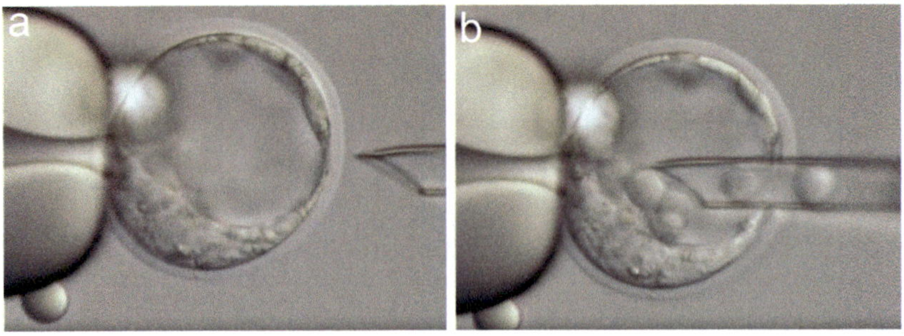

FIGURE 8.4.3 Blastocyst injection of ES cell. (A) A mouse embryo at blastocyst stage is held on the left by the holding capillary and on the right the injection capillary contains the ES cells. (B) The injection capillary gently penetrates the zona pellucida and delivers 10–20 ES cells.

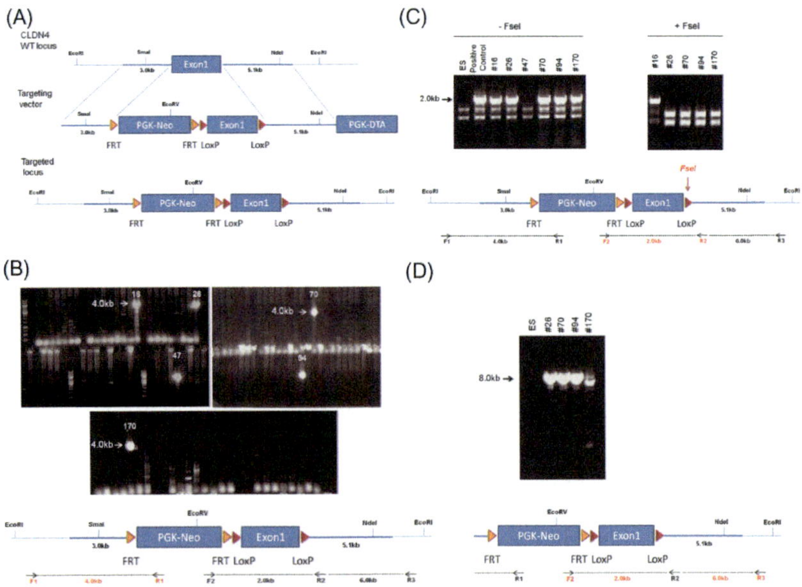

FIGURE 8.4.4 Generation of floxed claudin-4 allele. (A) A diagram shows the wild-type CLDN4 locus, the targeting construct, and the targeted locus. (B) PCR verification of 5′-arm recombination using primer pair F1/R1. (C) PCR verification of integration of exon1 and two loxP sites using primer pair F2/R2, followed by Fse1 digestion. (D) PCR verification of 3′-arm recombination using primer pair F2/R3. *Reproduced with permission from Gong, Y., Yu, M., Yang, J., Gonzales, E., Perez, R., Hou, M., Tripathi, P., Hering-Smith, K. S., Hamm, L. L., & Hou, J. (2014). The Cap1-claudin-4 regulatory pathway is important for renal chloride reabsorption and blood pressure regulation.* Proceedings of the National Academy of Sciences of the United States of America, 111, *E3766–3774.*

incorporated into the second loxP site to facilitate diagnosis (Fig. 8.4.4.C). F2 and R2 amplified a 2-kb band, which was absent in clone #47, suggesting the exon1 was lost in this clone. Fse1 site was missing in clone #16, suggesting its recombination has prematurely terminated prior to the second loxP site (Fig. 8.4.4.C). The primer (R3) is located outside the 3′-arm homologous region (Fig. 8.4.4D). When used with the primer (F2), R3 allows verification of the 3′-arm recombination. F2 and R3 amplified an 8 kb band in clones #26, #70 and #94. The amplicon was significantly shorter in clone #170, indicating it had a premature ending at the 3′-arm (Fig. 8.4.4D).

8.4.4.2 Breeding strategy for mutant mice

The clones #26, #70, and #94 were injected into the C57BL/6 blastocysts to generate chimeric founders. The founder mice were bred to wild-type C57BL/6 mice to generate F1 offspring. The presence of neo gene in F1 generation indicates germline transmission. Cleavage of the PGK-neo expression cassette was carried out by crossing F1 mouse with a transgenic mouse (Gt (ROSA)26Sor$^{tm1(FLP1)}$; the Jackson Lab) that expresses the FLP recombinase in the ROSA-26 locus. Gene expression in the ROSA-26 locus is ubiquitous, including the germ cells (Farley, Soriano, Steffen, & Dymecki, 2000). The offspring is heterozygous for the floxed allele of claudin-4. Homozygosity is achieved by intercross of heterozygous claudin-4 floxed mice (Fig. 8.4.5).

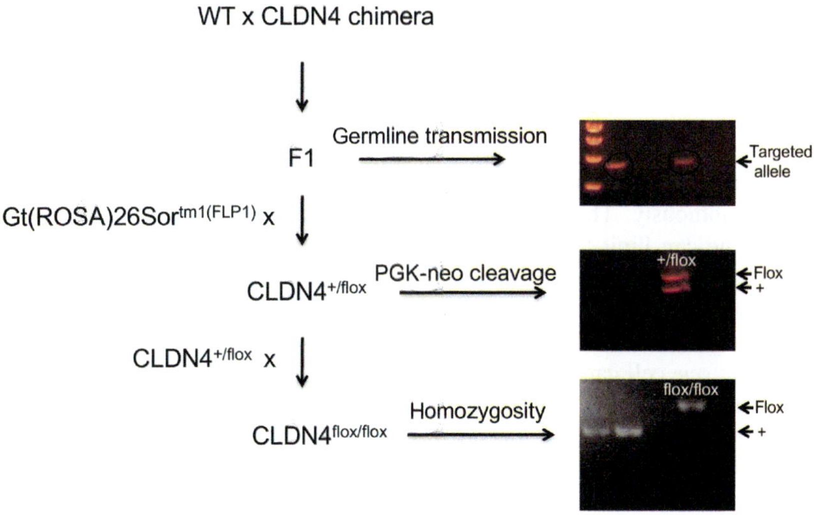

FIGURE 8.4.5 Breeding strategy for claudin-4 floxed mice. A flowchart indicates the stepwise breeding strategy to generate homozygous claudin-4 floxed mice. flox: flanked by loxP sites.

8.4.5 Troubleshooting

8.4.5.1 Issues related to targeting vector

Insertion of loxP sites into intronless genes, as most claudin genes, can be difficult. Care must be taken to avoid disrupting the regulatory sequences in 5′- and 3′-untranslated regions (UTRs). The PGK-neo expression cassette must be removed by the FLP recombinase. The neo gene in the targeted locus has been shown to influence the expression of neighboring genes, sometimes at a distance of 100 kb apart (Pham, MacIvor, Hug, Heusel, & Ley, 1996). Even short DNA sequences such as loxP and FRT may have off-target effects. If these sequences are accidentally placed into regulatory elements, they can alter the expression of not only the targeted gene, but also the neighboring genes (Meier et al., 2010).

8.4.5.2 Issues related to Cre expression

The Cre enzyme needs to be expressed at a sufficiently high level to get close to 100% recombination. The rate of Cre-mediated recombination can be improved by altering the Cre cDNA to include more mammalian codons and a nuclear localization signal to increase Cre protein abundance and trafficking to the nucleus. Despite these modifications, it is important to bear in mind that, since Cre is delivered as a transgene by pronuclear DNA injection, its expression might be limited by the factors elaborated in Chapter 8.2.5, including silencing, mosaicism, and nonspecificity. When Cre expression varies in a target organ, the efficacy of Cre-mediated gene knockout will vary accordingly.

8.4.5.3 Cell autonomy

Tissue-specific knockout strategy is best suited to study the genes that act cell-autonomously. The half-life of the mRNA and protein encoded by the target gene can limit the effectiveness of this approach. If the gene product is very stable, as many tight junction proteins, the reduction will lag behind the induced genetic ablation, potentially masking the mutant effect. Because tight junction is made by two neighboring cells, the deletion of a tight junction gene from one cell can be sufficient to affect the tight junction function even if the gene is intact in the partner cell. For example, the collecting duct of the mouse kidney is a heterogeneous epithelium, made of the principal cell and the intercalated cell. When the claudin-4 gene is knocked out selectively from the principal cell, the abundance of claudin-4 protein in the intercalated cell is concomitantly reduced, most likely due to the loss of *trans* claudin-4 interactions in the tight junction (Fig. 8.4.6) (Gong et al., 2014).

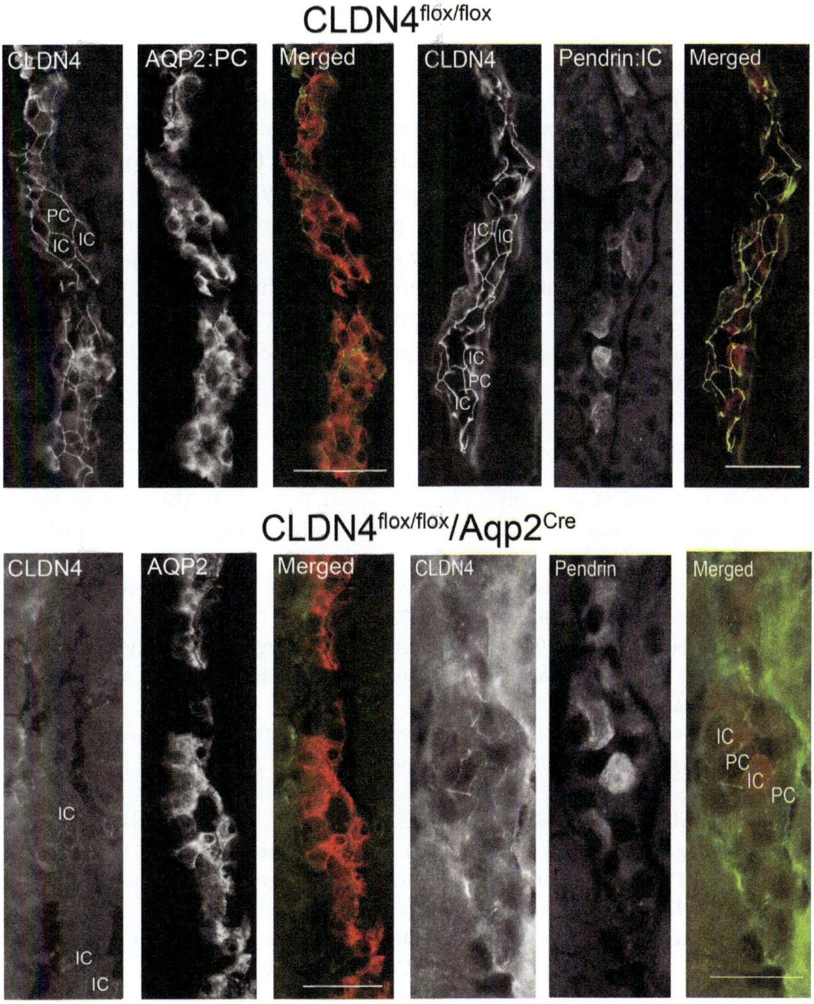

FIGURE 8.4.6 Collecting duct specific deletion of claudin-4 gene in the mouse kidney. Double immunostaining of claudin-4 protein with a principal cell marker, Aqp2, or with an intercalated cell marker, pendrin, in the collecting duct of the cortex from control ($CLDN4^{flox/flox}$) and knockout ($CLDN4^{flox/flox}/Aqp2^{Cre}$) mouse kidneys. IC, intercalated cell; PC, principal cell. In the control mouse kidney, the majority of cell-cell junctions made of claudin-4 are between the PCs, that is, the homotypic junction. Claudin-4 can also be found in cell junctions between the ICs and between PC and IC cells, that is, the heterotypic junction. In knockout mouse kidney, PCs are depleted with claudin-4 proteins. ICs retain claudin-4 protein expression in the tight junction but at significantly reduced levels. Bar: 20 μm. *Reproduced with permission from Gong, Y., Yu, M., Yang, J., Gonzales, E., Perez, R., Hou, M., Tripathi, P., Hering-Smith, K. S., Hamm, L. L., & Hou, J. (2014). The Cap1-claudin-4 regulatory pathway is important for renal chloride reabsorption and blood pressure regulation.* Proceedings of the National Academy of Sciences of the United States of America, 111, *E3766–3774.*

8.4.6 Concluding remarks

The use of a site-specific recombination system has made it possible to inactivate tight junction genes in a spatially and temporally restricted manner in transgenic mice. This approach is particularly useful to delineate the cell or tissue-specific function of a tight junction protein. Compared to the global knockout strategy, the conditional knockout strategy can address the role of a tight junction gene in the absence of paracrine or endocrine complications. Because tight junction function requires two or more cells to implement, the interpretation of the cell-autonomous effect of a tight junction gene, brought about by the conditional knockout approach, must take into consideration that the biochemistry of tight junction proteins in the neighboring cells could be altered simultaneously.

References

Andrews, B. J., Proteau, G. A., Beatty, L. G., & Sadowski, P. D. (1985). The FLP recombinase of the 2 micron circle DNA of yeast: Interaction with its target sequences. *Cell*, *40*, 795–803.

Austin, S., Ziese, M., & Sternberg, N. (1981). A novel role for site-specific recombination in maintenance of bacterial replicons. *Cell*, *25*, 729–736.

Deng, C., & Capecchi, M. R. (1992). Reexamination of gene targeting frequency as a function of the extent of homology between the targeting vector and the target locus. *Molecular and Cellular Biology*, *12*, 3365–3371.

Farley, F. W., Soriano, P., Steffen, L. S., & Dymecki, S. M. (2000). *Widespread recombinase expression using FLPeR (flipper) mice, . Genesis (New York, NY: 2000)* (28, pp. 106–110).
.

Gong, Y., Yu, M., Yang, J., Gonzales, E., Perez, R., Hou, M., ... Hou, J. (2014). The Cap1-claudin-4 regulatory pathway is important for renal chloride reabsorption and blood pressure regulation. *Proceedings of the National Academy of Sciences of the United States of America*, *111*, E3766–E3774.

Hogan, B., Costantini, F., & Lacy, E. (1986). *Manipulating the mouse embryo: A laboratory manual* (Vol. 34). Cold Spring Harbor, NY: Cold Spring Harbor Laboratory.

McCarrick, J. W., 3rd, Parnes, J. R., Seong, R. H., Solter, D., & Knowles, B. B. (1993). Positive-negative selection gene targeting with the diphtheria toxin A-chain gene in mouse embryonic stem cells. *Transgenic Research*, *2*, 183–190.

Meier, I. D., Bernreuther, C., Tilling, T., Neidhardt, J., Wong, Y. W., Schulze, C., ... Schachner, M. (2010). Short DNA sequences inserted for gene targeting can accidentally interfere with off-target gene expression. *FASEB Journal: Official Publication of the Federation of American Societies for Experimental Biology*, *24*, 1714–1724.

Pham, C. T., MacIvor, D. M., Hug, B. A., Heusel, J. W., & Ley, T. J. (1996). Long-range disruption of gene expression by a selectable marker cassette. *Proceedings of the National Academy of Sciences of the United States of America*, *93*, 13090–13095.

Torres, R. M., & Kühn, R. (1997). *Laboratory protocols for conditional gene targeting*. Oxford University Press.

Chapter 9

Perspective

Chapter 9.1

Scanning ion conductance microscopy

9.1.1 Concept of conductance scanning

The most common way to measure paracellular conductance is by applying Ohm's law to a circuit in which the epithelium can be viewed as a conductor (Eq. 3.1.6). The ion conductive pathways in the epithelium can be described by the transepithelial conductance, that is, the inverse quantity of transepithelial resistance (TER). TER, in essence, is influenced by both transcellular and paracellular pathways. Therefore how to isolate the paracellular conductance from the transepithelial conductance is a major challenge to the field. Conductance scanning employs a scanning electrode to measure local variations in current density close to a cell surface and has been applied to characterize the underlying conductive pathways (Fromter & Diamond, 1972). In this type of recording, one pair of electrodes applies a transepithelial potential across an epithelium, whereas a second pair of electrodes measures the resultant changes in the electric field at two discrete depths above a conductive pathway (close to and far away from the surface). The conductance through the pathway can then be calculated according to the following equation (Cereijido, Meza, & Martinez-Palomo, 1981; Gitter, Bertog, Schulzke, & Fromm, 1997).

$$G = \frac{E}{\rho V_{te}} = \frac{(\Delta V_{\text{close point}} - \Delta V_{\text{distant point}})/\Delta z}{\rho V_{te}} \tag{9.1.1}$$

where E is the electrical field and equals the potential difference, ΔV measured at two discrete depths relative to the vertical displacement of the pipet, Δz; ρ is the bathing solution resistivity, and V_{te} is the applied transepithelial potential.

A Laboratory Guide to the Tight Junction. DOI: https://doi.org/10.1016/B978-0-12-818647-3.00009-X

9.1.2 Practical application

Scanning ion conductance microscopy (SICM) is a technique based on scanning probe microscopy (SPM), which raster-scans nonconductive samples to generate topographic images (Hansma, Drake, Marti, Gould, & Prater, 1989). To incorporate conductance scanning into SICM, a dual-barrel pipet is utilized, in which the first barrel is for topographic imaging and positioning of the pipet; the second barrel is for recording the electric field over the epithelium (Fig. 9.1.1A) (Chen, Zhou, & Baker, 2012). In a perfusion cell culture system (e.g., Transwell) where epithelial cells grow to form a monolayer, the dual-barrel pipet is placed into the upper chamber. The current electrode (PE) in one barrel detects current passing from the PE to a reference electrode (RE), serving as the conventional SICM probe to record the pipet-surface distance and to obtain the topographic information of the cell membrane. The potential electrode (UE) in the second barrel serves to measure localized potential differences at cell surfaces relative to the RE. A transepithelial potential (V_{te}) between the RE and a working electrode (WE, bottom chamber) is applied over heterogeneous conductive pathways in the epithelium (Fig. 9.1.1A). The equivalent electric circuit of the dual-barrel SICM configuration is shown in Fig. 9.1.1B. To evaluate local conductance changes with SICM, potential deflections at the pipet tip are measured at two fixed pipet-surface distances (D_{ps}), which is precisely controlled by a piezoelectric positioner. One position is close to the surface

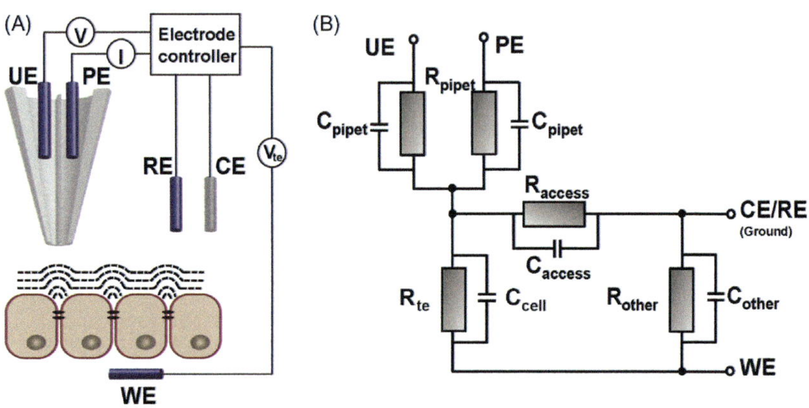

FIGURE 9.1.1 Schematic diagram and equivalent electric circuit of SICM. (A) A dual-barrel pipette is designed to obtain topographic information and measure local changes in transepithelial conductance. CE, Counter electrode; D_{ps}, pipet-surface distance; PE, current electrode; RE, reference electrode; UE, potential electrode; WE, working electrode. (B) The electric circuit from PE to RE monitors the access resistance (R_{access}); the electric circuit from WE to RE applies the transepithelial potential; the electric circuit from UE to WE records the transepithelial conductance. *Reproduced with permission from Hou, J. (2018).* The paracellular channel – biology, physiology and disease. *Academic Press.*

(0.2 μm), which is controlled through the robust feedback signal. A second position is 12.5 μm above the sample, which is half of the range of vertical piezo movement. The transepithelial conductance can be described by the following equation.

$$G = \frac{E}{\rho V^e} = \frac{(\Delta V_{0.2\mu m} - \Delta V_{12.5\mu m})/\Delta z}{\rho V^e} \quad (9.1.2)$$

in which E is the electrical field is obtained by measuring potential deflection (ΔV) induced by transepithelial potential (V_{te}) at two pipet-surface distances (D_{ps}); V^e is the range of transepithelial potential sweeping ($V^e = 120$ mV, -60 mV to $+60$ mV); ρ is the cell medium resistivity; and Δz is the vertical displacement of the pipet. To avoid depolarization of the cell plasma membrane, an alternating V_{te} is applied, the frequency of which is determined from impedance measurement to minimize the capacitive contribution. The magnitude of the transepithelial potential is less than 60 mV to avoid damage to the cell membrane.

9.1.3 Instrumentation

Commercial SICM is available from Ionscope (https://www.ionscope.com). The sample and the recording pipet are enclosed in a Faraday cage known as the scan head (Fig. 9.1.2). The scan head is built on top of an inverted light microscope and consists of two positioning manipulators for feedback scan control. The *Z* translational assembly moves the pipette in the *Z* or vertical direction normal to the sample surface, and the *XY* translational assembly, onto which the sample is placed, scans the sample in the *XY* plane. A commercial current amplifier (Axon Instruments Axopatch 200B) is used to monitor pipette ion current with a gain of 1 mV/1 pA and 2 kHz low-pass filter. Potentiometric signals are recorded using a commercial differential amplifier (Axon Instruments MultiClamp 700B) with 100 × gain and 20 Hz low-pass filter. All channel data (*XYZ* piezo sensor voltages, ion current, V^e, and potentiometric signals) are recorded in real time with Axon Digidata 1440 A and pClampex 10.6 software.

9.1.4 Tight junction conductance measurement

First, a topographic image is obtained with SICM imaging mode in the absence of transepithelial potential (Fig. 9.1.3A). The locations of the cell body (CB, transcellular pathway) and cell junctions (CJ, paracellular pathway) can then be pinpointed from the image to extract their spatial coordinates. The recording pipet is positioned over CB or CJ according to these coordinates to record the potential deflection (ΔV) induced by the applied transepithelial potential (V_{te}) at two discrete pipet-surface distances (D_{ps}, 0.2 and 12.5 μm), respectively. The recorded conductance over CBs (transcellular) or CJs (paracellular) displays

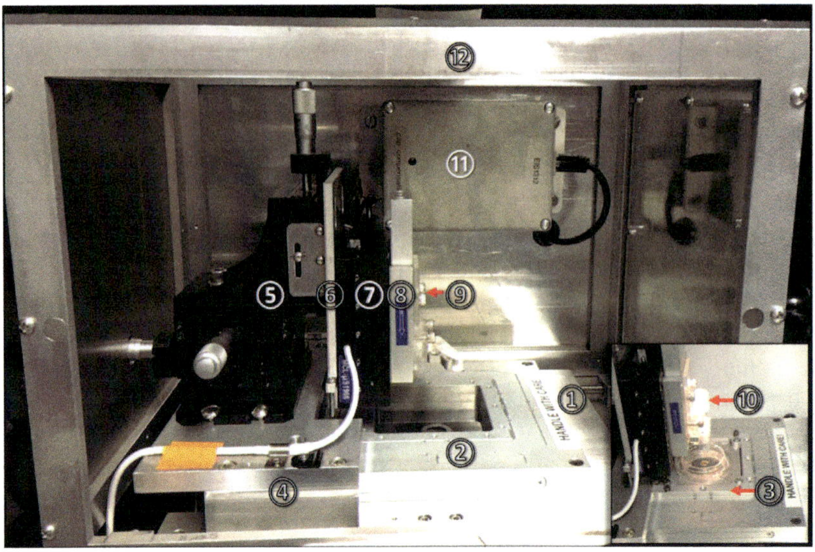

FIGURE 9.1.2 SICM scan head. The scan head is composed of three major components: (i) XY translation system (①–③): coarse positioning stage and fine positioning piezo; (ii) Z translation system (④–⑩): coarse positioning manipulator and stepper motor, fine positioning piezo and accessory parts for mounting; (iii) electronics system (⑪–⑫): amplifier headstage and Faraday cage. ①: XY coarse translation stage; ②: XY nanopositioner; ③: sample holder; ④: Z bridge; ⑤: 3-axis manipulator; ⑥: adapter plate; ⑦: Z stepper motor; ⑧: Z nanopositioner; ⑨: pipet mount; ⑩: pipet holder; ⑪: headstage; ⑫: Faraday cage (front door removed).

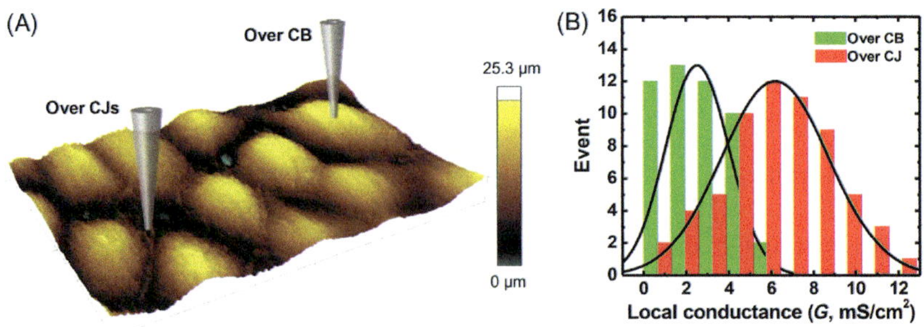

FIGURE 9.1.3 Characterization of epithelial cell monolayer with SICM. (A) A topographic image of the apical surface of the MDCK cell monolayer shows the location of the cell body (CB) and cell junction (CJ). (B) Histograms of conductance measurements obtained over cell bodies ($N = 49$) and cell junctions ($N = 62$). *Reproduced with permission from Hou, J. (2018).* The paracellular channel – biology, physiology and disease. *Academic Press.*

Gaussian distributions with means of 2.53 ± 1.49 and 6.20 ± 2.54 mS/cm^2, respectively (Fig. 9.1.3B) (Chen, Zhou, Morris, Hou, & Baker, 2013). The waveform of the potentiometric signal at $D_{ps} = 0.2$ μm is shown in Fig. 9.1.4. The paracellular conductance significantly differs from the transcellular

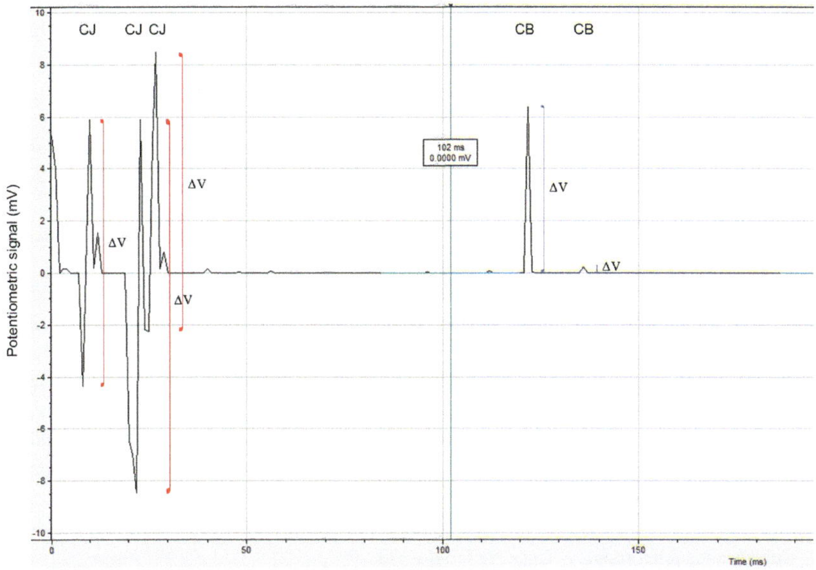

FIGURE 9.1.4 Waveform of the potentiometric signal. The potentiometric signals above cell junction (CJ) and cell body (CB) are shown in response to a linear sweeping potential of −60 mV to +60 mV at 5 Hz. The signals are recorded at $D_{ps} = 0.2\ \mu m$ and yield the value of $\Delta V_{0.2\mu m}$. $\Delta V_{12.5\mu m}$ approaches zero.

conductance. The spatial resolution for topographic imaging is under 100 nm and the lateral distribution of ΔV displays submicron spatial resolution (Chen et al., 2013).

9.1.5 Limitation and future direction

While the location of CJ or CB can be pinpointed by SICM, the continuous distribution of the multitude of competing conductive pathways surrounding CJ or CB, in fact, prevents the isolation of paracellular- or transcellular-specific conductance. Resolution in potentiometric measurement is determined largely by the scanning parameters employed, especially the geometry of the pipet tip, the probe-surface distance (D_{ps}), and the sweeping potential applied across the sample (V^e). The reduction of D_{ps} is a practical way to improve the resolution for SICM. To achieve it, the size of the pipet tip must be forged as small as possible. An extremely small pipet, however, has a high pipet resistance, which may reduce the spatial resolution in the topographic image. The potentiometric signals reflect an aggregate of paracellular channel opening and closing events on the time scale of 200 ms in response to a linear sweeping potential at 5 Hz. In the theory, the ion channel gating constant (τ) is less than 1 ms. A stimulation frequency of >1 kHz is therefore needed to resolve the gating behavior of the paracellular channel.

References

Cereijido, M., Meza, I., & Martinez-Palomo, A. (1981). Occluding junctions in cultured epithelial monolayers. *American Journal of Physiology Cell Physiology, 240*, C96–C102.

Chen, C. C., Zhou, Y., & Baker, L. A. (2012). Scanning ion conductance microscopy. *Annual Review of Analytical Chemistry (Palo Alto Calif), 5*, 207–228.

Chen, C. C., Zhou, Y., Morris, C. A., Hou, J., & Baker, L. A. (2013). Scanning ion conductance microscopy measurement of paracellular channel conductance in tight junctions. *Analytical Chemistry, 85*, 3621–3628.

Fromter, E., & Diamond, J. (1972). Route of passive ion permeation in epithelia. *Nature: New Biology, 235*, 9–13.

Gitter, A. H., Bertog, M., Schulzke, J. D., & Fromm, M. (1997). Measurement of paracellular epithelial conductivity by conductance scanning. *Pflügers Archiv – European Journal of Physiology, 434*, 830–840.

Hansma, P. K., Drake, B., Marti, O., Gould, S. A. C., & Prater, C. B. (1989). The scanning ion-conductance microscope. *Science, 243*, 641–643.

Chapter 9.2

Cryo-electron microscopy

9.2.1 Single-particle cryo-electron microscopy

9.2.1.1 Structural determination without crystallization

Unlike crystallographic approaches, single-particle cryo-electron microscopy (cryo-EM) determines the structures of biological macromolecules by averaging electron microscopic images of target molecules or particles embedded in a thin layer of vitreous ice in random orientations (Fig. 9.2.1) (Cheng, Grigorieff, Penczek, & Walz, 2015). Because it does not require

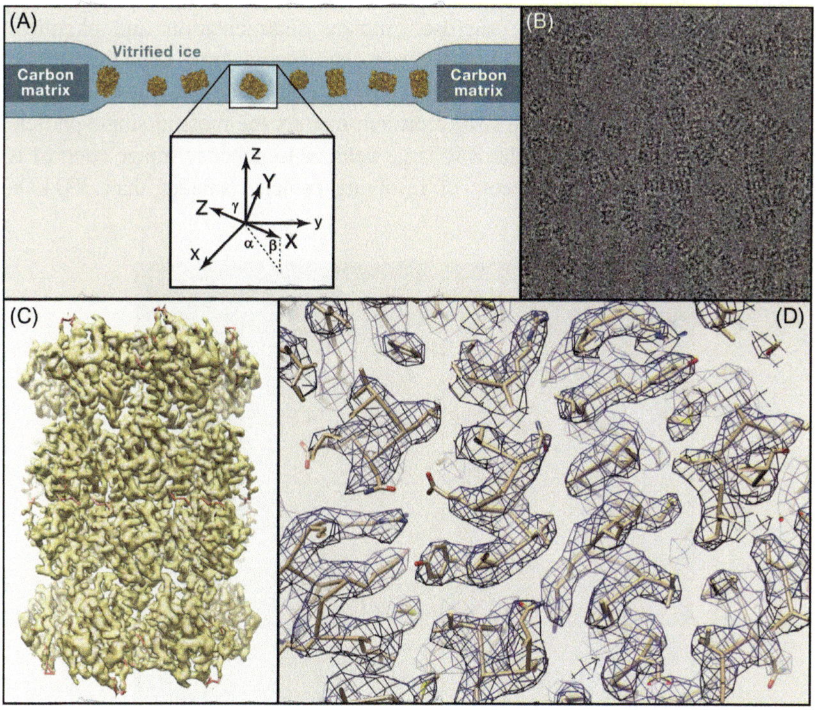

FIGURE 9.2.1 Single-particle cryo-EM. (A) Purified biological molecules are embedded in a thin layer of vitreous ice, in which they adopt random orientations. The orientations are specified by five in-plane position parameters, x and y, and three Euler angles α, β, and γ. The defocus value of the image defines the z position along the direction of the electron beam, which is determined separately. (B) The typical image of frozen-hydrated archaeal 20S proteasomes. (C) 3D reconstruction of the 20S proteasome at 3.3-Å resolution. (D) Side-chain densities of the map shown in (B) are comparable to those determined by X-ray crystallography at a similar resolution. *Reproduced with permission from Cheng, Y. (2015). Single-particle cryo-EM at crystallographic resolution.* Cell, 161, *450–457.*

crystallization or absolute sample homogeneity, this technique is particularly suited to the study of integral membrane proteins that can be maintained in an aqueous environment. Each particle image is a 2D projection of a molecule, whose spatial orientation and position are defined by six geometric parameters. These include three Euler angles and two in-plane positional parameters. The sixth parameter is the defocus that defines the z position along the direction of the electron beam. After further correction for aberrational errors of the microscope, a 3D structure can be reconstructed by combining images of many molecules that have been aligned to each other.

9.2.1.2 Protein quality and size

Although crystallization is no longer an absolute necessity for structural determination in single-particle cryo-EM, obtaining biochemically homogeneous proteins is still an essential requirement. In the case of purifying claudin-16 and claudin-19 heterodimers, sucrose gradient sedimentation and chemical crosslinking have to be used to separate the dimers from claudin-16 and claudin-19 monomers (Gong et al., 2015). The purified dimers show a variety of orientations (Fig. 9.2.2). The size of protein matters the most in single-particle cryo-EM. The need to use a relatively large defocus to generate image contrast is the major obstacle to the success of resolving proteins smaller than 300 kDa

FIGURE 9.2.2 Electron micrograph of claudin-16 and claudin-19 heterodimer. Purified claudin-16 and claudin-19 dimers are viewed by negative staining electron microscopy.

without high symmetry (Henderson, 2013). Claudin monomers or dimers are considered too small to be studied directly by single-particle cryo-EM. It is possible that adding some ordered soluble mass with recognizable structural features, for example, binding a conformation-specific Fab domain of an antibody, can provide a fiducial marker with which to facilitate the image alignment of claudin proteins and make 3D structural reconstruction feasible (Wu et al., 2012).

9.2.2 Cryo-electron microscopy of vitreous section

Cryo-electron microscopy of vitreous section (CEMOVIS) is a new method for visualizing the ultrastructure of subcellular organelle with nanometer resolution and in close to the native state. Similar to single-particle cryo-EM, CEMOVIS relies upon the target macromolecules being embedded in a thin layer of vitreous ice that is now part of the tissue section. CEMOVIS has been used to reveal the architecture of desmosome in the human epidermis (Al-Amoudi, Dubochet, & Norlen, 2005). The molecular interfaces of desmosomal cadherin assembly are seen as electron-dense, transverse lines with ~5 nm periodicity (Fig. 9.2.3A). These transverse lines are molecular clusters of cadherins undergoing *trans* W-like and *cis* V-like interactions (Fig. 9.2.3B) (Al-Amoudi, Diez, Betts, & Frangakis, 2007). Because the crystal structure of claudin is known, it is possible to apply CEMOVIS to tight junctions in vitrified tissue sections and extract the bonding information of the extracellular loops in claudin based upon the density map of the crystal structure of claudin. It will even be possible to visualize the structural alterations in claudin, which may correspond to the open or closed state of the paracellular channel that claudin makes, or reflect a physiological or pathological response to external stimulus.

9.2.3 Limitation and future direction

The tight junction is a heterogeneous subcellular organelle composed of lipids, integral membrane proteins, and peripheral binding proteins. To visualize the structure of tight junction at atomic resolution, some sort of biochemical purification is needed to break the tight junction into simpler components, such as junctional membrane and plaque. CEMOVIS is useful to visualize the extracellular interactions important for mediating cell adhesion and making paracellular channel pore. However, CEMOVIS is unable to resolve the protein structure embedded in the lipid bilayer. Integral membrane proteins will need to be solubilized in detergents or detergent-like amphipols and then visualized by single-particle cryo-EM. Single-particle cryo-EM is particularly useful to resolve large protein assemblies and dynamic complexes that are difficult or may even be impossible to crystallize. Smaller proteins with no symmetry

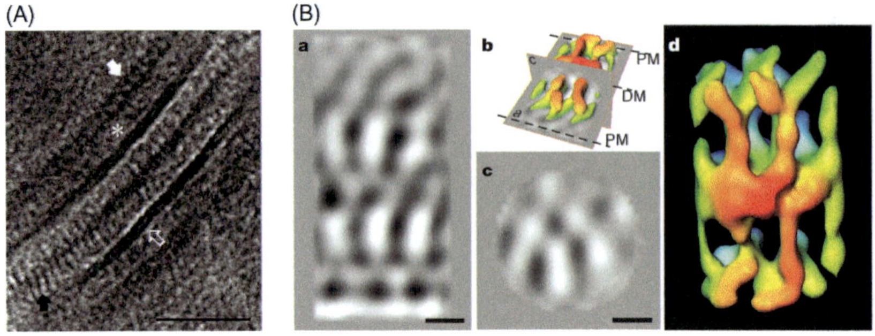

FIGURE 9.2.3 Cryo-EM of vitreous section. (A) In human epidermis, the desmosome is ~33 nm wide and contains transverse electron-dense lines with ~5 nm periodicity (*black arrow*). On the cytoplasmic side, an opaque zone of medium electron density (*white asterisk*) separates the electron-dense plasma membrane (*hollow white arrow*) from the electron-dense layer of desmosomal plaque (*solid white arrow*). Bar: 50 nm. (B) Visualization of the average of the subtomograms of desmosome. (a). Coronal slice through the average of the subtomograms, which are extracted from the extracellular space only. The plasma membranes are not shown but their position is indicated in (b). The elongated curved densities arrange in a periodic manner and their dimensions compare remarkably well with the W-like conformation observed in the C-cadherin X-ray structure. (b) Cartoon image showing the relative orientations of the slices visualized in a and c and the isosurface in (d). The position of the coronal slice visualized in (a), the axial slice in (c), and their location with respect to the isosurface shown in (d) are indicated. The dashed lines indicate the location of the dense midline (DM) as well as the position of the plasma membrane (PM). The latter is removed from the subtomograms and thus not visible in the average structure or shown in (a). (c) Axial slice through the averaged subtomograms. Distinct cadherin molecules can be recognized as dense blobs of approximately 3 nm in diameter. (d) Isosurface visualization of the organization of the cadherin molecules. The threshold is chosen so that the thickness of the central cadherin molecule is approximately 3 nm. The color coding varies as a function of the depth from red (close) to blue (far). Two cadherin dimers arranged in a *trans* W-like manner are shown in the foreground in orange, with their concave side facing left. One layer deeper, shown in green, four cadherin *trans* W-like dimers have their concave side oriented to the right. The orange and the green molecules emanating from the same cell surface interact to form V-like *cis* dimers. Two additional W-like cadherin dimers have their concave side oriented to the left (blue) and interact with the green molecules to form V-like *cis* dimers. Bar: 7 nm. *Reproduced with permission from Al-Amoudi, A., Diez, D.C., Betts, M.J., & Frangakis, A.S. (2007). The molecular architecture of cadherins in native epidermal desmosomes. Nature, 450, 832–837; Al-Amoudi, A., Dubochet, J., & Norlen, L. (2005). Nanostructure of the epidermal extracellular space as observed by cryo-electron microscopy of vitreous sections of human skin. The Journal of Investigative Dermatology, 124, 764–777.*

pose a challenge to single-particle cryo-EM due to its lack of resolution on the *z*-axis. Because tight junctions often consist of different claudin species that associate to form oligomers of different sizes, the biochemical heterogeneity of claudin oligomerization represents another obstacle to single-particle cryo-EM. More definitive knowledge of tight junction biochemistry and the ability to obtain more homogenous fraction of claudin oligomer will be prerequisites to successful structural study.

References

Al-Amoudi, A., Diez, D. C., Betts, M. J., & Frangakis, A. S. (2007). The molecular architecture of cadherins in native epidermal desmosomes. *Nature*, *450*, 832–837.

Al-Amoudi, A., Dubochet, J., & Norlen, L. (2005). Nanostructure of the epidermal extracellular space as observed by cryo-electron microscopy of vitreous sections of human skin. *The Journal of Investigative Dermatology*, *124*, 764–777.

Cheng, Y., Grigorieff, N., Penczek, P. A., & Walz, T. (2015). A primer to single-particle cryo-electron microscopy. *Cell*, *161*, 438–449.

Gong, Y., Renigunta, V., Zhou, Y., Sunq, A., Wang, J., Yang, J., ... Hou, J. (2015). Biochemical and biophysical analyses of tight junction permeability made of claudin-16 and claudin-19 dimerization. *Molecular Biology of the Cell*, *26*, 4333–4346.

Henderson, R. (2013). Avoiding the pitfalls of single particle cryo-electron microscopy: Einstein from noise. *Proceedings of the National Academy of Sciences of the United States of America*, *110*, 18037–18041.

Wu, S., Avila-Sakar, A., Kim, J., Booth, D. S., Greenberg, C. H., Rossi, A., ... Griner, S. L. (2012). *Fabs enable single particle cryoEM studies of small proteins*, . *Structure (London, England: 1993)* (20, pp. 582–592).

Chapter 9.3

Super-resolution microscopy for tight junction

9.3.1 Super-resolution microscopy

The resolution of a light microscope is defined by Abbe's diffraction limit equation (Chapter 5.1, Eq. 5.1.1). Conventional light microscopy operates in the resolution range of 200–300 nm laterally and 500–800 nm axially, limited by the wavelength of light (λ) and the numerical aperture (NA) of the objective lens. Super-resolution microscopy refers to a class of new light microscopic techniques that aim to break the diffraction barrier (Table 9.3.1). Among them, structured illumination microscopy (SIM) uses interference generated light patterns to create a Moiré effect from which higher-resolution information can be extracted (Guerra, 1995). Stimulated emission depletion microscopy (STED) reduces the effective excitation volume with a depletion laser, which restricts spontaneous fluorescence emission and shapes the effective scanning region to subdiffraction scales (Hell & Wichmann, 1994). Stochastic optical reconstruction microscopy (STORM) or photoactivated localization microscopy (PALM) stochastically activates a subset of photoactivatable probes at a time and then localizes individual photoactivatable fluorophores through fitting their point-spread functions (Betzig et al., 2006; Rust, Bates, & Zhuang, 2006).

9.3.2 Spatial separation of tight junction components

Endogenous tight junctions contain hundreds of proteins (Tang, 2006). The integral membrane proteins in the TJ are bundled with many other TJ proteins such as those in the junctional plaque and performing signaling functions. Super-resolution microscopic tools are essential for revealing the spatial relationship between integral membrane proteins and peripheral plaque proteins in the TJ. The TJ plaque contains a large number of adaptor proteins that form a protein network via multiple protein interactions to anchor the TJ onto the cytoskeleton. An important group of TJ plaque proteins is characterized by the presence of the so-called PDZ domain, which mediates the interaction of TJ plaque proteins with TJ integral membrane proteins including claudins (Fanning & Anderson, 1996). ZO-1 is an important TJ plaque protein with three PDZ domains (Stevenson, Siliciano, Mooseker, & Goodenough, 1986). Super-resolution microscopy reveals that the localization of ZO-1 protein is separated from that of claudin-2 protein in the mouse kidney tubule, which is compatible with the electron microscopic observation that TJ consists of two

TABLE 9.3.1 Super-resolution microscopic techniques.

	Structured illumination microscopy (SIM)	Stimulated emission depletion microscopy (STED)	Photoactivated localization microscopy (PALM) and stochastic optical reconstruction microscopy (STORM)
Microscopy type	Wide field	Laser scanning confocal	Wide field
Lateral resolution	100–130 nm	20–70 nm	10–30 nm
Axial resolution	300 nm	40–150 nm	10–75 nm
Probes	**Common photostable organic dyes** and fluorescent proteins	**Particular photostable organic dyes and** fluorescent proteins	**PALM: photoswitchable fluorescent** proteins STORM: photoswitchable organic dyes
Temporal resolution	Milliseconds to seconds	Milliseconds to seconds	Seconds to minutes
Maximum number of simultaneous colors	4	3	2
Live imaging	Yes	Yes	Yes
Considerations	Straightforward multicolor experiments and sample preparation. Reconstruction algorithm may cause artifacts	Best temporal resolution at the highest spatial resolution; however, maximal in-plane can be at the expense of axial resolution	Highest spatial resolution, however, sensitive to labeling density. Crosstalk between fluorophores may be an issue

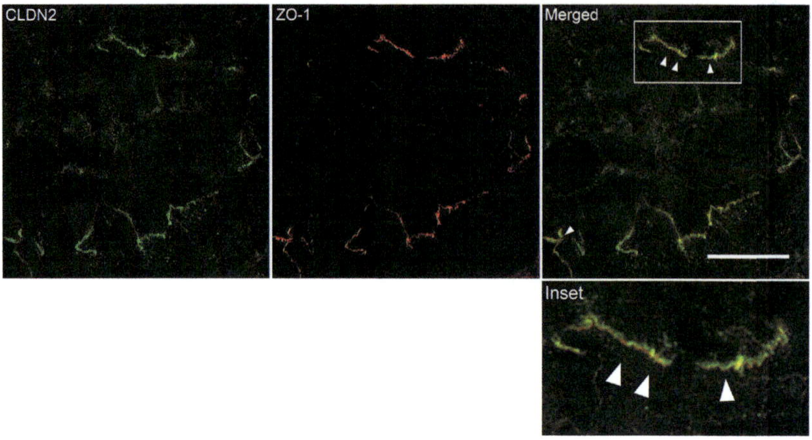

FIGURE 9.3.1 SIM reveals the separation of TJ membrane protein from TJ plaque protein. The proximal tubule from the mouse kidney was labeled with anti-claudin-2 (CLDN2) and anti-ZO-1 antibodies and imaged by SIM. Claudin-2 signals indicate the location of the TJ membrane whereas ZO-1 signals indicate the location of TJ plaque. Bar: 5 μm.

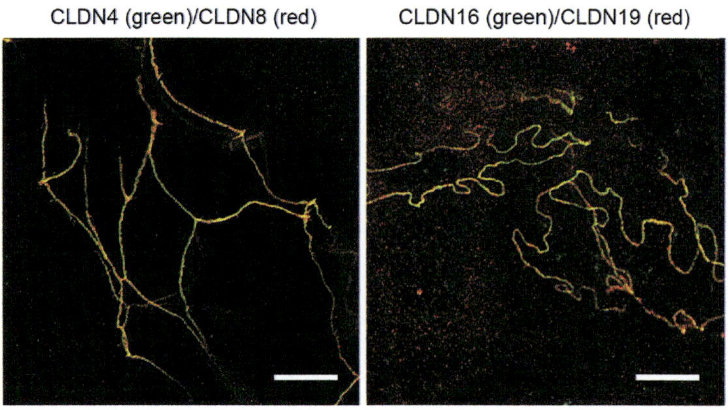

FIGURE 9.3.2 Colocalization of claudins in tight junction by SIM. The renal tubules from the mouse kidney were labeled with pairs of anti-claudin antibodies (CLDN4/CLDN8 and CLDN16/CLDN19), and imaged by SIM. Note that claudins always colocalize in the TJ strands. Bar: 5 μm.

anatomic layers—membrane and plaque (Fig. 9.3.1). Proteins within the same layer are separable by super-resolution microscopy (Fig. 9.3.2).

9.3.3 Architectural alteration in tricellular tight junction

The tricellular tight junction is ultrastructurally different from the bicellular tight junction. Freeze-fracture replica electron microscopy reveals that TJ strands are not continuous at the tricellular junction, where they make a 90

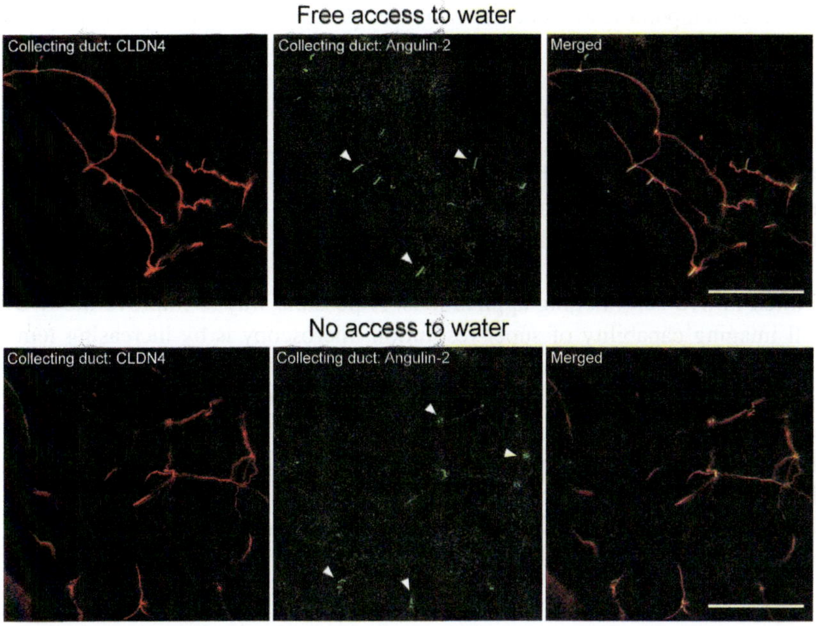

FIGURE 9.3.3 SIM reveals ultrastructural alteration in tricellular tight junction. The collecting ducts from the mouse kidney were labeled with anti-CLDN4 and anti-angulin-2 antibodies and imaged by SIM. Claudin-4 signals indicate the location of bicellular tight junction whereas angulin-2 signals indicate the location of the tricellular tight junction. Bar: 5 μm.

degrees turn to extend toward the basal direction (Staehelin, 1973). As a result, tricellular tight junction consists of three pairs of TJ strands that are arranged vertically and known as the central sealing elements (Hou, 2018). The tricellular tight junction is made of two classes of integral membrane proteins—tricellulin and angulins (Higashi & Miller, 2017). Super-resolution microscopic imaging of the proteins making tricellular tight junction may reveal the ultrastructural changes in this unique type of cell junction. Functionally, the tricellular tight junction is purported to contain a water-permeable pathway. The deletion of angulin-2 from tricellular tight junction can render the renal tubules more permeable to water in the mouse kidney (Gong et al., 2017). When the mouse kidney is placed under a water-deprived condition, the architecture of tricellular tight junction changes from the shape of the particle to the shape of a dumbbell, as if the junction is split open by an external force (Fig. 9.3.3). There appears to be a gap of angulin protein localization in tricellular tight junction, which can only be revealed by super-resolution microscopy.

9.3.4 Limitation and future direction

In general, every increase in optical resolution comes at the expense of more exposure, longer acquisition time, or higher energy load, which conversely decreases temporal resolution and increases photobleaching and phototoxicity. Expanding the information content by adding more dimensions, such as multicolor, 3D volume and time lapse, is often essential to address the spatial changes in various TJ components. Such spatial information can only suggest an architectural alteration of the tight junction but not serve as a direct proof. A correlative study combining super-resolution microscopy with electron microscopy may yield the much-needed confirmation that the structure of tight junction can be modeled by a defined group of molecules arranged with a certain biochemical order. Super-resolution microscopic approaches are often limited in live-cell imaging applications. A possible way to improve the live-cell imaging capability of super-resolution microscopy is by increasing temporal resolution and lowering photon burden. Challenges include optimized sample preparation and labeling, reduction in phototoxicity, and adaptation to imaging depth inside tissue. A major handicap of all super-resolution microscopic methods is their susceptibility to aberrations, in particular when the imaging depth is larger than 10 μm, which impacts contrast and resolution. Adaptive optics (AO), which includes deformable mirror devices to compensate for refractive index changes within a specimen, might provide a solution to the problem of imaging depth in super-resolution microscopy.

References

Betzig, E., Patterson, G. H., Sougrat, R., Lindwasser, O. W., Olenych, S., Bonifacino, J. S., ... Hess, H. F. (2006). Imaging intracellular fluorescent proteins at nanometer resolution. *Science (New York, NY)*, *313*, 1642–1645.

Fanning, A. S., & Anderson, J. M. (1996). Protein-protein interactions: PDZ domain networks. *Current Biology: CB*, *6*, 1385–1388.

Gong, Y., Himmerkus, N., Sunq, A., Milatz, S., Merkel, C., Bleich, M., & Hou, J. (2017). ILDR1 is important for paracellular water transport and urine concentration mechanism. *Proceedings of the National Academy of Sciences of the United States of America*.

Guerra, J. M. (1995). Super-resolution through illumination by diffraction-born evanescent waves. *Applied Physics Letters*, *66*, 3555–3557.

Hell, S. W., & Wichmann, J. (1994). Breaking the diffraction resolution limit by stimulated emission: Stimulated-emission-depletion fluorescence microscopy. *Optics Letters*, *19*, 780–782.

Higashi, T., & Miller, A. L. (2017). Tricellular junctions: How to build junctions at the TRICkiest points of epithelial cells. *Molecular Biology of the Cell*, *28*, 2023–2034.

Hou, J. (2018). *The paracellular channel – biology, physiology and disease*. Academic Press.

Rust, M. J., Bates, M., & Zhuang, X. (2006). Sub-diffraction-limit imaging by stochastic optical reconstruction microscopy (STORM). *Nature Methods*, *3*, 793–795.

Staehelin, L. A. (1973). Further observations on the fine structure of freeze-cleaved tight junctions. *Journal of Cell Science*, *13*, 763–786.

Stevenson, B. R., Siliciano, J. D., Mooseker, M. S., & Goodenough, D. A. (1986). Identification of ZO-1: A high molecular weight polypeptide associated with the tight junction (zonula occludens) in a variety of epithelia. *The Journal of Cell Biology*, *103*, 755–766.

Tang, V. W. (2006). Proteomic and bioinformatic analysis of epithelial tight junction reveals an unexpected cluster of synaptic molecules. *Biology Direct*, *1*, 37.

Chapter 9.4

Novel binders to tight junction

9.4.1 *Clostridium perfringens* enterotoxin

CPE is a 35-kDa protein produced by the bacterium *Clostridium perfringens* and causes intestinal tissue necrosis (Freedman, Shrestha, & McClane, 2016). Claudin-3 and claudin-4 are the receptor proteins for CPE (Katahira, Sugiyama et al., 1997; Katahira, Inoue, Horiguchi, Matsuda, & Sugimoto, 1997). The carboxyl-terminal domain of CPE (C-CPE) binds to claudins (Sonoda et al., 1999). The crystal structures of C-CPE bound claudin-4 and claudin-19 have been resolved (Saitoh et al., 2015; Shinoda et al., 2016). Both structures reveal conformational changes in the extracellular loop domains of claudins upon C-CPE binding. The C-CPE bound claudin-19 structure offers critical insights into the disassembly of the TJ strand made of claudin polymer (Fig. 9.4.1) (Saitoh et al., 2015). When applied to dog kidney epithelial cells, C-CPE removes claudin-3 and claudin-4 from the TJ, disintegrates the TJ strands, and increases the paracellular permeabilities to ions and solutes with sizes up to 10 kDa (Sonoda et al., 1999).

9.4.2 TJ modulating peptidomimetics

9.4.2.1 Occludin peptidomimetics

Occludin is a TJ integral membrane protein. It consists of four transmembrane domains and two extracellular loops (ECL1 and ECL2). Various forms of occludin peptidomimetics have been developed, all deriving from amino acid sequences in the ECL1 or ECL2 domain (Table 9.4.1). Among them, a peptidomimetic ($OCLN_{90-103}$) corresponding to amino acids 90–103 in the ECL1 domain of human occludin protein increases the paracellular permeabilities to ions and solutes with a wide range of molecular sizes, for example, mannitol (182 Da), 70-kDa or 2000-kDa dextran in human intestinal and airway epithelial cells (Everett, Vanhook, Barozzi, Toth, & Johnson, 2006; Tavelin et al., 2003). In *Xenopus* kidney epithelial cells, an ECL2-based peptidomimetic, corresponding to amino acids 184–227 and termed $OCLN_{184-227}$, increases the paracellular permeabilities to a variety of solutes, including mannitol, insulin (5.8 kDa), 3-kDa dextran, and 40-kDa dextran. Mechanistically, $OCLN_{184-227}$ delocalizes the occludin protein from the TJ, whereas the localization of ZO-1, ZO-2, or E-cadherin remains unaffected (Wong & Gumbiner, 1997).

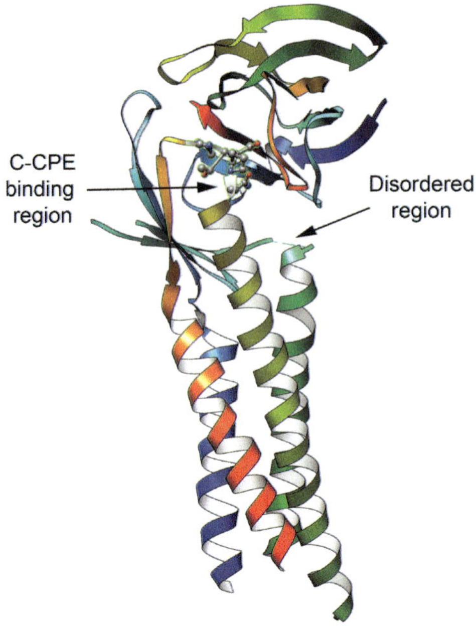

FIGURE 9.4.1 Molecular structure of C-CPE–claudin-19 complex. Crystal structures of mouse claudin-19 bound to C-CPE in ribbon representation. The side chains of residues (N150, P151, S152, T153, and P154) in the C-CPE binding region of claudin-19 are shown in ball and stick representation. The disordered region in claudin-19 corresponds to the extracellular helical structure in claudin-15. C-CPE directly binds to a region in the second extracellular loop of claudin-19, which is close to the hydrophobic pocket structure. The binding of C-CPE may distort the hydrophobic pocket structure required for claudin *cis*-interaction. The color changes gradually from the N-terminus (*blue*) to the C terminus (*red*) in each moiety. The depicted structure is based upon the X-ray analysis by Saitoh et al. (2015).

9.4.2.2 Claudin peptidomimetics

The two extracellular loops (ECL1 and ECL2) in claudins mediate their *cis* and *trans* interactions. Several peptidomimetics derived from the ECL1 or ECL2 domain in claudin show potent regulatory effects on the paracellular pathway (Table 9.4.1). A peptidomimetic ($CLDN1_{53-80}$) corresponding to amino acids 53–80 in the ECL1 domain of mouse claudin-1 protein increases the paracellular permeabilities to ions and solutes with sizes of <3 kDa in human intestinal epithelial cells (Mrsny et al., 2008). When fed to experimental rats, $CLDN1_{53-80}$ can enhance the gastric absorption of sucrose (Fig. 9.4.2) (Mrsny et al., 2008). A rat version of claudin-1 peptidomimetic ($CLDN1_{53-81}$) has been shown to open the perineurial barrier to facilitate the paracellular permeation of antinociceptive agents (Zwanziger et al., 2012). The $CLDN1_{53-81}$ effect is, however, not specific to claudin-1.

TABLE 9.4.1 **TJ modulating peptidomimetics.**

Peptide derived from	Motif sequence	Peptide sequence	Reference
Mouse claudin-1	ECL1:53-80	SCVSQSTGQIQCKVFDSLLNLNSTLQAT	Mrsny et al. (2008)
Rat claudin-1	ECL1:53-81	SSVSQSTGQIQSKVFDSLLNLNSTLQATR[a]	Zwanziger et al. (2012)
Mouse claudin-3	ECL2:145-149	DFYNP[b]	Baumgartner, Beeman, Hodges, and Neville (2011)
Mouse claudin-5	ECL2: 146–150	EFYDP[b]	Schlingmann et al. (2016)
Human occludin	ECL1:90-103	C_{14}-DRGYGTSLLGGSVG[c]	Everett et al. (2006) and Tavelin et al. (2003)
Human occludin	ECL1: 90–112	DRGYGTSLLGGSVGYPYGGSGFG	Van Itallie and Anderson (1997)
Chick occludin	ECL2: 184–227	GVNPQAQMSSGYYYSPLLAMCSQ AYGSTYLNQYIYHYCTVDPQE[d]	Vietor, Bader, Paiha, and Huber, (2001) and Wong and Gumbiner (1997)
Rat occludin	ECL2: 209–230	GSQIYTICSQFYTPGGTGLYVD	Chung, Mruk, Mo, Lee, and Cheng (2001)
Human occludin	ECL2: 210–228	SQIYALCNQFYTPAATGLYVD	Nusrat et al. (2005)

ECL1, First extracellular loop; *ECL2*, second extracellular loop.
[a]*The two underlined serines substitute cysteines, which prevent the formation of intrachain* disulfide bond.
[b]*The peptide is synthesized in the* D*-amino acid form.*
[c]*A lipophilic amino acid moiety [H_2N-CH(C12H25)-COOH] is conjugated to the N-terminus of the peptide to prevent its enzymatic degradation.*
[d]*The two underlined cysteines are modified by covalent linkage to acetamidomethyl group to prevent the formation of an intrachain disulfide bond.*

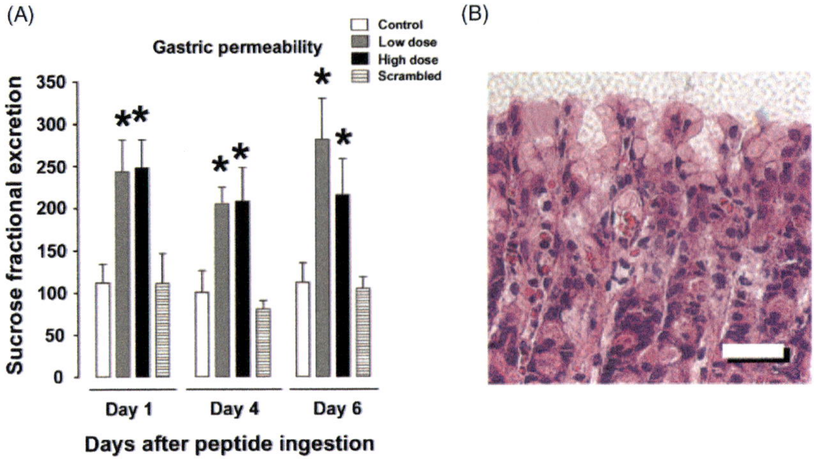

FIGURE 9.4.2 Claudin-1 peptidomimetic effect on gastric permeability in rats. (A) Rats were administered sucrose, lactulose, and mannitol alone (control, *white bars*) or along with claudin-1 peptidomimetic ($CLDN1_{53-80}$) at either 0.1 mg/kg body weight (low dose, *gray bars*), 1 mg/kg body weight (high dose, *black bars*), or with 1 mg/kg scrambled peptide (*stippled bars*) by oral gavage. Permeabilities of the stomach were assessed by measuring urinary disaccharide content. *, $P < .05$. (B) H&E-stained section of gastric tissue from rats treated with high-dose $CLDN1_{53-80}$. Peptide administration caused no epithelial cell loss or altered mucosal architecture. Bar: 100 μm. *Reproduced with permission from Mrsny, R. J., Brown, G. T., Gerner-Smidt, K., Buret, A. G., Meddings, J. B., Quan, C., Koval, M., & Nusrat, A. (2008). A key claudin extracellular loop domain is critical for epithelial barrier integrity.* The American Journal of Pathology, 172, *905–915.*

It can alter the localization of claudin-1 to claudin-5, occludin, JAM-A, and ZO-1 and cause TJ disintegration (Staat et al., 2015). There are two claudin peptidomimetics targeting the ECL2 domain: the claudin-3 peptidomimetic comprising the motif—DFYNP (aa. 145–149) in the ECL2 domain of mouse claudin-3 protein (Baumgartner et al., 2011); and the claudin-5 peptidomimetic comprising the motif—EFYDP (aa. 146–150) in the ECL2 domain of mouse claudin-5 protein (Schlingmann et al., 2016). Apart from regulating the paracellular permeability, $CLDN3_{145-149}$ can induce apoptosis in mammary epithelial cells (Baumgartner et al., 2011). $CLDN5_{146-150}$ protects the alveolar barrier function by destabilizing endogenous claudin-5 proteins, whose expression levels are increased in response to chronic alcohol treatment (Schlingmann et al., 2016).

9.4.3 Anti-claudin antibodies

The potential of developing monoclonal claudin antibodies as cancer therapeutics has been recognized for the past 10 years. Currently, two anti-claudin monoclonal antibodies (mAbs)—the anticlaudin-18.2 IMAB362 (Zolbetuximab) and the anti-claudin-6 IMAB027 have entered clinical trials (Sahin et al., 2008, 2015; Singh, Toom, & Huang, 2017). Several more claudin

TABLE 9.4.2 List of anticlaudin monoclonal antibodies and their applications.

Target	Name of mAb	Application	Reference
CLDN1	7A5	Drug delivery	Nakajima et al. (2015)
	OM-7D3-B3	Block HCV entry	Fofana et al. (2010)
	3A2	Treat cancer	Hashimoto et al. (2016)
CLDN2	1A2	Treat inflammatory bowel diseases	Takigawa et al. (2017)
	1A2	Treat cancer	Hashimoto et al. (2018)
CLDN3	KM3953	Treat cancer	Ando et al. (2015)
CLDN4	KM3934	Treat cancer	Suzuki et al. (2009)
	HKH-189	Cell sorting	Kawai et al. (2011)
CLDN5	R9	Drug delivery	Hashimoto et al. (2017)
CLDN6	IMAB027	Treat cancer	Micke et al. (2014)
	342927	Cell sorting	Ben-David, Nudel, and Benvenisty (2013)
	WU-9E1-G2	Block HCV entry	Fofana et al. (2013)
CLDN18.2	IMAB362	Treat cancer	Singh et al. (2017)

mAbs have been generated, which include claudin-1, claudin-2, claudin-3, claudin-4, and claudin-5 mAbs (Table 9.4.2). These antibodies target the extracellular loops in the claudin proteins. Among them, claudin-5 mAbs can enhance the permeation of solutes across the blood–brain barrier in a mammalian cell model (Hashimoto et al., 2017). The efficacy of claudin-5 mAb is determined by the epitope region in claudin-5 protein it binds to. The mAbs that bind to the ECL1 in claudin-5 only increase the paracellular permeability of small molecules, for example, fluorescein (376 Da), whereas the mAbs that bind to the ECL2 in claudin-5 increase the paracellular permeability of both small and large molecules, for example, fluorescein and 4 kDa dextran, at a much higher rate (Fig. 9.4.3) (Hashimoto et al., 2017). Mechanistically, claudin-5 proteins are delocalized from tight junctions to intracellular vesicles by these mAbs.

9.4.4 Limitation and future direction

Binding promiscuity is a major problem of current TJ binders. One solution is to utilize a single-domain antibody (sdAb), also known as nanobody.

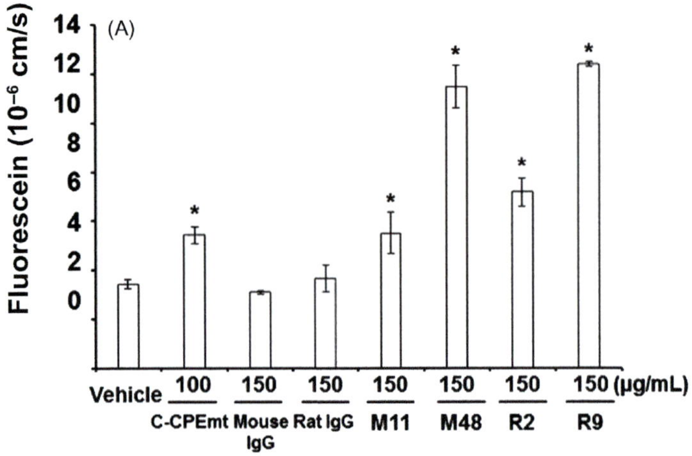

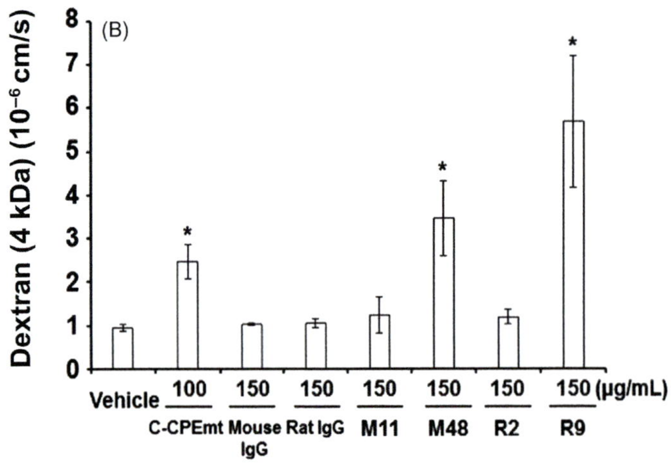

FIGURE 9.4.3 Effects of claudin-5 mAbs on blood–brain barrier. Cynomolgus monkey brain microvasculature endothelial cells were treated with anti-CLDN-5 mAbs or C-terminal half of *Clostridium perfringens* enterotoxin mutant (C-CPEmt) at the indicated concentrations for 12 hours. The cells were then incubated on the apical side with fluorescein (m.w. 376 Da) (A) or with 4 kDa dextran (B), and the permeabilities were measured from the apical side to the basal side. M11 and M48 are mouse antibodies; R2 and R9 are rat antibodies. M11 and R2 target the ECL1 in claudin-5; M48 and R9 target the ECL2 in claudin-5. *, $P < 0.05$ versus vehicle treatment. *Reproduced with permission from Hashimoto, Y., Shirakura, K., Okada, Y., Takeda, H., Endo, K., Tamura, M., Watari, A., Sadamura, Y., Sawasaki, T., Doi, T., et al. (2017). Claudin-5-binders enhance permeation of solutes across the blood–brain barrier in a mammalian model.* The Journal of Pharmacology and Experimental Therapeutics, 363, *275–283.*

Nanobody, engineered from the heavy-chain antibody found in camelids, consists of a single monomeric variable antibody domain of 12–15 kDa. The main advantages of nanobody over conventional antibodies include small size, thermostability, and the ability to recognize conformation-specific epitopes (Roovers, van Dongen, & van Bergen en Henegouwen, 2007). In the theory, a nanobody may recognize a domain important for either *cis* or *trans* claudin interaction, thereby allowing the delineation of the mode of claudin interaction in the tight junction. The claudin proteins assembled into the tight junction must have undergone conformational changes that, if can be recognized by nanobodies, will allow differentiating the localization of claudin proteins in the tight junction from the plasma membrane or the subcellular vesicle. Notably, claudins in tumor cells are not forming the tight junction, but they diffuse into the plasma membrane. A nanobody that binds specifically to claudins in the plasma membrane of tumor cells will have minimal adverse effects on normal TJ functions that require conformational changes during the claudin assembly process.

References

Ando, H., Suzuki, M., Kato-Nakano, M., Kawamoto, S., Misaka, H., Kimoto, N., ... Nakamura, K. (2015). Generation of specific monoclonal antibodies against the extracellular loops of human claudin-3 by immunizing mice with target-expressing cells. *Bioscience, Biotechnology, and Biochemistry*, *79*, 1272–1279.

Baumgartner, H. K., Beeman, N., Hodges, R. S., & Neville, M. C. (2011). A D-peptide analog of the second extracellular loop of claudin-3 and -4 leads to mislocalized claudin and cellular apoptosis in mammary epithelial cells. *Chemical Biology & Drug Design*, *77*, 124–136.

Ben-David, U., Nudel, N., & Benvenisty, N. (2013). Immunologic and chemical targeting of the tight-junction protein claudin-6 eliminates tumorigenic human pluripotent stem cells. *Nature Communications*, *4*, 1992.

Chung, N. P., Mruk, D., Mo, M. Y., Lee, W. M., & Cheng, C. Y. (2001). A 22-amino acid synthetic peptide corresponding to the second extracellular loop of rat occludin perturbs the blood-testis barrier and disrupts spermatogenesis reversibly in vivo. *Biology of Reproduction*, *65*, 1340–1351.

Everett, R. S., Vanhook, M. K., Barozzi, N., Toth, I., & Johnson, L. G. (2006). Specific modulation of airway epithelial tight junctions by apical application of an occludin peptide. *Molecular Pharmacology*, *69*, 492–500.

Fofana, I., Krieger, S. E., Grunert, F., Glauben, S., Xiao, F., Fafi-Kremer, S., ... Mee, C. J. (2010). Monoclonal anti-claudin 1 antibodies prevent hepatitis C virus infection of primary human hepatocytes. *Gastroenterology*, *139*, 953–964, 964.e951-954.

Fofana, I., Zona, L., Thumann, C., Heydmann, L., Durand, S. C., Lupberger, J., ... Reynolds, G. M. (2013). Functional analysis of claudin-6 and claudin-9 as entry factors for hepatitis C virus infection of human hepatocytes by using monoclonal antibodies. *Journal of Virology*, *87*, 10405–10410.

Freedman, J. C., Shrestha, A., & McClane, B. A. (2016). Clostridium perfringens *enterotoxin: Action, genetics, and translational applications*, . *Toxins (Basel)* (8). .

Hashimoto, Y., Hata, T., Tada, M., Iida, M., Watari, A., Okada, Y., ... Kondoh, M. (2018). Safety evaluation of a human chimeric monoclonal antibody that recognizes the extracellular loop domain of claudin-2. *European Journal of Pharmaceutical Sciences: Official Journal of the European Federation for Pharmaceutical Sciences*, *117*, 161–167.

Hashimoto, Y., Shirakura, K., Okada, Y., Takeda, H., Endo, K., Tamura, M., ... Doi, T. (2017). Claudin-5-binders enhance permeation of solutes across the blood–brain barrier in a mammalian model. *The Journal of Pharmacology and Experimental Therapeutics*, *363*, 275–283.

Hashimoto, Y., Tada, M., Iida, M., Nagase, S., Hata, T., Watari, A., ... Yagi, K. (2016). Generation and characterization of a human-mouse chimeric antibody against the extracellular domain of claudin-1 for cancer therapy using a mouse model. *Biochemical and Biophysical research Communications*, *477*, 91–95.

Katahira, J., Inoue, N., Horiguchi, Y., Matsuda, M., & Sugimoto, N. (1997). Molecular cloning and functional characterization of the receptor for *Clostridium perfringens* enterotoxin. *The Journal of Cell Biology*, *136*, 1239–1247.

Katahira, J., Sugiyama, H., Inoue, N., Horiguchi, Y., Matsuda, M., & Sugimoto, N. (1997). *Clostridium perfringens* enterotoxin utilizes two structurally related membrane proteins as functional receptors in vivo. *The Journal of Biological Chemistry*, *272*, 26652–26658.

Kawai, Y., Hamazaki, Y., Fujita, H., Fujita, A., Sato, T., Furuse, M., ... Minato, N. (2011). Claudin-4 induction by E-protein activity in later stages of CD4/8 double-positive thymocytes to increase positive selection efficiency. *Proceedings of the National Academy of Sciences of the United States of America*, *108*, 4075–4080.

Micke, P., Mattsson, J. S., Edlund, K., Lohr, M., Jirstrom, K., Berglund, A., ... Ponten, F. (2014). Aberrantly activated claudin 6 and 18.2 as potential therapy targets in non-small-cell lung cancer. *International Journal of Cancer*, *135*, 2206–2214.

Mrsny, R. J., Brown, G. T., Gerner-Smidt, K., Buret, A. G., Meddings, J. B., Quan, C., ... Nusrat, A. (2008). A key claudin extracellular loop domain is critical for epithelial barrier integrity. *The American Journal of Pathology*, *172*, 905–915.

Nakajima, M., Nagase, S., Iida, M., Takeda, S., Yamashita, M., Watari, A., ... Sawasaki, T. (2015). Claudin-1 binder enhances epidermal permeability in a human keratinocyte model. *The Journal of Pharmacology and Experimental Therapeutics*, *354*, 440–447.

Nusrat, A., Brown, G. T., Tom, J., Drake, A., Bui, T. T., Quan, C., & Mrsny, R. J. (2005). Multiple protein interactions involving proposed extracellular loop domains of the tight junction protein occludin. *Molecular Biology of the Cell*, *16*, 1725–1734.

Roovers, R. C., van Dongen, G. A., & van Bergen en Henegouwen, P. M. (2007). Nanobodies in therapeutic applications. *Current Opinion in Molecular Therapeutics*, *9*, 327–335.

Sahin, U., Jaeger, D., Marme, F., Mavratzas, A., Krauss, J., De Greve, J., ... Tureci, O. (2015). First-in-human phase I/II dose-escalation study of IMAB027 in patients with recurrent advanced ovarian cancer (OVAR): Preliminary data of phase I part. *Journal of Clinical Oncology*, *33*, 5537.

Sahin, U., Koslowski, M., Dhaene, K., Usener, D., Brandenburg, G., Seitz, G., ... Tureci, O. (2008). Claudin-18 splice variant 2 is a pan-cancer target suitable for therapeutic antibody development. *Clinical Cancer Research*, *14*, 7624–7634.

Saitoh, Y., Suzuki, H., Tani, K., Nishikawa, K., Irie, K., Ogura, Y., ... Fujiyoshi, Y. (2015). Tight junctions. Structural insight into tight junction disassembly by *Clostridium perfringens* enterotoxin. *Science (New York, NY)*, *347*, 775–778.

Schlingmann, B., Overgaard, C. E., Molina, S. A., Lynn, K. S., Mitchell, L. A., Dorsainvil White, S., ... Koval, M. (2016). Regulation of claudin/zonula occludens-1 complexes by hetero-claudin interactions. *Nature Communications*, *7*, 12276.

Shinoda, T., Shinya, N., Ito, K., Ohsawa, N., Terada, T., Hirata, K., ... Yokoyama, S. (2016). Structural basis for disruption of claudin assembly in tight junctions by an enterotoxin. *Scientific Reports*, *6*, 33632.

Singh, P., Toom, S., & Huang, Y. (2017). Anti-claudin 18.2 antibody as new targeted therapy for advanced gastric cancer. *Journal of Hematology & Oncology*, *10*, 105.

Sonoda, N., Furuse, M., Sasaki, H., Yonemura, S., Katahira, J., Horiguchi, Y., & Tsukita, S. (1999). Clostridium perfringens enterotoxin fragment removes specific claudins from tight junction strands: Evidence for direct involvement of claudins in tight junction barrier. *The Journal of Cell Biology*, *147*, 195–204.

Staat, C., Coisne, C., Dabrowski, S., Stamatovic, S. M., Andjelkovic, A. V., Wolburg, H., ... Blasig, I. E. (2015). Mode of action of claudin peptidomimetics in the transient opening of cellular tight junction barriers. *Biomaterials*, *54*, 9–20.

Suzuki, M., Kato-Nakano, M., Kawamoto, S., Furuya, A., Abe, Y., Misaka, H., ... Ando, H. (2009). Therapeutic antitumor efficacy of monoclonal antibody against Claudin-4 for pancreatic and ovarian cancers. *Cancer Science*, *100*, 1623–1630.

Takigawa, M., Iida, M., Nagase, S., Suzuki, H., Watari, A., Tada, M., ... Yagi, K. (2017). Creation of a claudin-2 binder and its tight junction-modulating activity in a human intestinal model. *The Journal of Pharmacology and Experimental Therapeutics*, *363*, 444–451.

Tavelin, S., Hashimoto, K., Malkinson, J., Lazorova, L., Toth, I., & Artursson, P. (2003). A new principle for tight junction modulation based on occludin peptides. *Molecular Pharmacology*, *64*, 1530–1540.

Van Itallie, C. M., & Anderson, J. M. (1997). Occludin confers adhesiveness when expressed in fibroblasts. *Journal of Cell Science*, *110*(Pt 9), 1113–1121.

Vietor, I., Bader, T., Paiha, K., & Huber, L. A. (2001). Perturbation of the tight junction permeability barrier by occludin loop peptides activates beta-catenin/TCF/LEF-mediated transcription. *EMBO Reports*, *2*, 306–312.

Wong, V., & Gumbiner, B. M. (1997). A synthetic peptide corresponding to the extracellular domain of occludin perturbs the tight junction permeability barrier. *The Journal of Cell Biology*, *136*, 399–409.

Zwanziger, D., Hackel, D., Staat, C., Bocker, A., Brack, A., Beyermann, M., ... Blasig, I. E. (2012). A peptidomimetic tight junction modulator to improve regional analgesia. *Molecular Pharmaceutics*, *9*, 1785–1794.

Chapter 9.5

De novo assembly of tight junction

9.5.1 Concept of de novo assembly of subcellular organelle

Claudin polymerization is part of a sophisticated cellular process known as TJ assembly, which involves hundreds of proteins via several key steps such as cell adhesion, polarity establishment, and cytoskeletal anchorage. Cell-free systems such as artificial membranes may allow bypass the regulatory hierarchy required for TJ biogenesis. Peripheral myelin protein 22 (PMP22), a protein structurally related to claudin, spontaneously organized the membrane into myelin-like lamellae when reconstituted into lipid bilayers (Fig. 9.5.1A) (Mittendorf et al., 2017). VE-cadherin self-assembled into an artificial adherens junction when reconstituted into liposomes (Fig. 9.5.1B) (Taveau et al., 2008). The extracellular domains in PMP22 or VE-cadherin are vital to the respective assembly process. Because the extracellular

(A) (B)

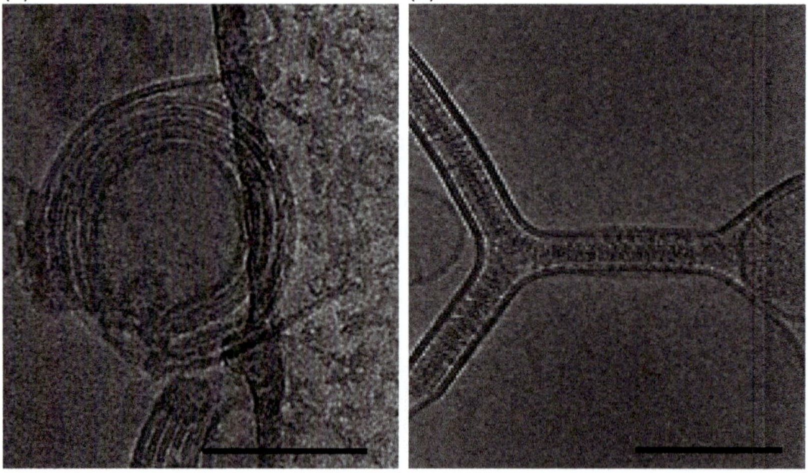

FIGURE 9.5.1 Cryo-electron micrographs of artificial myelin and adherens junction in artificial membranes. (A) Vitrified liposomes containing PMP22 proteins spontaneously form myelin-like structures. (B) Vitrified liposomes containing VE-cadherin proteins establish cell junctions resembling the adherens junction. Bar: 100 nm. *Reproduced with permission from Mittendorf, K. F., Marinko, J.T., Hampton, C.M., Ke, Z., Hadziselimovic, A., Schlebach, J.P., et al. (2017). Peripheral myelin protein 22 alters membrane architecture. Science Advances, 3, e1700220; Taveau, J.C., Dubois, M., Le Bihan, O., Trepout, S., Almagro, S., Hewat, E., et al. (2008). Structure of artificial and natural VEcadherin-based adherens junctions. Biochemical Society Transactions, 36, 189–193.*

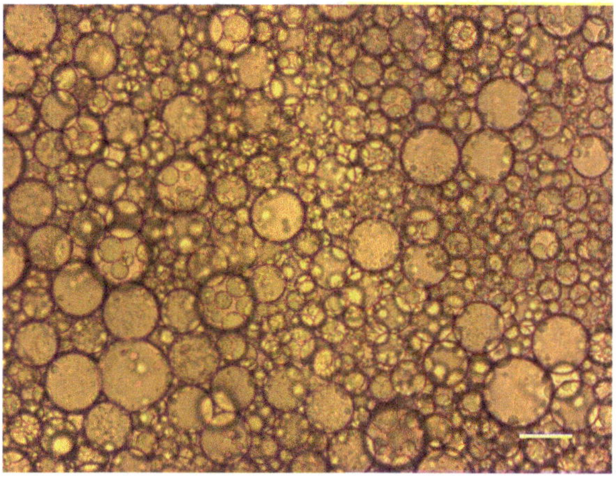

FIGURE 9.5.2 Bright-field image of GUVs. GUVs were made by rehydrating dried lipids in 100 mM sucrose buffer and imaged in a buffer containing 100 mM glucose, 100 mM KCl, and 5 mM HEPES (pH7.4). Bar: 50 μm.

domains in claudin also mediate cell adhesion (Kubota et al., 1999), liposomes reconstituted with claudins may form TJ-like structures.

9.5.2 Giant unilamellar vesicle

Giant unilamellar vesicle (GUV) has primarily been used to study the physical and chemical nature of lipid bilayers, including bilayer deformation, lateral phase coexistence (also known as rafts), and membrane fusion (Luisi & Walde, 2008). GUV resembles a cell by forming a spherical shell of the membrane surrounding an aqueous interior that can easily be made different from the aqueous exterior. The size of the GUV ranges from 1 to 100 μm in diameter. GUV can be imaged with a variety of light microscopic approaches (Dimova et al., 2006). GUV spontaneously forms by a swelling process when a dried lipid film is rehydrated (Fig. 9.5.2) (Manley & Gordon, 2008). Electroporation allows for preparing larger and more uniform GUVs (Politano, Froude, Jing, & Zhu, 2010). Ion channels can be reconstituted into GUVs to facilitate patch clamp studies (Riquelme, Lopez, Garcia-Segura, Ferragut, & Gonzalez-Ros, 1990). More importantly, GUV can be made taut using osmotic gradient or mechanical tension so that while being soft the membrane may retain different levels of curvature and stiffness (Aimon et al., 2014).

9.5.3 Protein incorporation and topological orientation in giant unilamellar vesicle

Incorporating membrane proteins into GUVs is a difficult task with limited success that applies only to a small subset of membrane proteins. One particular challenge is how to orientate protein topology during the process of protein incorporation. Microfluidic jetting may provide a solution by simultaneously creating and loading GUVs with controlled contents (Stachowiak et al., 2008). First, a membrane protein such as claudin is reconstituted into small proteoliposomes made from 1,2-diphytanoyl-*sn*-glycero-3-phosphocholine (DPhPC) (Rigaud, Pitard, & Levy, 1995). Then the claudin containing proteoliposomes is added to the inner DPhPC-stabilized droplet of an asymmetric two-droplet infinity chamber (Fig. 9.5.3A). After proteoliposomes have fused with the DPhPC monolayer, the central acrylic divider of the infinity chamber is removed and a lipid membrane forms spontaneously. Finally, GUVs are created by placing a 25-μm-diameter microfluidic nozzle close to the lipid membrane and jetting with a piezoelectric actuator (Fig. 9.5.3B). The orientation of claudin protein topology can be achieved by

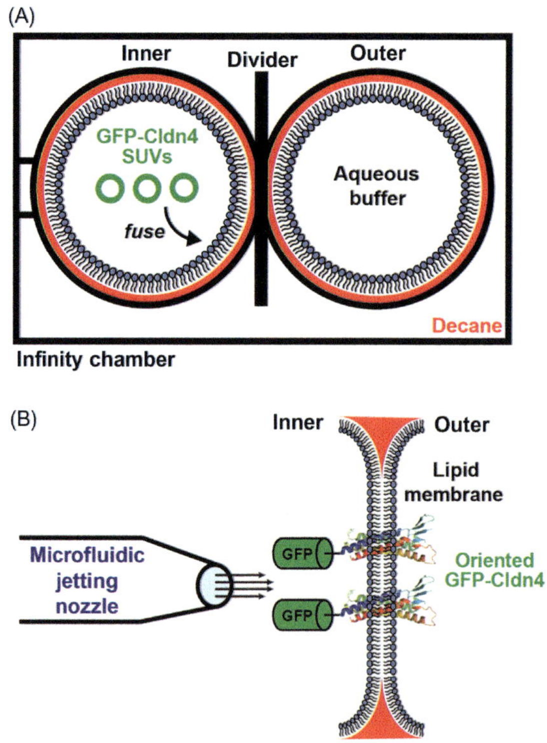

FIGURE 9.5.3 Reconstitution of claudin-4 in jetted GUV. (A) Infinity chamber configuration. (B) Microfluidic jetting scheme. A GFP tag is added to the N-terminus of claudin-4 protein to confer topological orientation. *SUV*, Small unilamellar vesicle. *Reproduced with permission from Belardi, B., Son, S., Vahey, M. D., Wang, J., Hou, J., & Fletcher, D. A. (2018). Claudin-4 reconstituted in unilamellar vesicles is sufficient to form tight interfaces that partition membrane proteins.* Journal of Cell Science, 132, jcs221556.

adding a large solubilization tag, for example, GFP on one side of the protein relative to the lipid membrane, which creates an energy barrier to prevent the protein from traversing the membrane (Belardi et al., 2018).

9.5.4 Probing claudin interactions in giant unilamellar vesicle

Claudins can interact in *cis* within the cell membrane or in *trans* between cell membranes (Furuse, Sasaki, & Tsukita, 1999). When incorporated into unpaired GUVs, claudins form large protein clusters as evidenced by the bright puncta along the jetted membrane, which is due to their *cis* interactions (Fig. 9.5.4A). In contrast, proteins not able to interact in *cis* are found uniformly distributed along the jetted membrane (Fig. 9.5.4B) (Belardi et al., 2018). In paired GUVs, claudins are enriched at interfaces (Fig. 9.5.5A). The mobility of claudin proteins at the interface is significantly lower than

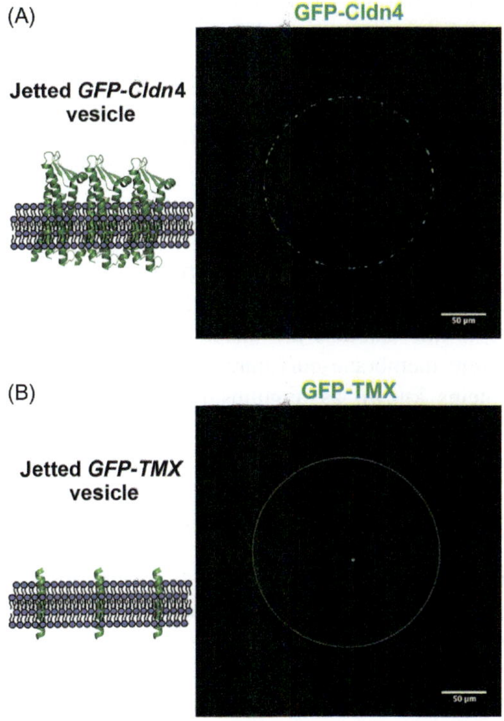

FIGURE 9.5.4 Claudin *cis*-interaction in GUV. Confocal micrograph of single jetted vesicle containing either GFP-Cldn4 (A) or GFP-TMX (B). TMX is a synthetic transmembrane protein that remains monomeric in lipid bilayers. Bar: 50 μm. *Reproduced with permission from Belardi, B., Son, S., Vahey, M. D., Wang, J., Hou, J., & Fletcher, D. A. (2018). Claudin-4 reconstituted in unilamellar vesicles is sufficient to form tight interfaces that partition membrane proteins.* Journal of Cell Science, 132, jcs221556.

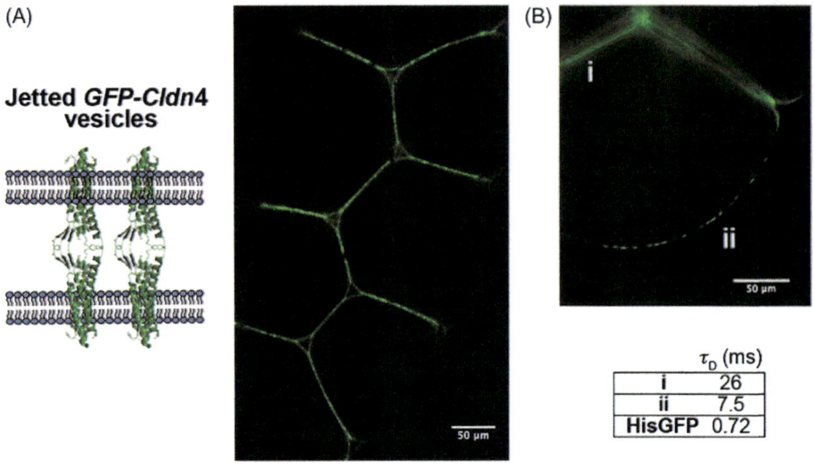

FIGURE 9.5.5 Claudin *trans*-interaction in GUV. (A) Confocal micrograph of paired jetted vesicles containing GFP-Cldn4. (B) Protein mobility is determined by fluorescence correlation spectroscopy as an inverse value of the diffusion time (τ_D). Region (i) denotes the interface; region (ii) denotes the free membrane. HisGFP indicates the mobility of an irrelevant protein not able to form vesicle junctions. Bar: 50 μm. *Reproduced with permission from Belardi, B., Son, S., Vahey, M. D., Wang, J., Hou, J., and Fletcher, D. A. (2018). Claudin-4 reconstituted in unilamellar vesicles is sufficient to form tight interfaces that partition membrane proteins.* Journal of Cell Science, 132, jcs221556.

that at the free membrane, suggesting that *trans* interactions interlock claudins in the interfacial membrane (Fig. 9.5.5B) (Belardi et al., 2018). An intriguing finding is that claudins are unable to colonize the tri-vesicle junctions, where different membrane curvature and intercellular spacing call for a new class of proteins known as angulins (Furuse, Izumi, Oda, Higashi, & Iwamoto, 2014).

9.5.5 Limitation and future direction

Claudin alone is unlikely to reconstitute the tight junction in vitro in GUV. Other molecules such as occludin, ZO-1, and JAM must play important roles. In particular, ZO-1 proteins have been found to self-organize into membrane-attached compartments via phase separation and the formation of condensed ZO-1 protein compartments can sequester key TJ membrane proteins such as claudins (Beutel, Maraspini, Pombo-Garcia, Martin-Lemaitre, & Honigmann, 2019). In the theory, ZO-1 proteins can be added to the aqueous interior of GUV to facilitate the assembly of claudin proteins at the vesicle junctions. The structural role of other TJ proteins may also be studied in GUV alone or in combination with claudin. A variety of light and electron microscopic approaches are available to allow directly visualizing the GUV junctions, which include confocal microscopy, freeze-fracture replica electron microscopy, and cryo-electron microscopy. Functional analyses of the GUV junctions, for instance, protein adhesion measurement and paracellular conductance recording, are challenging and will require new technologies such as atomic force microscopy and scanning ion conductance microscopy, respectively.

References

Aimon, S., Callan-Jones, A., Berthaud, A., Pinot, M., Toombes, G. E., & Bassereau, P. (2014). Membrane shape modulates transmembrane protein distribution. *Developmental Cell, 28*, 212–218.

Belardi, B., Son, S., Vahey, M. D., Wang, J., Hou, J., & Fletcher, D. A. (2018). Claudin-4 reconstituted in unilamellar vesicles is sufficient to form tight interfaces that partition membrane proteins. *Journal of Cell Science, 132*.

Beutel, O., Maraspini, R., Pombo-Garcia, K., Martin-Lemaitre, C., & Honigmann, A. (2019). Phase separation of zonula occludens proteins drives formation of tight junctions. *Cell, 179*, 923–936.e911.

Dimova, R., Aranda, S., Bezlyepkina, N., Nikolov, V., Riske, K. A., & Lipowsky, R. (2006). A practical guide to giant vesicles. Probing the membrane nanoregime via optical microscopy. *Journal of Physics: Condensed Matter, 18*, S1151–S1176.

Furuse, M., Izumi, Y., Oda, Y., Higashi, T., & Iwamoto, N. (2014). Molecular organization of tricellular tight junctions. *Tissue Barriers, 2*, e28960.

Furuse, M., Sasaki, H., & Tsukita, S. (1999). Manner of interaction of heterogeneous claudin species within and between tight junction strands. *The Journal of Cell Biology, 147*, 891–903.

Kubota, K., Furuse, M., Sasaki, H., Sonoda, N., Fujita, K., Nagafuchi, A., & Tsukita, S. (1999). Ca(2 +)-independent cell-adhesion activity of claudins, a family of integral membrane proteins localized at tight junctions. *Current Biology: CB, 9*, 1035–1038.

Luisi, P. L., & Walde, P. (2008). *Giant vesicles* (Vol. 22). John Wiley & Sons.

Manley, S., & Gordon, V. D. (2008). Making giant unilamellar vesicles via hydration of a lipid film. *Current Protocols in Cell Biology, Chapter 24*, Unit 24.23.

Mittendorf, K. F., Marinko, J. T., Hampton, C. M., Ke, Z., Hadziselimovic, A., Schlebach, J. P., ... Sanders, C. R. (2017). Peripheral myelin protein 22 alters membrane architecture. *Science Advances, 3*, e1700220.

Politano, T. J., Froude, V. E., Jing, B., & Zhu, Y. (2010). AC-electric field dependent electroformation of giant lipid vesicles. *Colloids and Surfaces B: Biointerfaces, 79*, 75–82.

Rigaud, J. L., Pitard, B., & Levy, D. (1995). Reconstitution of membrane proteins into liposomes: Application to energy-transducing membrane proteins. *Biochimica et Biophysica Acta, 1231*, 223–246.

Riquelme, G., Lopez, E., Garcia-Segura, L. M., Ferragut, J. A., & Gonzalez-Ros, J. M. (1990). Giant liposomes: A model system in which to obtain patch-clamp recordings of ionic channels. *Biochemistry, 29*, 11215–11222.

Stachowiak, J. C., Richmond, D. L., Li, T. H., Liu, A. P., Parekh, S. H., & Fletcher, D. A. (2008). Unilamellar vesicle formation and encapsulation by microfluidic jetting. *Proceedings of the National Academy of Sciences of the United States of America, 105*, 4697–4702.

Taveau, J. C., Dubois, M., Le Bihan, O., Trepout, S., Almagro, S., Hewat, E., ... ambert, O. (2008). Structure of artificial and natural VE-cadherin-based adherens junctions. *Biochemical Society Transactions, 36*, 189–193.

Chapter 9.6

Organoid model of tight junction biology

9.6.1 Organoid culture

Directed differentiation of human pluripotent stem cells (hPSC) to specific cellular endpoints may allow regeneration for a variety of human tissue types. The organotypic organization of hPSC cells to form "organoids" has provided complex multicellular models of developing human tissues including optic cup, intestine, cerebral cortex, stomach, lung, and kidney (Eiraku et al., 2011; Lancaster et al., 2013; McCracken et al., 2014; Miller et al., 2019; Spence et al., 2011; Takasato et al., 2015). Compared to primary or immortalized cell culture models, organoid culture models retain the 3D architecture of tissues and the proper supplementation of their supporting stromata. The kidney organoid, taken as an example, consists of nephrons associated with a tubular network surrounded by renal interstitia and endothelial cells. Within the network, individual nephrons segment into proximal and distal tubules, loops of Henle, and glomeruli composed of podocytes and rudimentary blood vessels (Fig. 9.6.1) (Takasato et al., 2015). Tight junctions are made by the tubular epithelial cells in the organoid (Fig. 9.6.2A). Claudins are expressed and localized to the tight junctions (Fig. 9.6.2B).

9.6.2 Organ on a chip

Although organoid cultures can support the formation of high-order tissue structures, they often lack the features that are critical for organ function, such as vascular perfusion, mechanical cues (e.g., breathing motions in the lungs and beating motions in the heart), circulating immune cells, and the ability to coculture normal microbiome for extended periods. Organ on a chip is a microfluidic cell culture device that contains hollow microchannels lined by living cells cultured on extracellular matrix (ECM)-coated, porous membrane and allows continuous perfusion with life-sustaining culture medium (Bhatia & Ingber, 2014). Mechanical forces can be applied to the device to mimic physiological movements (Kim, Li, Collins, & Ingber, 2016); air–liquid interfaces can be created in the device by introducing air into the hollow microchannels (Huh et al., 2010); electrical signals can be recorded from the device by integrating electrodes into the perfusion chamber (Douville et al., 2010). The use of organ-on-a-chip device has been shown to improve tight junction maturation in blood–brain barrier models

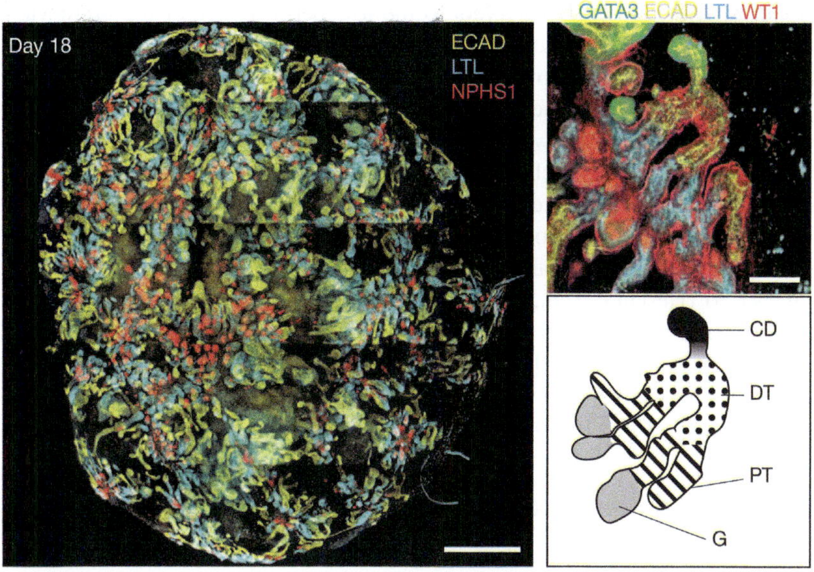

FIGURE 9.6.1 Kidney organoid culture. Left: Tile scan immunofluorescence micrograph of a whole kidney organoid displays structural complexity. Bar: 1 mm. Right: High-magnification immunofluorescence micrograph shows a nephron segmented into four compartments, including the collecting duct (CD, GATA3^{+}ECAD^{+}), the distal tubule (DT, GATA3^{-}ECAD^{+}LTL^{-}), the proximal tubule (PT, ECAD^{-}LTL^{+}), and the glomerulus (G, WT1^{+}). Bar: 100 μm. *Reproduced with permission from Takasato, M., Er, P. X., Chiu, H. S., Maier, B., Baillie, G. J., Ferguson, C., Parton, R. G., Wolvetang, E. J., Roost, M. S., Chuva de Sousa Lopes, S. M., et al. (2015). Kidney organoids from human iPS cells contain multiple lineages and model human nephrogenesis.* Nature, 526, *564–568.*

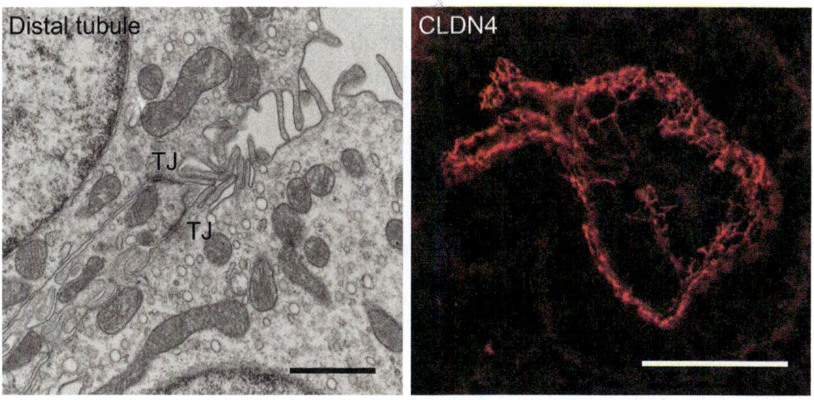

FIGURE 9.6.2 Tight junction in kidney organoid. Left: Electron micrograph reveals tight junctions in the kidney organoid distal tubule. Bar: 1 μm. Right: Immunofluorescence micrograph shows the claudin-4 (CLDN4) protein localization in tight junctions of the kidney organoid distal tubule. Bar: 50 μm.

(Booth & Kim, 2012). BBB refers to the low permeability of the brain vasculature, which is contributed to by three ways: (1) the highly specialized TJ restricts the paracellular pathway; (2) the rate of transcytosis is extremely low; and (3) the lack of membrane fenestration limits the direct exchange of solutes and water. Multiple cell types coordinate the development of BBB, which includes the endothelial cell, the pericyte, and the astrocyte. Fluid sheer stress is also important for the function of BBB (Desai et al., 2002). The organ-on-a-chip device can allow coculture of the three types of BBB-making cells under constant fluid sheer stress (Fig. 9.6.3) (Herland et al., 2016). The BBB developed on the chip exhibits tight junction permeability similar to that observed in vivo in animal BBB models (Yeon et al., 2012). Various TJ proteins including ZO-1 are expressed and localized to the tight junction in BBB on a chip (Griep et al., 2013; Herland et al., 2016).

9.6.3 Bioprinting of organ

Current organ-on-a-chip models lack the geometry, complexity, and functionality important for recapitulating the organ physiology. Bioprinting, with which living cells can be precisely deposited together with hydrogel-base scaffolds, allows fabricating the 3D tissue structures resembling those in naturally developed organs (Ozbolat & Yu, 2013). A bioprinted model of human vascularized proximal tubule consists of two adjacent conduits that are lined with confluent epithelium and endothelium, respectively, embedded in a permeable ECM and can be independently accessed using a closed-loop perfusion system (Fig. 9.6.4) (Lin et al., 2019). The bioprinted tissue exhibits active tubular-vascular exchange of solutes, for example, glucose and albumin, akin to native kidney proximal tubules. Because each bioprinted tubule can be independently perfused and recorded by external electrodes, the transport properties, including both transcellular and paracellular permeabilities, are interrogated on the single-tubule level, that is the gold standard for renal physiology, and was made possible only by the ex vivo tubule perfusion technique (Malnic, Klose, & Giebisch, 1966).

9.6.4 Limitation and future direction

The organotypic culture models provide an important alternative resource to animal models. Many drugs including TJ binders can be screened in vitro in cultured organs. These surrogates, however, only resemble human fetal organs. It is widely recognized that the expression profile of TJ proteins including claudins changes in an organ as it matures. Perhaps, the research value in organ culture is to allow genes to be manipulated or edited in vitro. Functional analyses of the genes will have to take into account that the phenotypes may be incomplete. The most promising approach to promote

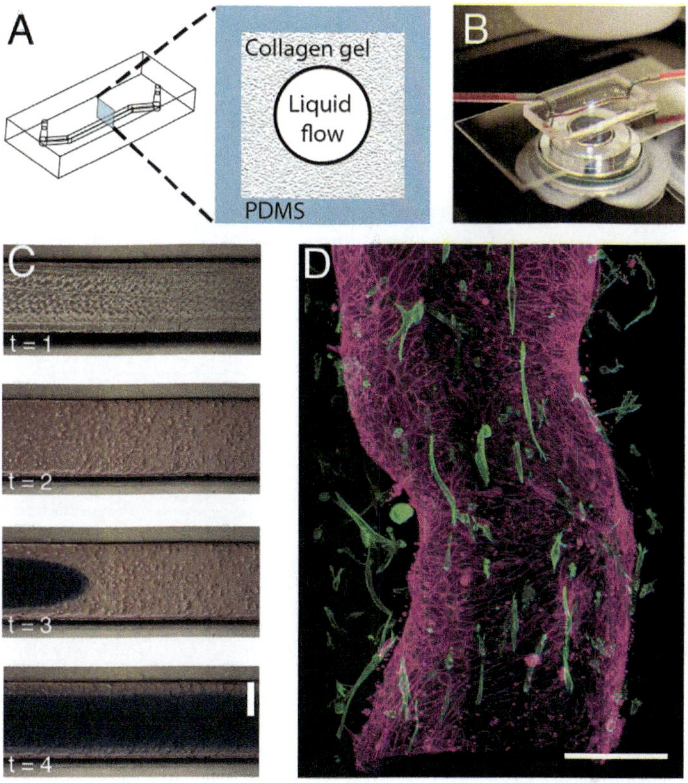

FIGURE 9.6.3 BBB on a chip. (A) Schematic diagram of the polydimethylsiloxane structure used to generate the BBB chip (left) and an illustration of a cross-section through the chip showing the microchannel containing the collagen gel (right). (B) Photograph of the BBB chip on the stage of an inverted microscope. (C) Time-lapse images of the microchannel before ($t = 1$) and after infusion of a neutralized collagen gel containing dispersed human astrocytes ($t = 2$), which was then followed by injection of a low viscosity liquid (dyed blue) driven by hydrostatic pressure to initiate "finger" formation in the center of the gel ($t = 3$), and eventually a continuous hollow cylindrical lumen throughout the length of the device ($t = 4$). Bar: 500 μm. (D) Immunofluorescence micrograph showing cell distribution in the BBB chip comprising endothelial cells lining the lumen and astrocytes embedded in the surrounding gel. Green indicates F-actin staining and magenta corresponds to VE-Cadherin staining. Bar: 200 μm. *Reproduced with permission from Herland, A., van der Meer, A. D., FitzGerald, E. A., Park, T. E., Sleeboom, J. J., & Ingber, D. E. (2016). Distinct contributions of astrocytes and pericytes to neuroinflammation identified in a 3D human blood–brain barrier on a chip.* PloS one, 11, *e0150360.*

organotypic maturation is by transplantation in vivo into host animals. In this setting, host vasculature invades the transplanted organ and exposes it to a variety of circulating factors, including growth factors, hormones, erythrocytes, leukocytes and more. Mechanical stress, nerve signal and microbiome in the host will then fine-tune the developmental process for the transplanted organ.

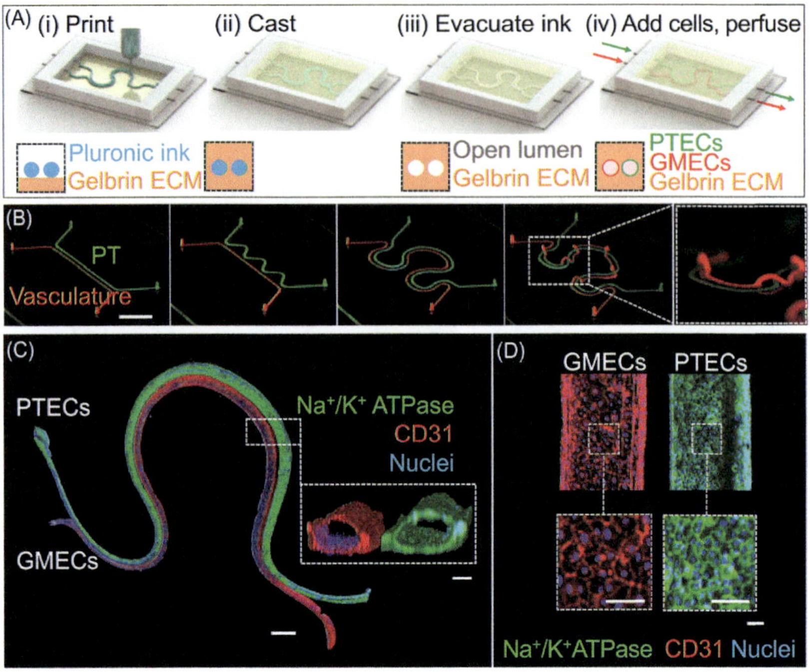

FIGURE 9.6.4 Bioprinted human proximal tubule. (A) Schematic view of the 3D bioprinting process. (B) Simple and complex 3D proximal tubules and blood vessels can be bioprinted. Bar: 10 mm. (C) Whole-mount immunofluorescence staining of the bioprinted tissue, in which Na + /K + ATPase labels the proximal tubule, CD31 labels the blood vessel. Bar: 1 mm. The gap between the tubular and vascular conduits is ~70 μm. Inset: cross-sectional image of the two open lumens. Bars: 100 μm. (D) High-magnification images of the bioprinted tissue after staining. Bar: 100 μm. PT: proximal tubule; PTEC: proximal tubule epithelial cell; GMEC: glomerular microvascular endothelial cell. *Reproduced with permission from Lin, N. Y. C., Homan, K. A., Robinson, S. S., Kolesky, D. B., Duarte, N., Moisan, A., & Lewis, J. A. (2019). Renal reabsorption in 3D vascularized proximal tubule models.* Proceedings of the National Academy of Sciences of the United States of America, 116, *5399–5404.*

References

Bhatia, S. N., & Ingber, D. E. (2014). Microfluidic organs-on-chips. *Nature Biotechnology*, *32*, 760–772.

Booth, R., & Kim, H. (2012). Characterization of a microfluidic in vitro model of the blood-brain barrier (muBBB). *Lab on a Chip*, *12*, 1784–1792.

Desai, S. Y., Marroni, M., Cucullo, L., Krizanac-Bengez, L., Mayberg, M. R., Hossain, M. T., ... Janigro, D. (2002). Mechanisms of endothelial survival under shear stress. *Endothelium: Journal of Endothelial Cell Research*, *9*, 89–102.

Douville, N. J., Tung, Y. C., Li, R., Wang, J. D., El-Sayed, M. E., & Takayama, S. (2010). Fabrication of two-layered channel system with embedded electrodes to measure resistance across epithelial and endothelial barriers. *Analytical Chemistry*, *82*, 2505–2511.

Eiraku, M., Takata, N., Ishibashi, H., Kawada, M., Sakakura, E., Okuda, S., ... Sasai, Y. (2011). Self-organizing optic-cup morphogenesis in three-dimensional culture. *Nature*, *472*, 51–56.

Griep, L. M., Wolbers, F., de Wagenaar, B., ter Braak, P. M., Weksler, B. B., Romero, I. A., ... van den Berg, A. (2013). BBB on chip: Microfluidic platform to mechanically and biochemically modulate blood-brain barrier function. *Biomedical Microdevices*, *15*, 145–150.

Herland, A., van der Meer, A. D., FitzGerald, E. A., Park, T. E., Sleeboom, J. J., & Ingber, D. E. (2016). Distinct contributions of astrocytes and pericytes to neuroinflammation identified in a 3D human blood-brain barrier on a chip. *PLoS One*, *11*, e0150360.

Huh, D., Matthews, B. D., Mammoto, A., Montoya-Zavala, M., Hsin, H. Y., & Ingber, D. E. (2010). Reconstituting organ-level lung functions on a chip. *Science (New York, NY)*, *328*, 1662–1668.

Kim, H. J., Li, H., Collins, J. J., & Ingber, D. E. (2016). Contributions of microbiome and mechanical deformation to intestinal bacterial overgrowth and inflammation in a human gut-on-a-chip. *Proceedings of the National Academy of Sciences of the United States of America*, *113*, E7–E15.

Lancaster, M. A., Renner, M., Martin, C. A., Wenzel, D., Bicknell, L. S., Hurles, M. E., ... Knoblich, J. A. (2013). Cerebral organoids model human brain development and microcephaly. *Nature*, *501*, 373–379.

Lin, N. Y. C., Homan, K. A., Robinson, S. S., Kolesky, D. B., Duarte, N., Moisan, A., & Lewis, J. A. (2019). Renal reabsorption in 3D vascularized proximal tubule models. *Proceedings of the National Academy of Sciences of the United States of America*, *116*, 5399–5404.

Malnic, G., Klose, R. M., & Giebisch, G. (1966). Microperfusion study of distal tubular potassium and sodium transfer in rat kidney. *American Journal of Physiology*, *211*, 548–559.

McCracken, K. W., Cata, E. M., Crawford, C. M., Sinagoga, K. L., Schumacher, M., Rockich, B. E., ... Zavros, Y. (2014). Modelling human development and disease in pluripotent stem-cell-derived gastric organoids. *Nature*, *516*, 400–404.

Miller, A. J., Dye, B. R., Ferrer-Torres, D., Hill, D. R., Overeem, A. W., Shea, L. D., & Spence, J. R. (2019). Generation of lung organoids from human pluripotent stem cells in vitro. *Nature Protocols*, *14*, 518–540.

Ozbolat, I. T., & Yu, Y. (2013). Bioprinting toward organ fabrication: Challenges and future trends. *IEEE Transactions on Biomedical Engineering*, *60*, 691–699.

Spence, J. R., Mayhew, C. N., Rankin, S. A., Kuhar, M. F., Vallance, J. E., Tolle, K., ... Zorn, A. M. (2011). Directed differentiation of human pluripotent stem cells into intestinal tissue in vitro. *Nature*, *470*, 105–109.

Takasato, M., Er, P. X., Chiu, H. S., Maier, B., Baillie, G. J., Ferguson, C., ... Chuva de Sousa Lopes, S. M. (2015). Kidney organoids from human iPS cells contain multiple lineages and model human nephrogenesis. *Nature*, *526*, 564–568.

Yeon, J. H., Na, D., Choi, K., Ryu, S. W., Choi, C., & Park, J. K. (2012). Reliable permeability assay system in a microfluidic device mimicking cerebral vasculatures. *Biomedical Microdevices*, *14*, 1141–1148.

Index

Note: The page numbers followed by "*f*" and "*t*" refer to figures and tables, respectively

A

A2E. *See* *N*-retinylidene-*N*-retinyl-ethanolamine (A2E)
Abbe's diffraction limit equation, 175, 213
AC. *See* Alternating current (AC)
Acetone, 217
Acrylic resins, 214–215, 240
Actin, 20
  arcs, 20–23
  bundle snapping, 20–23
  polymerization, 20, 21*f*
  reorganization, 20–23, 22*f*
Activators, 310
Actomyosin, 23–24
  assembly, 23*f*
  ring, 24
Adaptive optics (AO), 355–356
Additives in fixation, 233–234
Ade2
  gene, 70
  phenotype, 76
Adherens junction (AJ), 1, 2*t*, 3, 41
  contamination of, 47
  signaling, 27*f*
Aequorea victoria, 178–179
Affinity chromatography, 78–79
Ag/AgCl electrode, 94
Ageing cells, 179
Airy disc, 175, 176*f*, 199–200
AJ. *See* Adherens junction (AJ)
Alanine, 76–77
Aldehydes, 228
  crosslinks, 153–154
Aliasing, 97–99
α-actinin, 34–35
α-amine, 59
α-catenin, 24
Alternating current (AC), 96
Amino acids, 76–77
3-Aminotriazole (3-AT), 76
Amot. *See* Angiomotin (Amot)
Amphotropic virus, 280–282
Amplitude resolution, 97–99
Analog signals, 97–99
"Analog-to-digital" conversion, 97–99
Anchorage independence, 264–265
Angiomotin (Amot), 28–31
Angulin, 16–17, 369–370
  knockout mouse model, 16–17
Antibody, 57, 240, 244–245
Anti-claudin antibodies, 360–361, 361*t*
Antigen–antibody complex, 186
AO. *See* Adaptive optics (AO)
Apical membrane potential, 95–96
Aquaporins, 149
Artificial membranes, 366–367
Asp55, 35
Asp64, 35
3-AT. *See* 3-Aminotriazole (3-AT)
Autofluorescence, 179
Axial diffraction pattern, 176–177, 177*f*
Axial resolution, 176

B

BAC. *See* Bacterial artificial chromosome (BAC)
Bacterial artificial chromosome (BAC), 311
Band-pass filter, 97
Baseline, 112
Basolateral membrane potential, 95–96
BBB. *See* Blood–brain barrier (BBB)
β-catenin, 26, 28
β-galactosidase (β-gal), 75
β-sheet domain, 35
Bicellular tight junction, 16–17
Bile canaliculus, 41
Binding promiscuity, 361–363
Biochemical approaches for tight junction
  claudin oligomer isolation by chemical cross-linking, 59–000

Biochemical approaches for tight junction (*Continued*)
  immunoprecipitation of *cis* and *trans* claudin interactions, 49–000
  recombinant claudin protein production in *P. pastoris*, 78–000
  tight junction isolation by subcellular fractionation, 41–000
  yeast two-hybrid assay of claudin interaction, 70–000
Biochemistry of tight junction
  biochemical organization of tight junction, 33
  claudin interaction with tight junction plaque proteins, 37–38
  models of claudin interaction, 36–37, 37*f*
  molecular structure of claudin protein, 35, 36*f*
  tight junction anchorage onto cytoskeleton, 38–39
  tight junction enriched protein fraction, 33–34
  tight junction integral protein fraction, 34–35
Biophysical approaches for tight junction
  electrophysiology of epithelial transport, 89–000
  epithelial cell cultures in Ussing chamber, 100–000
  epithelial ohmmeter, 128–000
  epithelial tissues in Ussing chamber, 117–000
  flux assay in Ussing chamber, 140–000
  impedance measurement in Ussing chamber, 133–000
  water permeability measurement in Ussing chamber, 145–000
Bioprinting of organ, 374
Bits, 97–99
Blastocysts, 307
Blood–brain barrier (BBB), 243–244, 372–374, 375*f*
Bowman's capsule, 170
Branched actin polymerization, 20–23
Breeding strategy for mutant mice, 337

C

C57BL/6 blastocysts, 337
Ca^{++}-sensing receptor (CaSR), 314–315
Cadherin(s), 3, 6, 26
  cadherin-mediated cell adhesion, 28
  cadherin–β-catenin complex, 26
  compatibility, 8–9
  ectodomain, 6
  interaction, 6
  puncta, 20–23
  structure, 7*f*
Calcium chelator, 47
Calcium switch assay, 0–275, 277*f*
  data analysis, 277
  experimental procedure, 276–277
  materials and reagents, 275–276
    Ca^{++}-switch buffers, 275, 276*t*
    cell culture medium, 275
    cell model, 275
    equipment, 275
  troubleshooting, 278
Canonical Wnt signaling, 28, 29*f*
Carboxyl-terminal domain of CPE (C-CPE), 357
Cardiovascular system, 162
Casein kinase 1α (CK1α), 28
CaSR. *See* Ca^{++}-sensing receptor (CaSR)
Catenins, 3
  signaling, 26
CB. *See* Cell body (CB)
C-CPE. *See* Carboxyl-terminal domain of CPE (C-CPE)
CD-1, 304
Cdc42, 26
Cell adhesion
  cadherin compatibility and tissue morphogenesis, 8–9
  cadherin interaction, 6
  claudin compatibility and tissue barrier, 10
  claudin interaction, 6–7
  connexin compatibility and electric synapse, 10
  connexin interaction, 8
Cell body (CB), 343–345
Cell capacitance and underestimation, 131
Cell culture, 263–000
  anchorage independence, 264–265
  cell growth curve, 264*f*
  cloning and selection, 265
  contamination, 266–267
  cryopreservation, 265–266
  data analysis, 224
  electron microscopy for tight junction, 220
  experimental procedure, 222–224
    cell culture and fixation, 222–223
    embedding, 223–224
    sectioning and poststaining, 224
  fixation for electron microscopy, 220–221

gene transfer, 265
materials and reagents, 221–222
buffers, 221–222
cell culture medium, 221
cell model, 221
equipment, 221
mycoplasma, 267*f*
primary culture and cell transformation, 263
subculture and propagation, 263–264
troubleshooting, 225–226
fixation of membrane structure, 225
poststaining of tight junction, 225–226
Cell junctions (CJ), 343–345
adherens junction, 3
desmosome, 3–4
GJ, 4–5
junction category, 1
molecular mechanisms connecting, 30*f*
TJ, 1–3
types, 2*t*
Cell membrane electrophysiology principles, 91–94
current through capacitor, 93–94
electric current, 92
electrode, liquid junction potential, and salt bridge, 94
equivalent electrical circuit of cell membrane, 91
Ohm's law, 92
voltage divider, 92–93
Cell models
calcium switch assay, 275–000
cell culture, 263–000
epithelial cells, 269–000
retrovirus-mediated RNA interference, 288–000
retrovirus-mediated transgene expression, 279–000
Cell monolayer quality, 114
Cell transformation, 263
Cell-free systems, 366–367
Cells, wide-field fluorescence microscopy for
data analysis, 184
dome formation in MDCK cells, 185*f*
experimental procedure, 183–184
fixation and permeabilization, 181–182
immunofluorescence labeling, 181
double or triple, 186–187
materials and reagents, 182–183
buffers, 182–183
cell culture medium, 182
cell model, 182
equipment, 182
troubleshooting, 184–187
antibody avidity, 186
antibody specificity, 186
claudin-4 protein subcellular localization, 185*f*
fixation artifact, 184–186
limit of resolution, 187
Cells on transwell, confocal microscopy for, 199–200, 204*f*, 209*f*
apicobasal polarity, 200
data analysis, 203
experimental procedure, 202–203
fixation of tight junction, 201
materials and reagents, 201–202
buffers, 201–202
cell culture medium, 201
cell model, 201
equipment, 201
troubleshooting, 203–205
axial resolution, 205
live-cell imaging, 205
signal sensitivity, 203–204
CEMOVIS. *See* Cryo-electron microscopy of vitreous section (CEMOVIS)
Centrifugation, 45
CFTR. *See* Cystic fibrosis transmembrane conductance regulator (CFTR)
cfu. *See* Colony forming units (cfu)
Chemical cross-linkers, claudin oligomer isolation by, 59, 60*t*, 62
Chimeric pups, 307
"Chopstick" electrode system, 128
Chromatography, 78–79
Circumferential actin cable, 20–23
Cis claudin interaction, 8*f*
immunoprecipitation of, 49–53, 54*f*
Cis oligomer, 68
CJ. *See* Cell junctions (CJ)
CK1α. *See* Casein kinase 1α (CK1α)
Claudin oligomer isolation, 59–000
chemical cross-linkers, 59, 62
data analysis, 65–66
ectopic claudin expression model, 59
experimental procedure, 63–65
chemical cross-linking, 65
HEK293 cell transfection, 63–64
solubilization of claudin protein, 64
sucrose gradient centrifugation, 64–65, 65*f*
materials and reagents, 59–63
buffers, 60–62

Claudin oligomer isolation (*Continued*)
cell culture medium, 60
cell model for ectopic gene expression, 59
chemical cross-linkers, 62
equipment, 62–63
plasmids, 60
troubleshooting, 66–68
cis vs. trans oligomer, 68
molecular ruler and nearest neighbor approaches, 67
specificity of cross-linking, 66–67
in vitro *vs.* in vivo cross-linking, 67–68
Claudin(s), 1–3, 6, 35, 45, 49, 367–370
alignment of first extracellular loop domain in, 15*f*
claudin-3, 357
claudin-4, 35, 357
claudin-5, 243–244, 244*f*
claudin-14, 314–315, 314*f*
claudin-15, 35
claudin-16, 10, 322–323
claudin-19, 10, 35
$CLDN1_{53-80}$, 358–360
$CLDN5_{146-150}$, 358–360
compatibility, 10
interaction, 6–7
with tight junction plaque proteins, 37–38
models of claudin interaction, 36–37, 37*f*
molecular structure, 35, 36*f*
peptidomimetics, 358–360
permeabilities, 13*t*
polymerization, 366–367
proteins, 86
tight junction localization, 35*f*
Cloning and selection, 265
Clostridium perfringens enterotoxin (CPE), 357
Coimmunoprecipitation, 49
Colony forming units (cfu), 322
Concentration gradients, 89
Conditional gene knockout by homologous recombination
data analysis, 335–337
breeding strategy for mutant mice, 337
screening of targeted embryonic stem cell clones, 335–337
design of targeting vector, 330–331
experimental procedure, 332–335
DNA electroporation, 332–334
embryonic stem cell injection to blastocyst, 335
embryonic stem cell screening, 334–335
plating embryonic stem cells, 332
materials and reagents, 331–332
buffer, 332
cell culture medium, 331–332
cell line, 331
equipment, 331
site-specific recombination system, 329
troubleshooting, 338–339
cell autonomy, 338–339
issues related to Cre expression, 338
issues related to targeting vector, 338
Conductance scanning, 341
Confocal microscopy. *See also* Light microscopy
for cells on transwell, 199–000
for thick tissue sections, 207–000
Congenic mouse strain, 304
Connexin(s), 4–5, 8
compatibility, 10
interaction, 8
Connexon, 4–5
Constitutive heterochromatin, 315–316
Contractile apparatus, 23–24
Contractility, 23–24
Contrast, 199–200, 200*f*
Cover glass, 181, 184–186, 200. *See also* Cells, wide-field fluorescence microscopy for
CPE. *See Clostridium perfringens* enterotoxin (CPE)
Cre expression, issues related to, 338
Cre-loxP from bacteriophage P1, 329
Cross-link formation, 153–154
Cross-linking, 59, 155
Cryo-electron microscopy (Cryo-EM), 0–347, 348*f*, 350*f*
CEMOVIS of vitreous section, 349
limitation and future direction, 349–350
single-particle, 347–349, 347*f*
protein quality and size, 348–349
structural determination without crystallization, 347–348
Cryo-electron microscopy of vitreous section (CEMOVIS), 349
Cryo-EM. *See* Cryo-electron microscopy (Cryo-EM)
Cryofixation, 217
Cryopreservation, 265–266
Cryostat section, 189–191, 191*t*, 195*f*

freezing of fresh unfixed tissue, 189
microtome temperature, 190
postsectioning fixation, 190–191
sectioning technique, 190
Crystal structures of claudin proteins, 35
C-terminal half of ubiquitin (Cub), 70–71
C-terminal transcription activation domain, 70
Cub. *See* C-terminal half of ubiquitin (Cub)
Cub-LexAVP16 module, 74–75
Current through capacitor, 93–94
Cx45/Cx43 heterotypic junctions, 10
Cysteine, 76–77
Cystic fibrosis transmembrane conductance regulator (CFTR), 117–118
Cytochalasins, 38–39
Cytoskeleton, tight junction anchorage onto, 38–39

D

Data
acquisition, 97–99
analysis, 160
DC. *See* Direct current (DC)
DDM. *See n*-Dodecyl β-D-maltoside (DDM)
de Broglie equation, 213
De novo assembly of tight junction
GUV, 367
limitation and future direction, 370–371
probing claudin interactions in giant unilamellar vesicle, 369–370
protein incorporation and topological orientation in GUV, 367–369
subcellular organelle, 366–367
Dehydrants, 154
Denaturant extraction, 41
Denaturation, 153–154
Deoxycholate (DOC), 33–34, 44
Depolymerization, 20
Desmocollins, 3–4
Desmogleins, 3–4
Desmosomal cadherins, 3–4
Desmosome, 1, 2*t*, 3–4
desmosome-intermediate filament complex, 3–4
Detergent, 41
Deuterium oxide ([2]H_2O), 150
Diabetes insipidus, 16–17
Digital noise, 96
Digitization, 97–99
1,2-Diphytanoyl-*sn*-glycero-3-phosphocholine (DPhPC), 367–369
Direct current (DC), 133
DMEM. *See* Dulbecco's Modified Eagle's medium (DMEM)
DMSO, 66–67
DNA electroporation, 332–334
DNA-activation domain (DNA-AD), 70
DNA-binding domain (DNA-BD), 70
DOC. *See* Deoxycholate (DOC)
Domain swapping theory, 6
Double-strand breaks (DSB), 307–308
Downstream core promoter element, 311
DPhPC. *See* 1,2-Diphytanoyl-*sn*-glycero-3-phosphocholine (DPhPC)
DSB. *See* Double-strand breaks (DSB)
DTA toxin gene, 330–331
Dual transformation and reporter gene expression assay, 74
Dulbecco's Modified Eagle's medium (DMEM), 50, 60, 250, 256–257, 319

E

EA domain. *See* Extracellular anchor domain (EA domain)
EC domains. *See* Extracellular cadherin domains (EC domains)
E-cadherin, 8–9, 26–27
ECLs. *See* Extracellular loops (ECLs)
ECM. *See* Extracellular matrix (ECM)
Ectopic claudin expression model, 59
Edge damage, 125–126
EDTA, 47
in IMAC Ni-charged affinity chromatography, 87
E-face pit, 260
EGF, 26–27
EGTA, 47, 275
Electric
conductance of paracellular channel, 12–13
current, 92
potential, resistance, and capacitance of cell membrane, 89–91
pulse, 115, 126
shunting process, 96
synapse, 10
Electrode, 94
quality of, 114–115
Electrogenic ionic current, 100
Electrolytes, 233–234
Electromagnetic lens, 213–214
Electromagnetic radiation, 213–214
Electron microscopy for tight junction
electromagnetic lens, 213–214
FF-EM, 246–000

Electron microscopy for tight junction (*Continued*)
  FRIL, 256–000
  immunolabeling techniques, 217–218
  low temperature methods, 217
  negative staining, 216–217
  positive staining, 216
  specimen preparation, 214–215
  transmission electron microscopy
    for cell culture, 220–000
    for immunolabeling application, 240–000
    for tissue section, 228–000
    for tracer assay, 235–000
  ultramicrotomy, 215–216
  wave-particle duality of electron, 213
Electrophysiological rig, 105, 118, 136
Embryonic stem cells (ES cells), 303, 307
  injection to blastocyst, 335
  screening, 334–335
En bloc staining, 232–233
ENaC. *See* Epithelial Na^+ channel (ENaC)
End-joining, 315
ENU. *See* Ethyl nitrosourea (ENU)
Eph4 cells, 37–38
Epithelial cells, 0–269, 270*t*
  cultures in Ussing chamber, 100–000
  data analysis, 272
  experimental procedure, 271
    growing MDCK cells on plastic substrate, 271
    seeding MDCK cells on permeable Transwell filter, 271
  lines, 269–270
  materials and reagents, 270–271
    cell culture medium, 271
    cell model, 270
    equipment, 270
  troubleshooting, 272–273
Epithelial Na^+ channel (ENaC), 117–118
Epithelial ohmmeter, 0–128, 128*f*
  "chopstick" electrode system, 128
  current clamp and Ohm's law, 128
  data analysis, 129–130
  experimental procedure, 129
  materials and instrumentation, 128–129
  troubleshooting, 130–131
    cell capacitance and underestimation, 131
    nonuniform electric field, 130
Epithelial tissues in Ussing chamber, 117–000
  data analysis, 123–125
    electrical contribution from multiple tissue layers, 125
    MATLAB functions, 123–124
  experimental procedure, 121–123
    measuring paracellualr conductance, 121–123
    measuring short-circuit current, 121
    setting up Ussing chamber, 121
  intestinal epithelium, 117–118
  materials and instrumentation, 118–120
    electrophysiological rig, 118
    superfusate, 118
    Ussing chamber assembly, 118
  renal epithelium, 118
  troubleshooting, 125–126
    edge damage, 125–126
    electric pulse, 126
    tissue viability and variability, 126
Epithelial transport, electrophysiology of, 89–000
  basic principles of cell membrane electrophysiology, 91–94
  data acquisition and digitization, 97–99
  electric potential, resistance, and capacitance of cell membrane, 89–91
  electrophysiology of epithelium, 94–96
  noise prevention and signal conditioning, 96–97
Epithelium
  electrophysiology of, 94–96
  equivalent electrical circuit of, 94
  TER, 94–95
  transepithelial potential, 95–96
Epoxy resins, 214–215
ε-amine, 59
Equivalent electrical circuit of cell membrane, 91
Equivalent short-circuit current, 101
ES cells. *See* Embryonic stem cells (ES cells)
Escherichia coli, 86
Etching, 248
Ethanol, 155, 217
Ethyl nitrosourea (ENU), 305
Euchromatin, 315–316
Eukaryotes, 78
Eukaryotic cells, 20
External noise, 96
Extracellular anchor domain (EA domain), 3–4
Extracellular cadherin domains (EC domains), 3, 6

Extracellular loops (ECLs), 1–6, 14–16, 35, 357
Extracellular matrix (ECM), 372–374

F

F1 hybrids, 304
F1/R1 (primer pair), 335–337
F2 hybrids, 304
F2/R2 (primer pair), 335–337
F-actin, 20, 23–24
False-positive interaction, 76
Faraday cage, 96
FBS. *See* Fetal bovine serum (FBS)
Ferritin, 217–218
Fetal bovine serum (FBS), 50, 60
FF-EM. *See* Freeze-fracture electron microscopy (FF-EM)
FGF, 26–27
Fibrils, 1–3, 33
Fick's law, 140–141
Filtering, 97, 98*f*
FITC. *See* Fluorescein isothiocyanate (FITC)
Fixation, 0–153, 214–215
  data analysis, 160
  duration, 157
    diffusion coefficient, 156*t*
  for electron microscopy, 220–221
  experimental procedure, 158–160
    immersion fixation, 160
    perfusion fixation through heart, 158–159
    perfusion fixation through the abdominal aorta, 159
  materials and reagents, 158
  of membrane structure, 225
  and permeabilization, 181–182
  troubleshooting, 160
Fixation artifacts, 157
Fixatives
  classification of, 153
  concentration of, 155
  fixation artifacts, 157
  mechanism of, 153–155
    lipid oxidization and cross-linking, 155
    protein cross-linking and denaturation, 153–154
    reaction of fixatives with nucleic acids, 155
  osmolality of fixative solution, 155–156
  penetration of, 156
  temperature of, 156
FLP-FRT from *S. cerevisiae*, 329
Fluorescein isothiocyanate (FITC), 178–179, 184, 194, 202
Fluorescence
  intensity, 146
  labels, 178–179
  microscopy, 177–178
    Jablonski energy diagram, 178*f*
Flux assay in Ussing chamber, 140–000
  data analysis, 142
  experimental procedure, 142
  Fick's law, 140–141
  materials and instrumentation, 141–142
    buffer, 141
    cell culture, 142
    liquid scintillation counter, 142
    radioisotope, 142
  radioisotope, 141
  troubleshooting, 143
    differentiation of paracellular from transcellular pathway, 143
    radiation safety, 143
Flux-ratio equation, 143
Foreign DNA, 307–308
Formaldehyde, 179, 228, 260
Fractionation, 41
Fracturing, 246–247
Freeze etching, 217
Freeze fracturing, 217
Freeze substitution, 217
Freeze-fracture electron microscopy (FF-EM), 246–000
  data analysis, 253
  experimental procedure, 251–253
    cell culture and fixation, 251
    freeze fracturing, 251–253
    perfusion and fixation, 251
  materials and reagents, 249–250
    buffers, 250
    cell culture medium, 250
    cell model, 249
    equipment, 249
  principle, 246
  technical consideration, 246–249
  troubleshooting, 254–255
Freeze-fracture replica immunolabeling (FRIL), 256–000
  data analysis, 258–260
  experimental procedure, 258
  materials and reagents, 256–258
    buffers, 257–258
    cell culture medium, 256–257
    cell model, 256

Freeze-fracture replica immunolabeling (FRIL) (*Continued*)
 equipment, 256
 troubleshooting, 260
Freeze-fractured gap junction, 4–5
Freezing, 246
FRIL. *See* Freeze-fracture replica immunolabeling (FRIL)
FseI (restriction endonuclease site), 335–337
Full width at half maximum (FWHM), 199–200

G

G-actin, 20
Gal upstream activating sequences (Gal-UAS), 70
Gal4 protein, 70
γ-catenin, 26
Gap junction (GJ), 1, 2*t*, 4–5
 channel structure, 9*f*
 electron microscopy of, 4*f*
Gastrointestinal tract, 166–167, 167*f*
Gene knockout, 307–308
Gene targeting, 307–308
 in laboratory mouse, 307–308, 308*f*
  homologous recombination, 307–308
  manipulation of mouse embryonic stem cells, 307
Gene transfer, 265
Gene trapping, 318
GFP. *See* Green fluorescent protein (GFP)
Giant unilamellar vesicle (GUV), 367
 claudin *cis*-interaction in, 369*f*
 claudin *trans*-interaction in, 370*f*
 probing claudin interactions in, 369–370
 protein incorporation and topological orientation in, 367–369
GJ. *See* Gap junction (GJ)
Glomerulus, 170
Glucose, 276
Glutaraldehyde, 179, 217–218, 225, 228
Glycine, 76–77
Glycosylphosphatidylinositol (GPI), 46
Goldman–Hodgkin–Katz equation, 89–90, 101
GPI. *See* Glycosylphosphatidylinositol (GPI)
Green fluorescent protein (GFP), 178–179
GSK3β, 28, 29*f*
Guanidine, 49
Guanidine-HCl, 34–35
GUV. *See* Giant unilamellar vesicle (GUV)

H

Heavy metals, 216
HEK293 cell transfection, 63–64
Hepatocyte, 41
Herpes simplex VP16 transactivator protein, 70–71
Heterochromatin, 315–316
High-pass filter, 97
High-pressure freezing, 217
Hippo signaling, 28–31
*His*3 gene, 70
*His*3 reporter stringency test, 76
HisGFP-claudin purification from *Pichia pastoris*, 83–85
 affinity chromatography, 83–84
 lysis and extraction, 83
 size-exclusion chromatography, 84–85
Histidine-tagged proteins (His_{10} proteins), 87
HIV. *See* Human immunodeficiency virus (HIV)
Homologous arms, 330–331
Homologous DNA, 330–331
Homologous recombination, 307–308
Homophilic cadherin interactions, 24, 26
Homozygosity, 337
Horseradish peroxidase, 217–218
hPSC. *See* Human pluripotent stem cells (hPSC)
Hum noise, 96
Human embryonic HEK293 cells, 49
Human immunodeficiency virus (HIV), 318
Human pluripotent stem cells (hPSC), 372
Hybrid mouse strain, 304
Hydrophobic interaction chromatography, 78–79
Hydrostatic pressure, cancelation of, 148

I

IBs. *See* Inclusion bodies (IBs)
ICR (outbred strain), 304
Ideal electrical ground, 96
IMAB027 (anticlaudin-6), 360–361
IMAB362 (anticlaudin-18.2), 360–361
IMAC Ni-charged affinity chromatography
 imidazole concentration optimization for, 87
 pH and EDTA levels in, 87
 reducing nonspecific protein interaction in, 87
Imidazole concentration optimization for IMAC Ni-charged affinity chromatography, 87

Immunoelectron microscopy, 240
Immunofluorescence, 181, 196*f*
  mouse kidney section for, 195*f*
Immunolabeling, 217–218
  data analysis, 243–244
  experimental procedure, 242–243
    embedding, 242
    perfusion and fixation, 242
    sectioning and poststaining, 242–243
  immunoelectron microscopy, 240
  low temperature embedding, 240
  materials and reagents, 240–242
    buffers, 241–242
    equipment, 240–241
  transmission electron microscopy for, 240–000
  troubleshooting, 244–245
    negative result, 244–245
    nonspecific binding, 245
Immunoprecipitation
  of *cis* and *trans* claudin interactions, 49–50
  coimmunoprecipitation, 49
  data analysis, 56–57
  experimental procedure, 51–56
    cis claudin interactions, 51–53, 54*f*
    trans claudin interactions, 53–56, 56*f*
  materials and reagents, 50–51
    buffers, 50–51
    cell culture medium, 50
    cell model for ectopic gene expression, 50
    plasmids, 50
  preservation of protein interaction, 49
  troubleshooting, 57–58
    choice of antibody, 57
    lysis condition, 57
    nonspecific protein interactions, 57–58
Impedance measurement
  data analysis, 137
  equivalent electric circuit for, 133*f*
  experimental procedure, 136–137
  impedance of resistor and capacitor, 134–135
  materials and instrumentation, 136
    buffer, 136
    cell culture, 136
    electrophysiological rig, 136
  Nyquist plot, 135
  sinusoidal current waveform, 134
  troubleshooting, 137–138
    paracellular *vs.* transcellular pathway, 138
    phase shift, 138
    sample-electrode distance, 137–138
  in Ussing chamber, 133
Impedance of resistor and capacitor, 134–135
In vitro cross-linking, 67–68
In vivo cross-linking, 67–68
Inbred mouse strain, 303–304
Inclusion bodies (IBs), 78
Infinite-time potential. *See* Steady-state potential
Initiator element, 311
Insertional mutation, 318
Intercellular junction systems, 1
  adherens junction, 3
  desmosome, 3–4
  electron micrograph of, 2*f*
  GJ, 4–5
  TJ, 1–3
Intestinal epithelium, 117–118
Intracellular membrane, contamination of, 45
Intrinsic noise, 96–97
Intron–exon boundaries, 311–312
Ion
  channel in tight junction, 12
  ion-exchange chromatography, 78–79
  selectivity of paracellular channel, 14–16
I–V curve, 112–113

J

JAM (molecule), 370–371
Junction signaling
  canonical Wnt signaling, 28
  catenin signaling, 26
  hippo signaling, 28–31
  RTKs, 26–27
  small GTPases, 26

K

Kidney, 170, 173*f*
Kimizuka–Koketsu equation, 101–102
Knife fracture, 246
Knockdown phenotype, 325–326
Knockout (KO), 307–308
  mouse kidney tubules, 16–17
  phenotype, 325–326
Kozak sequence, 311–312

L

Laboratory mouse (*Mus musculus*), 303
lacZ
  gene, 70
  reporter, 75

Lag phase, 263–264
Lanthanum, 225–226, 235
Lead, 216
Lead citrate, 225–226
Lectins, 45
Lentivirus-mediated gene knockdown, 318–000
  data analysis, 322–323
  experimental procedure, 320–322
    lentivirus production, 320–321
    perivitelline injection of lentivirus and production of transgenic mice, 322
  lentivirus-mediated transgenesis, 318
  materials and reagents, 318–319
    buffers, 319
    cell culture medium, 319
    cell model for ectopic gene expression, 319
    equipment, 318
    plasmids, 319
  RNAi in live mice, 318
  troubleshooting, 323–326
    knockdown *vs.* knockout phenotype, 325–326
    silencing of recombinant lentivirus, 323–324
    toxicity of siRNA expression, 324
Leukemia inhibitory factor (LIF), 307
LexA-DNA-binding domain, 70–71
LIF. *See* Leukemia inhibitory factor (LIF)
Light elements, 216
Light microscopy. *See also* Confocal microscopy
  autofluorescence, 179
  axial resolution in, 176
  depth of field in, 177
  fluorescence microscopy, 177–178
  fluorescent labels, 178–179
    natural fluorescent proteins, 179*t*
  lateral resolution in, 175
  photobleaching, 180
Lipid microdomains. *See* Lipid rafts
Lipid oxidization, 155, 155*f*
Lipid rafts, 46
  contamination of lipid-raft microdomains, 46
Liquid junction potential, 94
Liquid scintillation counter, 142
Liver, 168–169, 169*f*
London resins (LR), 240
Loop current, 95
Low temperature embedding, 240
Low temperature methods, 217
Low-pass filter, 97
Lowicryl resins, 240
LoxP, 329
LR. *See* London resins (LR)
Lung, 164–166

M

mAbs. *See* Monoclonal antibodies (mAbs)
Macula adherens. *See* Desmosome
Macula communicans. *See* Gap junction (GJ)
Madin-Darby canine kidney (MDCK) cells, 269–270
  claudin gene silencing in, 292*f*
  strain II cells, 107
  time course of TER recording in, 272*f*
Mammalian gene regulation, 310
Mammalian liver, 41
MATLAB functions, 113, 123–124
MbYTH system. *See* Membrane yeast two-hybrid system (MbYTH system)
Mechanosensitive signal transduction, 24
MEF feeder cells. *See* Mouse embryonic fibroblast feeder cells (MEF feeder cells)
Membrane
  capacitance, 90–91
  conductance, 90
  potential, 89–90
  resistance, 90
Membrane yeast two-hybrid system (MbYTH system), 70–71
Mendelian diseases, 325–326
Methanol, 155, 184–186, 217
Methionine, 76–77
Methylation of gene, 315–316
Microbial contaminations, 266–267, 266*f*
Microbial decontamination, 266–267
Microfluidic jetting, 367–369
Microtome temperature, 190
Microtubule-disrupting agents, 38–39
Millman equation, 90
Missense mutations, 325–326
MLCK. *See* Myosin light-chain kinase (MLCK)
MLCP. *See* Myosin light-chain phosphatase (MLCP)
Molecular ruler, 67
Moloney murine leukemia virus, 318
Monoclonal antibodies (mAbs), 360–361
Mouse colon preparation, 118–120

Mouse embryo, pronuclear injection of, 312, 312*f*
Mouse embryonic fibroblast feeder cells (MEF feeder cells), 331
Mouse embryonic stem cells, manipulation of, 307
Mouse genetics and transgenics, 303–000
  gene targeting in laboratory mouse, 307–308
  laboratory mouse, 303
  mouse genome, 305
  mouse strain, 303–304
    congenic mouse strain, 304
    hybrid mouse strain, 304
    inbred mouse strain, 303–304
    outbred mouse strain, 304
  random mutagenesis in laboratory mouse, 305
  transgenesis in laboratory mouse, 305–307
Mouse L fibroblast cells, 49
Mouse mating, 312
Mouse models of tight junction physiology
  conditional gene knockout by homologous recombination, 329–000
  lentivirus-mediated gene knockdown, 318–000
  mouse genetics and transgenics, 303–000
  transgenic overexpression by DNA injection, 310–000
Multiple tissue layers, electrical contribution from, 125
Mus musculus. *See* Laboratory mouse (*Mus musculus*)
Mycoplasmal infections, 266–267
Myelin, 170, 174*f*
MyoII. *See* Myosin II (MyoII)
Myosin, 20–23
  myosin-mediated tension, 20–23
Myosin II (MyoII), 23*f*
  activation, 23–24
  regulatory light-chain phosphorylation, 24
Myosin light-chain kinase (MLCK), 23–24
Myosin light-chain phosphatase (MLCP), 23–24

N

NA. *See* Numerical aperture (NA)
Na^+/glucose cotransporter 1 (SGLT1), 117–118
Nanobody. *See* Single-domain antibody (sdAb)
N-cadherin, 8–9
n-Dodecyl β-D-maltoside (DDM), 83
Nearest neighbor approaches, 67
Negative selection, 330–331
Negative staining, 216–217
Neo gene, 335–337
Neomycin, 330–331
Nernst equation, 89
Nervous system, 170–171
NMY51 yeast strain, transformation of bait and prey constructs into, 72–73
Noise prevention, 96–97
Nonspecific antibody binding, 197, 210
Nonuniform electric field, 130
Notch filter, 97
Novel binders to tight junction
  CPE, 357
  TJ modulating peptidomimetics, 357–360
NP-40 (nonionic detergent), 33–34
N-retinylidene-*N*-retinyl-ethanolamine (A2E), 179
N-terminal DNA-binding domain, 70
N-terminal half of ubiquitin (Nub), 70–71
Nub. *See* N-terminal half of ubiquitin (Nub)
Nuclear karyopherin exportin-5, 324
Nucleic acids, fixatives reaction with, 155
Numerical aperture (NA), 175, 187, 199–200, 352
Nyquist plot, 135
Nyquist theorem, 97–99

O

Occludin, 34–35, 45, 370–371
  $OCLN_{184-227}$, 357
  peptidomimetics, 357
  tight junction localization of, 35*f*
Oct3/4 gene promoter, 307
OECTs. *See* Organic electrochemical transistors (OECTs)
Ohm's law, 92, 128, 341
Open reading frame (ORF), 314–315
ORF. *See* Open reading frame (ORF)
Organ on a chip, 372–374
Organ systems, survey of tight junction in, 162–171
  cardiovascular system, 162
  gastrointestinal tract, 166–167
    pulmonary alveolus, 165*f*
  kidney, 170
    renal epithelium, 171*f*
  liver, 168–169
    hepatic lobule, 169*f*
  lung, 164–166

Organ systems, survey of tight junction in (*Continued*)
nerve, 170–171
skin, 162–164
epidermis, 164*f*
tight junction in blood, 163*f*
Organic electrochemical transistors (OECTs), 131–132
Organoid model of tight junction biology
bioprinting of organ, 374
limitation and future direction, 374–375
organ on chip, 372–374
organoid culture, 372
Osmium, 216–218, 225
Osmium tetroxide, 155
Osmolality, 233–234
of fixative solution, 155–156
Outbred mouse strain, 304

P

p120-catenin subfamily, 26
Packaging of retrovirus, 279–280, 281*f*
PALM. *See* Photoactivated localization microscopy (PALM)
Paracellular
channel
electric conductance of, 12–13
ion channel in tight junction, 12
ion selectivity of, 14–16
paracellular water channel, 16–17
size selectivity of, 16
conductance, 101
measuring, 110–112, 121–123
measuring paracellular ion selectivity, 112
pathway, 94, 96, 138
permeability, 24
resistance, 100
water channel, 16–17
Paracellular differentiation from transcellular pathway, 143
Paraffin section, 189
Patch clamp technique, 115–116
PBS. *See* Phosphate-buffered saline (PBS)
PDZ. *See* Postsynaptic density 95/discs large/ZO-1 (PDZ)
Peak amplitude, 112
Penetration of fixatives, 156
Perijunctional
actin belt, 20–23
actomyosin ring, 38–39
cytoskeleton
actin polymerization, 20
actin reorganization, 20–23
contractile apparatus, 23–24
contractility and paracellular permeability, 24
mechanosensitive signal transduction, 24
Peripheral myelin protein 22 (PMP22), 366–367
Perivitelline injection of lentivirus, 322
Permeable Transwell filter, seeding MDCK cells on, 271
PGK-neo expression cassette, 338
pH levels in IMAC Ni-charged affinity chromatography, 87
Phase shift, 138
Phenylmethanesulfonyl fluoride (PMSF), 50–51, 61
Phosphate-buffered saline (PBS), 271
Photoactivated localization microscopy (PALM), 352
Photobleaching, 180
Pichia pastoris
EasyComp transformation kit, 81
induction, 82–83
recombinant claudin protein production in, 78–000
transformation, 81–82
prepare competent *P. pastoris* cells, 81
yeast strain, 79
Plakoglobin, 26
Plasmids, 50, 60, 72, 79, 319
Plastic substrate, growing MDCK cells on, 271
Plating embryonic stem cells, 332
PMP22. *See* Peripheral myelin protein 22 (PMP22)
PMSF. *See* Phenylmethanesulfonyl fluoride (PMSF)
Pol II. *See* Polymerase II (Pol II)
Poly(A) tail, 311–312
Polyadenylation, 311–312
Polymerase II (Pol II), 310
Polymerization of branched actin networks, 20–23
Polyvinyl alcohol adhesive (PVA adhesive), 249, 256
Positive selection, 330–331
Positive staining, 216
Postembedding labeling, 217–218
Postsectioning fixation, 190–191
Poststaining, 232–233
of tight junction, 225–226

Postsynaptic density 95/discs large/ZO-1 (PDZ), 1–3
Potential difference, 93
Preembedding labeling, 217–218
Prey protein, 70–71
Primary amines, 59
Primary antibody, 186
Primary culture, 263
Prokaryotes, 78
Promoter, 311
Pronuclear
  DNA injection, 313
  injection of mouse embryo, 312, 312*f*
Propagation, 263–264
Protein
  cross-linking, 153–154, 154*f*
    formation, 153–154
  incorporation and topological orientation in GUV, 367–369
  preservation of protein interaction, 49
  solubilization, 86
  stability in *S. cerevisiae*, 76–77
  topology, 74–75
Proteoliposomes, 367–369
Proton permeability, 150
Pseudotyping with VSV-G protein, 280–282
PVA adhesive. *See* Polyvinyl alcohol adhesive (PVA adhesive)

Q

qPCR3utr (primer pair), 314–315
qPCRorf (primer pair), 314–315
Quantitative β-galactosidase assay, 75

R

Rac1, 26
Radiation safety, 143
Radioisotope, 141
Rafts, 367
Random mutagenesis in laboratory mouse, 305
Rayleigh criterion, 175, 176*f*, 199–200
RE. *See* Reference electrode (RE)
Receptor tyrosine kinases (RTKs), 26–27
Recombinant claudin protein production in *Pichia pastoris*, 78–000
  chromatography, 78–79
  data analysis, 86
  experimental procedure, 81–85
    HisGFP-claudin purification, 83–85
    P. pastoris induction, 82–83
    P. pastoris transformation, 81–82
  materials and reagents, 79–81
    chromatography, 81
    P. pastoris EasyComp transformation kit, 81
    P. pastoris yeast strain, 79
    plasmids, 79
    yeast growth media, 79–80
  recombinant protein expression system, 78
    eukaryotes, 78
    prokaryotes, 78
  troubleshooting, 86–88
    molecular weight and shape in size-exclusion chromatography, 87–88
    optimization of imidazole concentration for IMAC Ni-charged affinity chromatography, 87
    pH and EDTA levels in IMAC Ni-charged affinity chromatography, 87
    protein solubilization, 86
    reducing nonspecific protein interaction in IMAC Ni-charged affinity chromatography, 87
    resolution in size-exclusion chromatography, 88
Recombinant protein expression system, 78
  eukaryotes, 78
  prokaryotes, 78
Recombinant retroviral vector, 279
Reference electrode (RE), 342–343
Regulatory elements, 311
Regulatory light chain (RLC), 23–24
  phosphorylation, 23–24
Renal epithelium, 118
Replication, 248–249
Resolution, 199–200
Resting membrane potential, 89–90
Retrovirus-mediated RNA interference, 0–288
  data analysis, 291
  experimental procedure, 290–291
  materials and reagents, 289–290
    buffers, 290
    cell culture medium, 290
    cell model for ectopic gene expression, 289
    plasmids, 289
  rational selection of siRNA sequence, 291*f*
  as tool to study loss of gene function, 288
  troubleshooting, 292–293
    expression level of siRNA, 292–293
    expression level of target gene, 293

Retrovirus-mediated transgene expression, 0–279, 280*f*, 286*f*
data analysis, 285
experimental procedure, 283–285
materials and reagents, 282–283
buffers, 282–283
cell culture medium, 282
cell model for ectopic gene expression, 282
plasmids, 282
packaging of retrovirus, 279–280
pseudotyping with VSV-G protein, 280–282
recombinant retroviral vector, 279
troubleshooting, 285–287
transduction efficiency, 286–287
viral titer, 285–286
Retroviruses, 318
Rho GTPase, 24, 26
Rho kinase (ROCK), 23–24
Rho signaling, 24
Ringer's solution, 228
RISC. *See* RNA-induced silencing complex (RISC)
RLC. *See* Regulatory light chain (RLC)
RNA interference (RNAi), 288, 318
in live mice, 318
molecular mechanism of, 289*f*
sequence selection for, 288–289
tool to study loss of gene function, 288
RNA-induced silencing complex (RISC), 288
RNAi. *See* RNA interference (RNAi)
ROCK. *See* Rho kinase (ROCK)
ROSA-26 locus, 337
RTKs. *See* Receptor tyrosine kinases (RTKs)

S

Saccharomyces cerevisiae, 70
FLP-FRT from, 329
protein stability in, 76–77
yeast strain, 72
Salt bridge, 94
Sample-electrode distance, 137–138
Sampling, 97–99
Scanning ion conductance microscopy (SICM), 0–343, 342*f*, 344*f*
characterization of epithelial cell monolayer with, 344*f*
conductance scanning, 341
instrumentation, 343
limitation and future direction, 345–346
practical application, 342–343
tight junction conductance measurement, 343–345
waveform of potentiometric signal, 345*f*
Scanning probe microscopy (SPM), 342–343
Scrambled siRNAs, 292–293
Screening of targeted embryonic stem cell clones, 335–337
sdAb. *See* Single-domain antibody (sdAb)
SDS. *See* Sodium dodecyl sulfate (SDS)
Sectioning technique, 190
Self-contained Ussing chamber system, 102, 118
Seromusculature layers, 125
Seromusculature stripping, 125
Sf21 cells, 49
Sf9 cells, 49
SGLT1. *See* Na^+/glucose cotransporter 1 (SGLT1)
Short-circuit current, measuring, 108–110, 121
SICM. *See* Scanning ion conductance microscopy (SICM)
Signal conditioning, 96–97
Silencing of recombinant lentivirus, 323–324
SIM. *See* Structured illumination microscopy (SIM)
Single-domain antibody (sdAb), 361–363
Sinusoidal current waveform, 134
siRNA. *See* Small interfering RNA (siRNA)
Site-specific recombination system, 329
Size selectivity of paracellular channel, 16
Size-exclusion chromatography, 78–79
molecular weight and shape in, 87–88
resolution in, 88
Skin, 162–164, 165*f*
Small GTPases, 26
Small interfering RNA (siRNA), 288, 318
toxicity of siRNA expression, 324
Sodium dodecyl sulfate (SDS), 256
Solenoid, 213–214
Solubilization of claudin protein, 64
Spatial separation of tight junction components, 352–354
Specificity of cross-linking, 66–67
Split-ubiquitin MbYTH system, 70–71, 71*f*, 75*f*
SPM. *See* Scanning probe microscopy (SPM)
Steady-state potential, 93–94
Stochastic optical reconstruction microscopy (STORM), 352
STORM. *See* Stochastic optical reconstruction microscopy (STORM)

Strands, 1–3, 33
Structured illumination microscopy (SIM), 352, 354*f*, 355*f*
Subcellular fractionation, 41
  data analysis, 45
  detergent and denaturant extraction, 41
  experimental procedure, 44–45
  materials and reagents, 42–44
    animals, 42
    buffers, 43–44
    centrifuge, 43
    homogenizer, 42
  tight junction isolation by, 41–000
  troubleshooting, 45–47
    contamination of adherens junction, 47
    contamination of intracellular membrane, 45
    contamination of lipid-raft microdomains, 46
    extrapolation to other organ systems, 47
Subcellular organelle, de novo assembly of, 366–367
Subculture, 263–264
Sucrose gradient centrifugation, 64–65, 65*f*
Super-resolution microscopy, 0–352, 353*t*
  architectural alteration in tricellular tight junction, 354–355
  colocalization of claudins in tight junction, 354*f*
  limitation and future direction, 355–356
  spatial separation of tight junction components, 352–354
  tools, 352–354
Superfusate, 106, 118, 147, 147*t*
Superresolution optical techniques, 38–39
Swiss Webster, 304
Switch noise, 96

T

TAL. *See* Thick ascending limb (TAL)
Tamm–Horsfall protein (THP), 314–315
Targeting vector
  design of, 330–331
  issues related to, 338
TATA motif, 310
Temperature of fixation, 156
Temporal resolution, 97–99
Tensile fracture, 246
TER. *See* Transepithelial resistance (TER)
Tetramethyl-rhodamine isothiocyanate (TRITC), 178–179, 184, 194, 202
TF-II recognition element, 311
TFs. *See* Transcription factors (TFs)
Thick ascending limb (TAL), 314–315
Thick tissue sections, confocal microscopy for, 207–000
  data analysis, 208–209
  experimental procedure, 208
  low-magnification confocal micrograph, 211*f*
  materials and reagents, 207–208
    animals, 207
    buffers, 207–208
    equipment, 207
    tissue-tek, 207
  troubleshooting, 210
Thin tissue section, wide-field fluorescence microscopy for, 189–000
  cryostat section *vs.* paraffin section, 189
  experimental procedure, 193–194
    cutting cryostat sections, 193–194
    data analysis, 194–195
    fixation and tight junction pattern, 196
    freezing tissues, 193
    immunolabeling cryostat sections, 194
    nonspecific antibody binding, 197
    troubleshooting, 196–197
  materials and reagents, 192–193
    animals, 192
    buffers, 192–193
    equipment, 192
    tissue-tek, 192
THP. *See* Tamm–Horsfall protein (THP)
Threonine, 76–77
Tight junction (TJ), 1–3, 2*t*, 33, 166, 181–182, 201, 225, 232, 269, 349–350
  anchorage onto cytoskeleton, 38–39
  architectural alteration in tricellular, 354–355
  assembly, 366–367
  atlas, 0–161, 162*f*
    survey of tight junction in organ systems, 162–171
  biochemical approaches
    claudin oligomer isolation by chemical cross-linking, 59–000
    immunoprecipitation of *cis* and *trans* claudin interactions, 49–000
    recombinant claudin protein production in *P. pastoris*, 78–000
    yeast two-hybrid assay of claudin interaction, 70–000
  biochemical organization, 33

Tight junction (TJ) (*Continued*)
biochemistry
biochemical organization, 33
claudin interaction with TJ plaque proteins, 37–38
models of claudin interaction, 36–37, 37*f*
molecular structure of claudin protein, 35, 36*f*
biophysical approaches
electrophysiology of epithelial transport, 89–000
epithelial cell cultures in Ussing chamber, 100–000
epithelial ohmmeter, 128–000
epithelial tissues in Ussing chamber, 117–000
flux assay in Ussing chamber, 140–000
impedance measurement in Ussing chamber, 133–000
water permeability measurement in Ussing chamber, 145–000
claudin interaction with tight junction plaque proteins, 37–38
colocalization of claudins in, 354*f*
in colon, 168*f*
conductance measurement, 343–345
de novo assembly
GUV, 367
limitation and future direction, 370–371
probing claudin interactions in giant unilamellar vesicle, 369–370
protein incorporation and topological orientation in GUV, 367–369
subcellular organelle, 366–367
in detergent extracted bile canaliculus membrane, 42*f*
enriched protein fraction, 33–34
epithelial and endothelial, 232*f*
formation, 200
freeze-fracture replica of, 3*f*
glomerular parietal epithelium, 172*f*
histological approaches
fixation, 0–158
fixatives, 153–000
tight junction atlas, 161–000
integral protein fraction, 34–35
ion channel in tight junction, 12
isolation by subcellular fractionation, 41–000
modulating peptidomimetics, 357–360, 359*t*
anti-claudin antibodies, 360–361, 361*t*
claudin peptidomimetics, 358–360
limitation and future direction, 361–363
occludin peptidomimetics, 357
poststaining of, 225–226
proteins, 181
in small intestines, 168*f*
spatial separation of tight junction components, 352–354
stepwise enrichment of TJ protein from liver cell membrane, 46*f*
three-dimensional reconstruction of, 205*f*, 210*f*
ultrastructure, 34*f*
Time constant, 94
Tissue
barrier, 10, 235
fixation, 228
morphogenesis, 8–9
perfusion, 228
viability and variability, 126
Tissue section, transmission electron microscopy for, 228–000
data analysis, 232
experimental procedure, 230–232
embedding, 230–231
perfusion and fixation, 230
sectioning and poststaining, 231–232
materials and reagents, 229–230
buffers, 229–230
equipment, 229
troubleshooting, 232–234
osmolality, electrolytes, and additives in fixation, 233–234
poststaining *vs*. en bloc staining, 232–233
TJ. *See* Tight junction (TJ)
TM domains. *See* Transmembrane domains (TM domains)
Toxicity of siRNA expression, 324
Tracer assay, transmission electron microscopy for, 235–000
data analysis, 237–238
experimental procedure, 237
lanthanum, 235
materials and reagents, 235–237
buffers, 236–237
equipment, 235
tissue barrier, 235
troubleshooting, 238–239
Trans claudin interactions, 6
immunoprecipitation of, 49–50, 53–56, 56*f*

Trans oligomer, 68
Transcellular
  pathway, 94, 138
  resistance, 100
Transcription factors (TFs), 310
Transcription regulation, 310–311
  core promoter architecture, 310–311
  general transcription machinery, 310
Transepithelial conductance, 342–343
Transepithelial potential, 95–96
Transepithelial resistance (TER), 94–95, 269–270, 277, 341
Transepithelial voltage, effect of, 149
Transepithelial water permeability, 145–000
  measuring, 148
Transgene
  design, 311–312
    intron–exon boundaries, Kozak sequence, and polyadenylation, 311–312
    promoter and regulatory elements, 311
  release, 313
  silencing, 315–316
Transgenesis in laboratory mouse, 305–307
Transgenic mice, 305–307
  production, 313, 322
Transgenic mosaicism, 316
Transgenic overexpression by DNA injection, 310–000
  data analysis, 314–315
  experimental procedure, 313
    pronuclear injection of DNA and production of transgenic mice, 313
    transgene release, 313
  materials and reagents, 313
    buffers, 313
    equipment, 313
  pronuclear injection of mouse embryo, 312, 312*f*
  transcription regulation, 310–311
  transgene design, 311–312
  troubleshooting, 315–316
    transgene silencing, 315–316
    transgenic mosaicism, 316
    unwanted transgene expression, 315
Transient transfections, 265
Transmembrane domains (TM domains), 8, 35
Transmission electron microscopy
  for cell culture, 220–000
  for immunolabeling application, 240–000
  for tissue section, 228–000
  for tracer assay, 235–000
Transwell, 220
  insert, 102–105
  membrane, 130
  permeable supports, 102–105
Tricellular tight junction
  architectural alteration in, 354–355
Tricellular tight junction (*vide infra*), 16–17, 354–355
Tricellulin, 16–17
Tris, 66–67
TRITC. *See* Tetramethyl-rhodamine isothiocyanate (TRITC)
Tritiated water ([3]H_2O), 150
Triton X-100, 49
Tungsten, 216–217
Two-step cross-linking process, 66–67

U

Ubiquitin-specific proteases (UBPs), 70–71
UBPs. *See* Ubiquitin-specific proteases (UBPs)
Ultramicrotome, 215–216, 215*f*, 216*t*
Ultramicrotomy, 215–216
Ultrastructure
  of tight junction, 240
  of tissue specimens, 217–218
5′-or 3′-Untranslated region (UTR), 314–315, 338
Unwanted transgene expression, 315
Uranium, 216–217
Uranyl acetate, 225–226
Urea, 49
Ussing chamber
  data analysis, 112–113
    baseline and peak amplitude, 112
    I–V curve, 112–113
    MATLAB functions, 113
  epithelial cell cultures in, 100–000
  epithelial tissues in, 117–000
  experimental procedure, 107–112
    measuring paracellualr conductance, 110–112
    measuring paracellular ion selectivity, 112
    measuring short-circuit current, 108–110
    setting up Ussing chamber, 107–108
  flux assay in, 140–000
  impedance measurement in, 133
  materials and instrumentation, 105–107
    cell culture, 107
    electrophysiological rig, 105
    superfusate, 106
    Ussing chamber assembly, 105–106
  perfusion rig, 146

Ussing chamber (*Continued*)
  practical applications, 102–105
    Classic Ussing chamber, 102
    self-contained Ussing chamber, 102
    Transwell permeable supports, 102–105
  theoretic considerations, 100–102
    recording of paracellular transport, 101–102
    recording of transcellular transport, 100–101
  troubleshooting, 114–115
    electric pulse, 115
    mouse colon preparation, 118–120
    quality of cell monolayer, 114
    quality of electrodes, 114–115
  water permeability measurement in, 145–000
Ussing chamber, classic, 102
UTR. *See* 5′-or 3′-Untranslated region (UTR)

V

Vascular perfusion, 228
VE-cadherin, 366–367
Vehicle osmolality, 155–156
Vesicular stomatitis virus G (VSV-G), 280–282
vide infra. *See* Tricellular tight junction (*vide infra*)
Vinculin, 34–35
Viral envelopes, 280–282
Viral transduction efficiency, 286–287
Vitrification, 217
Voltage divider, 92–93
Voltmeter, 93
VSV-G. *See* Vesicular stomatitis virus G (VSV-G)

W

Water permeability measurement in Ussing chamber, 145–000
  data analysis, 148–149
  experimental procedure, 148
    cancelation of hydrostatic pressure, 148
    measuring transepithelial water permeability, 148
  materials and instrumentation, 146–148
    cell culture, 148
    superfusate, 147
    Ussing chamber perfusion rig, 146
  simplified model, 146
  transepithelial water permeability, 145–000
  troubleshooting, 149–150
    differentiating paracellular from transcellular water pathway, 149
    limitation in direct measurement of volume, 150
    proton permeability, 150
    effect of transepithelial voltage, 149
Water-immiscible resins, 214–215
Water-miscible resins, 214–215
Wave-particle duality of electron, 213
Wavelength of electron, 213
Western blot, verification of bait and prey protein expression by, 73–74
White noise, 96–97
Wide-field fluorescence microscopy
  for cells on cover glass, 181–000
  for thin tissue section, 189–000

Y

Y2H assay of claudin interaction. *See* Yeast two-hybrid assay of claudin interaction (Y2H assay of claudin interaction)
Yeast artificial chromosome (YAC), 311
Yeast two-hybrid assay, classic, 70, 70*f*
Yeast two-hybrid assay of claudin interaction (Y2H assay of claudin interaction), 70–000
  classic yeast two-hybrid assay, 70, 70*f*
  data analysis, 74–76
    *Ade*2 phenotype, 76
    *His*3 reporter stringency test, 76
    protein topology, 74–75
    quantitative β-galactosidase assay, 75
  experimental procedure, 72–74
    dual transformation and reporter gene expression assay, 74
    transformation of bait and prey constructs into NMY51 yeast strain, 72–73
    verification of bait and prey protein expression by Western blot, 73–74
  materials and reagents, 72
    plasmids, 72
    S. cerevisiae yeast strain, 72
    yeast growth media, 72
  MbYTH system, 70–71
  troubleshooting, 76–77
    false-positive interaction, 76
    protein stability in *S. cerevisiae*, 76–77

Z

ZO-1 protein, 4–5, 24, 33–35, 37–39, 370–371
ZO-2 protein, 37–38
Zonula adherens. *See* Adherens junction (AJ)
Zonula occludens. *See* Tight junction (TJ)

9780128186473